Supplements to the 2nd Edition of

RODD'S CHEMISTRY OF CARBON COMPOUNDS

ELSEVIER SCIENTIFIC PUBLISHING COMPANY
1 Molenwerf
P.O. Box 211, Amsterdam, The Netherlands

ELSEVIER SCIENCE PUBLISHING COMPANY INC.
52 Vanderbilt Avenue
New York, New York, 10017

Library of Congress Card Number: 64-4605

ISBN 0-444-42183-1

Printed in The Netherlands

Supplements to the 2nd Edition of

RODD'S CHEMISTRY OF CARBON COMPOUNDS

VOLUME I

ALIPHATIC COMPOUNDS

★

VOLUME II

ALICYCLIC COMPOUNDS

★

VOLUME III

AROMATIC COMPOUNDS

★

VOLUME IV

HETEROCYCLIC COMPOUNDS

★

VOLUME V

MISCELLANEOUS
GENERAL INDEX

★

Supplements to the 2nd Edition (Editor S. Coffey) of

RODD'S CHEMISTRY OF CARBON COMPOUNDS

A modern comprehensive treatise

Edited by
MARTIN F. ANSELL
Ph.D., D.Sc. (London) F.R.S.C. C. Chem.
Department of Chemistry, Queen Mary College,
University of London (Great Britain)

Supplement to

VOLUME I ALIPHATIC COMPOUNDS

Part FG:
Penta- and Higher Polyhydric Alcohols, Their Oxidation Products and Derivatives; Saccharides; Tetrahydric Alcohols, Their Analogues, Derivatives and Oxidation Products

ELSEVIER SCIENTIFIC PUBLISHING COMPANY
Amsterdam – Oxford – New York 1983

CONTRIBUTORS TO THIS VOLUME

Robert J. Ferrier, Ph.D., D.Sc., F.R.S.N.Z.,
Department of Chemistry, Victoria University,
Wellington, New Zealand

Robert A. Hill, M.A., Ph.D.,
Department of Chemistry,
The University, Glasgow G12 8QQ, Scotland

Riaz Khan, M.Sc., Ph.D.,
Tate and Lyle Research Laboratory,
The University, Reading, England

Berit S. Paulsen, Ph.D.,
The Institute of Pharmacy,
The University of Oslo, Oslo 3, Norway

Jens K. Wold, Ph.D.,
The Institute of Pharmacy,
The University of Oslo, Oslo 3, Norway

PREFACE TO SUPPLEMENT IFG

This supplement covers chapters 22, 23, 24, 25 of volumes IF and IG of the second edition and through the efforts of an international group of contributors covers the progress in the carbohydrate and related fields. Although the chapters in this book stand on their own, it is intended that each one should be read in conjunction with the parent chapter in the second edition.

At a time when there are many specialist reviews, monographs and reports available, there is still in my view an important place for a book such as "Rodd", which gives a broader coverage of organic chemistry. One aspect of the value of this work is that it allows the expert in one field to quickly find out what is happening in other fields of chemistry. On the other hand a chemist looking for the way into a field of study will find in "Rodd" an outline of the important aspects of that area of chemistry together with leading references to other works to provide more detailed information.

As editor I wish to thank all the contributors for the very readable critical assessments they have made of the areas of chemistry covered by their respective chapters. As an organic chemist I have enjoyed reading the chapters and profited from learning of advances in fields of chemistry that are outside my own special interests.

This volume, like other recent supplements, has been produced by direct reproduction of manuscripts. I am most grateful to the contributors for all the care and effort which they and their secretaries have put into the production of their manuscripts, including in most cases the diagrams. I am confident that readers will find this presentation acceptable. I also wish to thank the staff at Elsevier for all the help they have given me and for seeing the transformation of authors' manuscripts to published work.

December 1982 Martin F. Ansell

CONTENTS

VOLUME I FG

Aliphatic Compounds; Penta- and Higher Polyhydric Alcohols, Their Oxidation Products and Derivatives; Saccharides; Tetrahydric Alcohols, Their Analogues, Derivatives and Oxidation Products

Chapter 24. Oligosacchardies, Polysaccharides and Related Compounds
by R.H. Khan, J.K. Wold and B.S. Paulsen

Chapter 25. Tetrahydric Alcohols, Their Analogues, Derivatives and Oxidation Products
by R.A. Hill

OFFICIAL PUBLICATIONS

B.P.	British (United Kingdom) Patent
F.P.	French Patent
G.P.	German Patent
Sw.P.	Swiss Patent
U.S.P.	United States Patent
U.S.S.R.P.	Russian Patent
B.I.O.S.	British Intelligence Objectives Sub-Committee Reports
F.I.A.T.	Field Information Agency, Technical Reports of U.S. Group Control Council for Germany
B.S.	British Standards Specification
A.S.T.M.	American Society for Testing and Materials
A.P.I.	American Petroleum Institute Projects
C.I.	Colour Index Number of Dyestuffs and Pigments

SCIENTIFIC JOURNALS AND PERIODICALS

With few obvious and self-explanatory modifications the abbreviations used in references to journals and periodicals comprising the extensive literature on organic chemistry, are those used in the World List of Scientific Periodicals.

LIST OF COMMON ABBREVIATIONS AND SYMBOLS USED

A	acid
Å	Ångström units
Ac	acetyl
a	axial; antarafacial
as, asymm.	asymmetrical
at	atmosphere
B	base
Bu	butyl
b.p.	boiling point
C, mC and μC	curie, millicurie and microcurie
c, C	concentration
C.D.	circular dichroism
conc.	concentrated
crit.	critical
D	Debye unit, 1×10^{-18} e.s.u.
D	dissociation energy
D	dextro-rotatory; dextro configuration
DL	optically inactive (externally compensated)
d	density
dec. or decomp.	with decomposition
deriv.	derivative
E	energy; extinction; electromeric effect; Entgegen (opposite) configuration
E1, E2	uni- and bi-molecular elimination mechanisms
E1cB	unimolecular elimination in conjugate base
e.s.r.	electron spin resonance
Et	ethyl
e	nuclear charge; equatorial
f	oscillator strength
f.p.	freezing point
G	free energy
g.l.c.	gas liquid chromatography
g	spectroscopic splitting factor, 2.0023
H	applied magnetic field; heat content
h	Planck's constant
Hz	hertz
I	spin quantum number; intensity; inductive effect
i.r.	infrared
J	coupling constant in n.m.r. spectra; joule
K	dissociation constant
kJ	kilojoule

LIST OF COMMON ABBREVIATIONS

k	Boltzmann constant; velocity constant
kcal	kilocalories
L	laevorotatory; laevo configuration
M	molecular weight; molar; mesomeric effect
Me	methyl
m	mass; mole; molecule; *meta*-
ml	millilitre
m.p.	melting point
Ms	mesyl (methanesulphonyl)
[M]	molecular rotation
N	Avogadro number; normal
nm	nanometre (10^{-9} metre)
n.m.r.	nuclear magnetic resonance
n	normal; refractive index; principal quantum number
o	*ortho*-
o.r.d.	optical rotatory dispersion
P	polarisation, probability; orbital state
Pr	propyl
Ph	phenyl
p	*para*-; orbital
p.m.r.	proton magnetic resonance
R	clockwise configuration
S	counterclockwise config.; entropy; net spin of incompleted electronic shells; orbital state
S_N1, S_N2	uni- and bi-molecular nucleophilic substitution mechanisms
S_Ni	internal nucleophilic substitution mechanisms
s	symmetrical; orbital; suprafacial
sec	secondary
soln.	solution
symm.	symmetrical
T	absolute temperature
Tosyl	*p*-toluenesulphonyl
Trityl	triphenylmethyl
t	time
temp.	temperature (in degrees centigrade)
tert.	tertiary
U	potential energy
u.v.	ultraviolet
v	velocity
Z	zusammen (together) configuration

LIST OF COMMON ABBREVIATIONS

α	optical rotation (in water unless otherwise stated)
$[\alpha]$	specific optical rotation
α_A	atomic susceptibility
α_E	electronic susceptibility
ε	dielectric constant; extinction coefficient
μ	microns (10^{-4} cm); dipole moment; magnetic moment
μB	Bohr magneton
μg	microgram (10^{-6} g)
λ	wavelength
ν	frequency; wave number
χ, χ_d, χ_μ	magnetic, diamagnetic and paramagnetic susceptibilities
~	about
(+)	dextrorotatory
(-)	laevorotatory
(±)	racemic
$\ominus$	negative charge
$\oplus$	positive charge

Chapter 22

POLYHYDRIC ALCOHOLS (ALDITOLS)

R.J. FERRIER

1. General introduction

The alditols, whose basic chemistry was covered fully in the Second Edition, are the products of reduction of free sugars, and consequently many of the developments noted in Chapter 23 have relevance here also. For these reasons this Chapter is brief, and covers very selectively newer developments relating specifically to the alditols. Some of the material is also relevant to a range of other acyclic derivatives such as the aldose 1,1-acetals or -dithioacetals, and also many heterocyclic compounds derived from aldoses and which have polyhydroxyalkyl substituents.

These polyols are of potential value as polyfunctional materials for the synthesis of polymers, and D-glucitol, in particular, has applicability as a humectant in several industries and also as a sweetening agent - particularly in diabetic foods. Xylitol has gained significance as an alternative sweetening agent to sucrose because of its comparable sweetness and lack of cariogenic properties ("Xylitol", ed. J.N. Counsell, Applied Science Publishers, London, 1978).

(a) Natural occurrence

A wide range of alditols occur in Nature: simpler compounds, for example, glycerol, erythritol, xylitol and arabinitol (undetermined absolute configuration) have been found in a *Cannabis* plant (G. Haustveit and J.K. Wold, Carbohydrate Res., 1973, 29, 325), the last two being rare and D-arabinitol also occurring in a lichen (W.D. Ollis *et al.*, J.chem.Soc.(C),

1971, 1318). It has also been found substituted at $O_{(5)}$ and both at $O_{(1)}$ and $O_{(5)}$ by β-D-fructofuranosyl units in a fungus which also contains D-mannitol with this ketose as substituent at both the primary positions (R.L. Mower, G.R. Gray and C.E. Ballou, Carbohydrate Res., 1973, 27, 119).

The following higher alditols appear to occur most commonly in plants: D-*glycero*-D-*gluco*-heptitol (β-sedoheptitol), D-*glycero*-D-*manno*-heptitol (D-volemitol), D-*glycero*-D-*galacto*-heptitol (D-perseitol) and D-*erythro*-D-*galacto*-octitol (N.K. Richtmyer, *ibid.*, 1970, 12, 135, 139, 233).

Branched-chain alditols to have been isolated are 2-*C*-methyl-D-erythritol (from a *Convolvulus*, S.W. Shah *et al.*, Acta chem. Scand., 1976, 30, 903) and 1-*O*-(α-D-galactopyranosyl)hamamelitol (from a *Primula*, E. Beck, Z. Pflanzenphysiol., 1969, 61, 360).

(b) Stereochemistry and complexing with inorganic ions

Relatively low energy barriers for rotation about the central bonds of alditols means that these compounds are flexible and continuously changing shape in solution, and this leads to their having low optical rotations relative to cyclic compounds. It, however, does not mean that all rotamer states are equally populated, and good generalisations are available regarding the favoured conformations. Since the same rotamers are found to be favoured in the solid state and in solution, it can be concluded that crystal forces are of less significance than intramolecular factors in determining the preferred conformations of such compounds in the crystal lattice.

Alditols favour conformations with planar zig-zag carbon chains unless, in such, there are eclipsing interactions (which are analogous to axial, axial interactions in pyranoid systems) between oxygen atoms on alternate carbon atoms. In such cases rotation occurs within the carbon chain to remove the interaction:

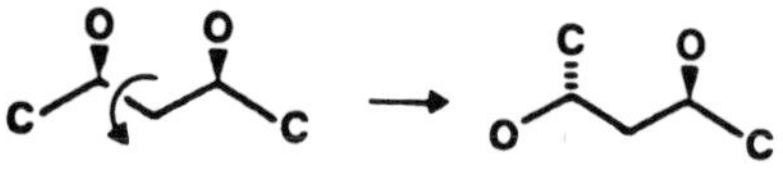

This leads to the generalisation that "the carbon chain adopts the extended, planar zig-zag when the configurations at alternate carbon centres are different (in the D or L sense), and is bent and non-planar when they are the same" (G.A. Jeffrey and H.S. Kim, Carbohydrate Res., 1970, 14, 207). The conformations of the three pentitols in the crystalline form illustrate this principle:

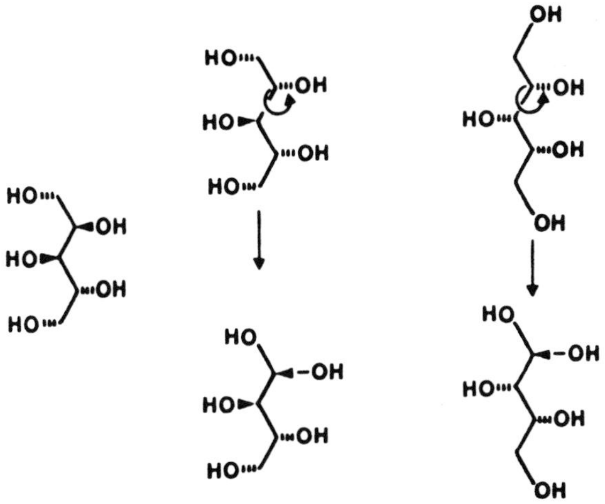

D-arabinitol ribitol xylitol

In the crystal the hexitols behave similarly, D-mannitol, which has the configurations LLDD at $C_{(2)}$ - $C_{(5)}$, has a planar carbon chain as does galactitol which also is without pairs of alternate carbon atoms with the same configurations (DLLD). Allitol (DDDD) suffers rotation around both the $C_{(2)}$, $C_{(3)}$ and $C_{(4)}$, $C_{(5)}$ bonds, D-altritol (LDDD) around bond $C_{(4)}$, $C_{(5)}$, D-glucitol (DLDD) around bond $C_{(2)}$, $C_{(3)}$, and D-iditol (LDLD) around bond $C_{(3)}$, $C_{(4)}$. This last rotation removes both unfavourable interactions concurrently.

^{1}H N.M.R. spectra of alditols in D_2O normally cannot be used to determine preferred conformations because of the overlapping of signals, but addition of lanthanide ions - especially europium - can remove this problem and lead to fully resolved spectra from which coupling constants can be measured. Because the equilibria established between the complexed and uncomplexed alditols are rapid on the N.M.R. time scale, weighted average spectra are obtained, and when the degree of complexing is small the spectra approximate to those of the uncomplexed species. By this method it was shown (S.J. Angyal, D. Greeves and J.A. Mills, Austral.J.Chem., 1974, 27, 1447) that the pentitols and hexitols adopt preferentially in aqueous solution the same conformations as are found in the crystalline forms. This must not depend to any great extent on specific hydrogen-bonding factors because the alditol acetates (with the exception of allitol hexa-acetate) adopt similar conformations in chloroform solution (S.J. Angyal, R.Le Fur and D. Gagnaire, Carbohydrate Res., 1972, 23, 121). The preferred conformations of higher alditols have been predicted (J.A. Mills, Austral.J.Chem., 1974, 27, 1433).

Polyhydroxy compounds complex with many metal cations in aqueous solution to extents which depend largely on the orientations of contiguous triols. Inositols and pyranoid compounds form strongest complexes with such ions when axial, equatorial, axial triols are present, and all alditols can adopt orientations with analogous triol structures (which contain destabilising 1,3-O/O interactions) by suitable rotations. The stabilities of the complexes formed depend upon the energies required to adopt these conformations and thus, compounds with contiguous *threo*, *threo*-triols develop no additional unfavourable intereactions and complex strongly, while *erythro*, *threo*-compounds develop an additional C/C *gauche* interaction and *erythro*, *erythro*-compounds are even less stable in the required conformation since eclipsed 1,3-C/C interactions are involved. These therefore form least stable complexes as evidence by, for example, the electrophoretic mobilities of

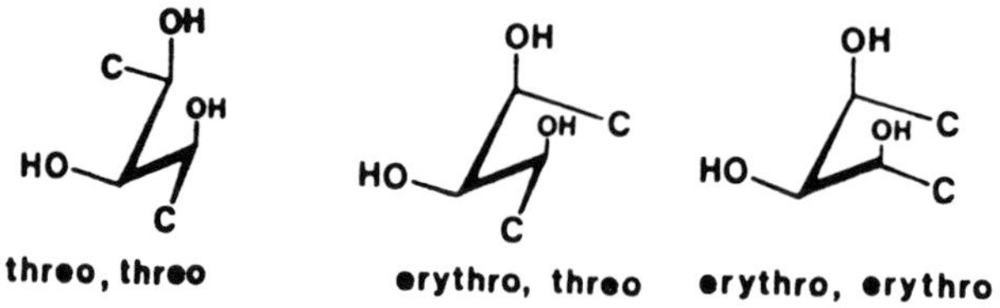

threo, threo **erythro, threo** **erythro, erythro**

ribitol and allitol in calcium acetate electrolyte. While these compounds have mobilities (relative to *cis*-inositol) of 0.08 and 0.09, xylitol and iditol (with "all-*trans*" structures) have mobilities of 0.18 and 0.24, respectively. Arabinitol and the other hexitols give intermediate values. Analogous conclusions regarding complexing are obtainable from

N.M.R. studies of europium ion complexes (Angyal, Greeves and Mills, *loc.cit.*; S.J. Angyal, Tetrahedron, 1974, 30, 1695).

Alditols can be separated on a preparative scale using their differential binding to cationic resins in the calcium, strontium or barium forms (J.K.N. Jones and R.A. Wall, Canad.J. Chem., 1960, 38, 2290).

Alditols are also separable as anionic complexes with polybasic acids. Stabilities of these complexes may also be dependent upon the above factors which lead to favoured reaction of compounds containing *threo*, *threo*-triols (Second Edition, p.41). The periodate ion forms complexes with *axial*, *equatorial*, *axial*-contiguous triols on 6-membered rings and, seemingly, also with allitol which can readily form stereochemically analogous systems. Some of the hexitol is recovered unchanged following treatment with 1 or 2 molar equivalents of periodate (L. Hough, P.N.S. Iyer and B.E. Stacey, Carbohydrate Res., 1976, 52, 228).

Alditols form complexes with molybdate which, in the cases of optically active compounds, give characteristic circular dichoism spectra with 3 or 4 bands in the 200-350 nm region (W. Voelter *et al.*, Ber., 1969, 102, 1005).

(c) Synthesis

Divinyl ketone offers access to all the pentitols (the ribitol and xylitol being *meso*-compounds, and the arabinitol being produced in the racemic form) (Scheme 1) (P. Chautemps, Compt.Rend., 1977, 284C, 807).

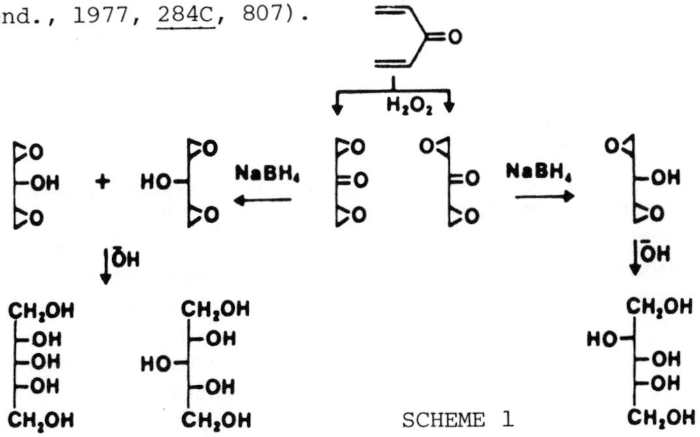

SCHEME 1

All of the methods applicable to the synthesis of free sugars in all of their modified forms have relevance since alditols result from reduction. Of special significance for the preparation of higher alditols are acetylenic compounds such as those which can be prepared by use of ethynyl Grignard reagents on dialdose derivatives (R. Hems, D. Horton and M. Nakadate, Carbohydrate Res., 1972, 25, 205). These can be appropriately converted into alditols having carbon chains extended by one or two atoms or into dimeric products (Scheme 2) (R.B. Roy and W.S. Chilton, J.org.Chem., 1971, 36, 3242).

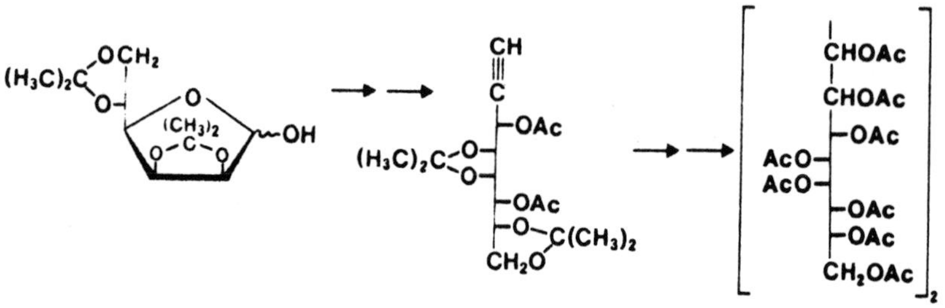

SCHEME 2

(d) Chromatographic separation

Since, on reduction, aldoses lose chirality at the anomeric centres and thus anomers give single products, alditols are frequently used for the chromatographic identification of sugars; in particular they are commonly applied in gas chromatographic and mass spectrometric studies - often as their trimethylsilyl or methyl ethers or acetyl or trifluoroacetyl esters. (G.G.S. Dutton, Adv.carbohydrate Chem.Biochem., 1973, 28, 11; 1974, 30, 9; J. Lönngren and S. Svensson, *ibid.*, 1974, 29, 41).

Unsubstituted alditols can be separated analytically or preparatively as cationic complexes (p. 4), and also as complexes of a variety of anionic species. Anionic resin columns allow the efficient separation of most alditols as borate complexes (K. Larsson and O. Samuelson, Carbohydrate Res., 1976, 50, 1).

2. Reactions and *O*-substituted derivatives

(a) Oxidation

Reaction of alditols with ferrous salts and excess of alkaline hydrogen peroxide (Fenton's reagent) causes almost complete conversion into formic acid following initial oxidation of the primary groups to give aldoses which then degrade to formic acid and the next lower aldoses by way of hydroperoxides (H.S. Isbell, H.L. Frush and E.W. Parks, *ibid.*, 1976, 51, C5). On the other hand, mercury(II) acetate in refluxing methanol causes preferential oxidation of the secondary groups, and from D-arabinitol, for example, the three possible pentuloses are produced together with products of further reaction (probably furan derivatives) (Scheme 3). Since ethylene glycol

SCHEME 3

oxidises to glycolaldehyde under these conditions the reaction is not altogether specific for secondary hydroxyl groups (L. Stankovič, K. Linek and M. Fedoroňko, *ibid.*, 1969, 10, 579).

Oxidations of alditols with bacterial enzyme systems can afford useful access to unusual sugars, L-psicose, for example, being obtainable from allitol by use of *Acetomonas oxydans* following the Bertrand-Hudson rule (L. Hough *et al.*, Phytochem., 1968, 7, 1). Some such reactions proceed further, and L-sorbose produced from D-glucitol by use of *Acetobactor suboxydans* is further oxidised with good specificity to D-*threo*-hexos-2,5-diulose (K. Sato *et al.*, Agric.biol.Chem. Japan, 1967, 31, 877). The catabolism of alditols by yeasts is discussed by J.A. Barnett, Adv.carbohydrate Chem.Biochem., 1976, 32, 125, and structural modifications which permit alditols to remain substrates for the dehydrogenase systems of *Acetobacter suboxydans* are noted by D.T. Williams and J.K.N. Jones (Canad.J.Chem., 1967, 45, 741). These may include the incorporation of single deoxy-, acetamidodeoxy-, S-ethylthio-, O-methyl- and *O*-acetyl-groups, and also 1,1-di(S-ethylthio)-groups (i.e. diethyl dithioacetals).

Attempts to carry out specific intramolecular oxidations (which would mimic both these biochemical reactions and D.M.S.O.-based alcohol oxidations) by use of 1-deoxy-1-ethylsulphinyl-L-galactitol were not successful - probably because steric factors prevented the basic centres in the ylide intermediates from abstracting the carbon-bonded protons of the alcohol groups (R.J. Ferrier and D.T. Williams, Carbohydrate Res., 1969, 10, 157).

(b) Cyclic acetals and ketals

J.F. Stoddart ("Stereochemistry of Carbohydrates", Wiley-Interscience, New York, 1971) has discussed at length the complex matters of stereochemistry arising with a range of alditol acetals, diacetals and triacetals.

A wide range of cyclic acetals containing 5-, 6- and 7-membered rings are formed by condensation of alditols with aldehydes or ketones, and their nature depends upon many factors including the stereochemistry of the alditol, the type of carbonyl compound used and the conditions of the reaction. In particular, products may vary markedly according to whether they are formed under kinetic or thermodynamic control. For example, D-glucitol reacts with aldehydes under acid conditions to give initially the 2,3-acetals which then rearrange to the 2,4-isomers, the products predicted by the Barker-Bourne rules. (T.G. Bonner *et al.*, J.chem.Soc.(B)., 1968, 827). 1-Deoxy-D-glucitol with butyraldehyde behaves similarly, the 2-deoxy-isomer gives the 1,3-substituted product and then the 3,4-isomer, and 3-*O*-methyl-D-glucitol reacts to give the 2,4-acetal throughout the reaction (*idem.*, *ibid.*, 1971, 957). In the case of 1-deoxy-D-galactitol, however, more substantial rearrangement occurs and the initially formed 2,3-*O*-butylidene acetals (pair of diastereoisomers) give way mainly to the 4,6-acetal (*idem.*, J.chem.Soc.,Perkin I, 1975, 1323). Rearrangement can also occur during further substitution reactions, and for example, 2,3:4,5-di-*O*-benzylidene-galactitol, on treatment with benzaldehyde and zinc chloride, affords the 1,3:2,4:5,6-triacetal (R. Bonn and I. Dyong, Ber., 1972, 105, 3833 who emphasise the thermodynamic basis of the Barker-Bourne rules).

Reactions can be very complex, the acetonation of D-glucitol giving mixtures of 1,2-, 3,4-, 1,2:3,4-, 3,4:5,6-, 1,2:5,6- and 1,2:3,4:5,6-substituted products (T.G. Bonner *et al.*,

Carbohydrate Res., 1972, 21, 29), and further complications arise in the case of products derived with unsymmetrical carbonyl compounds. The three isomeric forms of 1,2:5,6-bis-O-(trifluoromethyl)ethylidene-D-mannitol (1)-(3) have been separated and characterised by ^{1}H N.M.R. methods (T.D. Inch and R.V. Ley, *ibid.*, 1970, 14, 95), and related studies have

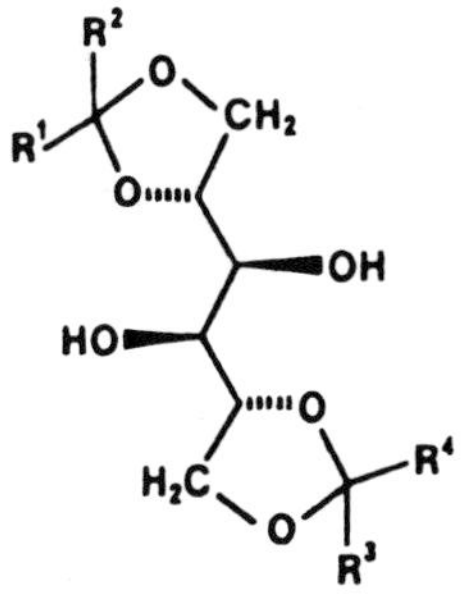

	CH_3	CF_3
(1)	R^1, R^3	R^2, R^4
(2)	R^1, R^4	R^2, R^3
(3)	R^2, R^4	R^1, R^3

been carried out on isomeric 1,2:5,6-di-*O*-bromoethylidene-D-mannitols (H.B. Sinclair, J.org.Chem., 1968, 33, 3714) and on the 1,2-*O*-substituted compound (*idem.ibid.*, 1969, 34, 3845).

Many developments have taken place in the synthesis of specific acetals, and derivatives of D-mannitol have received particular attention. Treatment of 1,6-disulphonate esters with potassium hydrogen carbonate in DMSO affords the 1,2:5,6-dicarbonate ester which, treated with acetone in the presence of acid, and then with base, gives the 3,4-*O*-isopropylidene compound (G. Hanisch and G. Henseke, Ber, 1967, 100, 3225). Direct acid-catalysed acetonation of the hexitol using both acetone and 2,2-dimethoxypropane affords a simplified synthesis of the 1,2:5,6-diacetal (G. Kohan and G. Just, Synthesis, 1974, 192), and 1,3:4,6-di-*O*-benzylidene-D-mannitol is obtainable (also in modest yield, but directly) using benzaldehyde in acidified DMSO (H.B. Sinclair, Carbohydrate Res., 1970, 12, 150).

Mixed acetals are of value for specific syntheses, the removal of the aromatic acetal of 3,4-*O*-benzylidene-1,6-dideoxy-1,6-difluoro-2,5-*O*-methylene-D-mannitol allowing a route to 1-deoxy-1-fluoro-L-glycerol (W.J. Lloyd and R. Harrison, *ibid.*, 1973, 26, 91).

Condensation between formaldehyde and erythritol illustrates further the potential complexity of these reactions: use of formaldehyde leads to the 1,3-acetal, the 1,2:3,4-, 1,3:2,4- and 1,4:2,3-diacetals and also to a dimer containing a 12-membered ring produced by intermolecular acetalation of two molecules of the 1,3-*O*-methylene compound (R.B. Jensen *et al.*, Acta chem.Scand., 1975, B29, 373). The relative thermodynamic stabilities of the 1,3:2,4- and 1,4:2,3-diacetals have been measured (the former is favoured) and discussed in terms of destabilising interactions between *gauche*-related vicinal oxygen atoms (I.J. Burden and J.F. Stoddart, J.chem.Soc.Perkin I, 1975, 666).

A new procedure for synthesising cyclic methylene acetals involves warming appropriate hydroxy compounds in DMSO in the presence of *N*-bromosuccinimide. Applied to 1,6-di-*O*-benzoyl-D-mannitol the method gives the 2,3:4,5-di-*O*-methylene acetal in good yield (S. Hanessian, P. Lavallee and A.G. Pernet, Carbohydrate Res., 1972, 26, 258).

Partially substituted acetal derivatives provide the normal routes to specifically substituted alditol ethers and esters. They can also be used in the synthesis of uloses, 1,2:4,5:6,7-tri-*O*-isopropylidene-D-*glycero*-D-*galacto*-heptitol yielding D-*gluco*-hept-3-ulose on oxidation of the alcohol group at $C_{(3)}$ followed by hydrolysis (T. Okuda, S. Saito and Y. Shiobara, *ibid.*, 1975, 39, 237).

(c) Esters

A range of boronate and borinate esters of alditols are now known (R.J. Ferrier, Adv.carbohydrate Chem.Biochem., 1978, 35, 31), and the former can be prepared by condensation of alditols with boronic acids or by heating with trialkyl- or triaryl-boranes or by thermolysis of per(dialkylborinyl) esters. In the case of D-mannitol, the hexakis(diethylborinyl) compound gives initially the 3,4-ethylboronate tetraborinate and then the 1,2:3,4:5,6-tris(ethylboronate) (W.V. Dahlhoff, W. Schüssler and R. Köster, Ann., 1976, 387). Selective de-esterification can be effected with such compounds which makes them of value for the preparation of specifically substituted compounds e.g. 1,6-diesters of galactitol (W.V. Dahlhoff and R. Köster, J.org.Chem., 1976, 41, 2316). The structures of alditol boronate esters have been assigned using mass spectrometric methods (C.J. Griffiths, I.R. McKinley and

H. Weigel, Carbohydrate Res., 1979, 72, 35).

(d) Ethers

Different possible approaches to the synthesis of specific alditol ethers is illustrated by the preparation of 1-*O*-methyl-D-mannitol from the 1,2-anhydro-3,4:5,6-di-*O*-isopropylidene derivative, the 2-ether from 1,2,6-tri-*O*-benzoyl-3,4-*O*-isopropylidene-D-mannitol and the 3-substituted isomer from 3-*O*-methyl-D-fructose by reduction (S. Bayne *et al.*, J.chem. Soc.(C), 1967, 114).

Reciprocal linking of the ends of pairs of acyclic compounds gives macrocyclic products. Asymmetric oxygenated precursors linked through ether bonds afford macrocycles which have similarities to the crown ethers, and their asymmetry affords them the possibility of exhibiting chiral recognition. Some alditol derivatives have the required characteristics, and Stoddart and collaborators have utilised these in notable studies of the selective co-ordination of resolvable ammonium salts (J.F. Stoddart, Chem.Soc.Rev., 1979, 8, 85). The synthesis of a chiral 18-crown-6 derivative from 1,2:5,6-di-*O*-isopropylidene-D-mannitol is illustrated in Scheme 4.

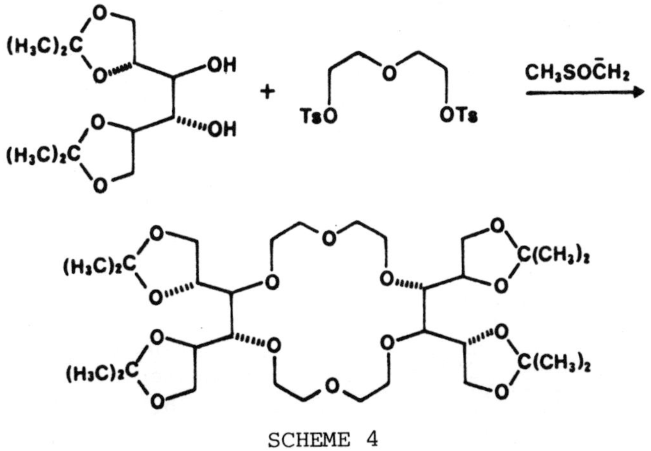

SCHEME 4

3. Anhydroalditols

The synthesis, the spectroscopic and other physical properties, the reactions and the industrial and biological uses of monodi- and tri-anhydroalditols have been reviewed (S. Soltzberg, Adv.carbohydrate Chem.Biochem., 1970, 25, 229).

The dehydration of the tetritols and pentitols which occurs on extended heating in acid solution gives 5-membered cyclic anhydrides as the chief products, and the rates of reaction of isomers are significantly different. Arabinitol, ribitol and xylitol have relative rates of 1:5:3 which are attributable to different interactions between C-O dipoles in the transition states. These involve normal non-bonded *cis*-interactions, but also an unfavourable dipolar effect exists when the hydroxyl group adjacent to the leaving group is *quasi*-equatorial. While ribitol and xylitol, because of their symmetry, each give one racemic 1,4-anhydride, D-arabinitol gives two related anhydrides, i.e. 1,4-anhydro-D-arabinitol and -D-lyxitol, and these constitute 70% of the reaction products (Scheme 5). A small proportion of the pyranoid 1,5-anhydride

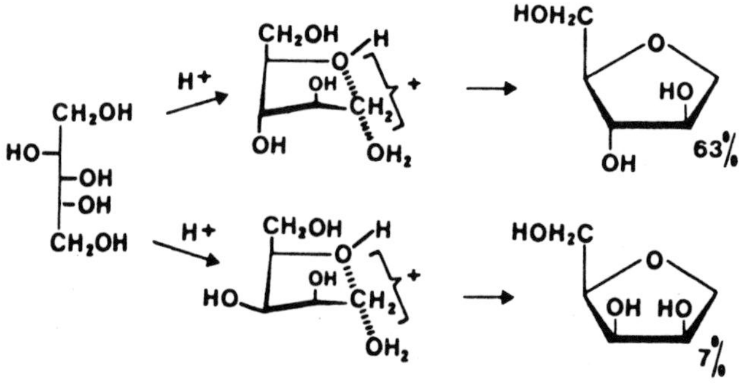

SCHEME 5

(5%) is also present as are isomeric 1,4-anhydrides which are formed by way of terminal epoxides (B.G. Hudson and R. Barker, J.org.Chem., 1967, 32, 3650). Likewise, the hexitols give mainly tetrahydrofuran compounds, the ease of formation of 1,4-anhydrides decreasing in the order allitol, talitol, iditol, glucitol, altritol, galactitol, gulitol and mannitol. Under the conditions of the dehydrations those anhydrides with the *cis*-relationship between the hydroxyl substituents at $C_{(3)}$ and the dihydroxyethyl substituents at $C_{(4)}$ react further to give 1,4:3,6-anhydrides, the iditol and gulitol compounds reacting 40 times faster than 1,4-anhydro-glucitol and -mannitol. This is because, with the latter pair, in the transition state leading to the bonding of $O_{(3)}$ to $C_{(6)}$, the

$C_{(5)}-O_{(5)}$, $C_{(4)}-O_{(4)}$ bonds become eclipsed, whereas, with the other pair, $O_{(5)}$ is far removed from the ring oxygen atom [$O_{(4)}$]. In the case of D-mannitol, which is the most resistant of the hexitols, considerable proportions of 2,5-anhydride are formed, and the 1,5-anhydride is also produced (R. Barker, J. org.Chem., 1970, 35, 461).

Erythritol, as well as giving the 1,4-anhydro-compound on heating in strong sulphuric acid, also gives an intermolecular dimeric anhydride [*trans*-2,5-di(1,2-dihydroxyethyl)-1,4-dioxane] (Scheme 6) (A.H. Haines and A.G. Wells, Carbohydrate Res., 1973, 27, 261).

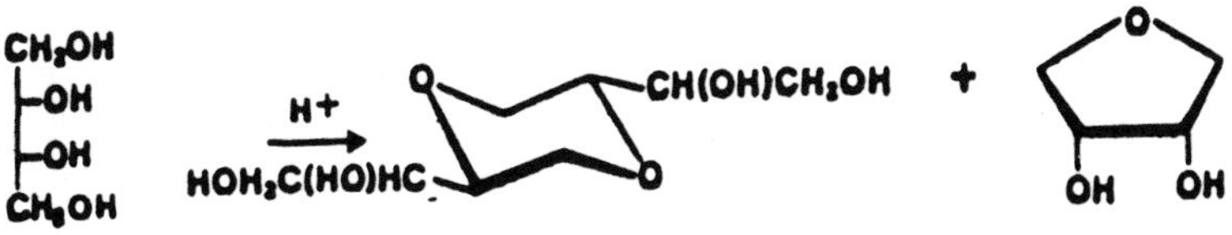

SCHEME 6

The use of fuming hydrochloric acid or hydrogen bromide leads to the production of halogenated products as well as direct anhydrides (Scheme 7) (J.M. Ballard and B.E. Stacey, *ibid.*, 1973, 30, 83).

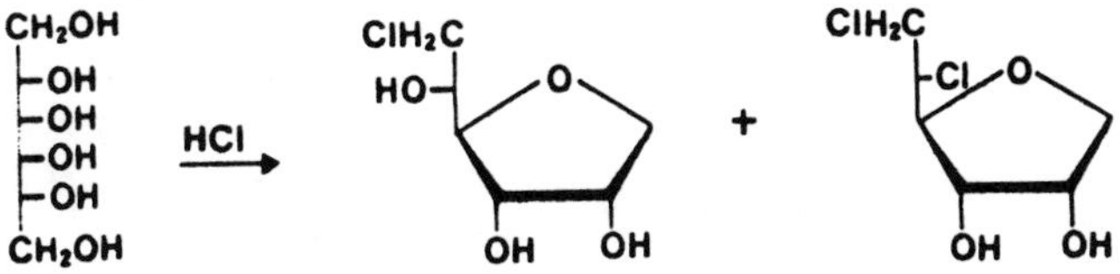

SCHEME 7

As in the case of dialkyl dithioacetals of aldoses selective sulphonylation of the primary hydroxyl groups of alditols can lead to anhydro-ring formation. Treatment of erythritol, L-arabinitol and D-mannitol with two equivalents of 2,4,6-tri-isopropylbenzenesulphonyl chloride in pyridine leads mainly to the terminal diesters, but ribitol, xylitol, D-glucitol and D-galactitol give large proportions of 1,4-anhydro-5- or 6-esters amongst their products (M. Mort, B.E. Stacey and B. Tierney, *ibid.*, 1975, 43, 183).

An alternative and specific route to 1,4- and 1,5-anhydroalditols involves the reduction of corresponding acylated glycosyl halides. If carried out with lithium aluminium hydride the reaction also causes reductive removal of the ester groups but the use of tributyltin hydride with 2,2-azobisisobutyronitrile as radical initiator, allows specific reduction at the anomeric centres (J. Augé and S. David, *ibid.*, 1977, 59, 255).

Recognition that 1,2:5,6-dianhydro-D-mannitol has potent inhibitory effects on tumour cells (M. Jarman and W.C.J. Ross, Chem. and Ind., 1967, 1789) has led to specific interest in alditol epoxides. Terminal diesters or dihalides, under very mild conditions, give 1,2:5,6-dianhydrides from hexitols. These may not, however, be the final products under stronger alkaline conditions (Scheme 8) (*idem.*, Carbohydrate Res., 1969, 9, 139; R.S. Tipson and A. Cohen, *ibid.*, 1968, 7, 232).

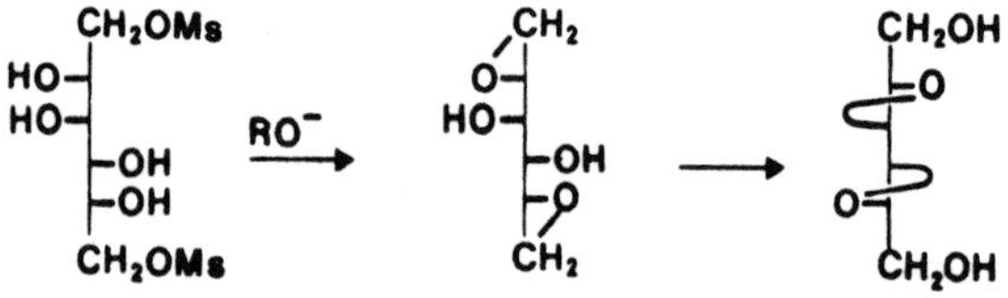

SCHEME 8

Nucleophilic halide opening of the terminal diepoxides gives terminal dihalides which, as well as the disulphonates, also display cytostatic activity because they revert to the epoxides under biological conditions (J. Kuszmann and L. Vargha, *ibid.*, 1971, 17, 309). When 1,6-dichloro-1,6-dideoxy-D-mannitol is treated with sodium methoxide a dianhydride is produced which has been characterised as the unusual 2,5:3,6-dianhydro-D-glucitol (*idem.*, *ibid.*, 1968, 8, 157). If it has this structure it could have been formed via the 1,2:5,6-diepoxide but not then via the 2,3:4,5-isomer.

2,3:4,5-Dianhydrides also result from base treatment of 3,4-disulphonates; 3,4-di-*O*-mesyl-D-mannitol in this way gives 2,3:4,5-dianhydro-D-iditol, but under acidic conditions the 1,4:3,6-dianhydride of this alditol is produced (Scheme 9) (D.R. Hicks and B. Fraser-Reid, Canad.J.Chem., 1974, 52, 3367).

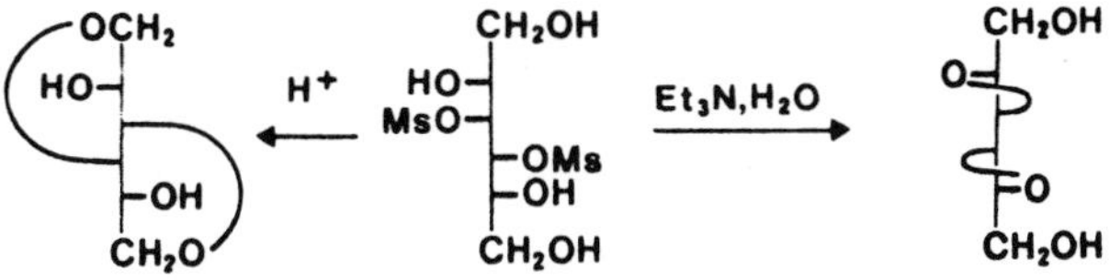

SCHEME 9

3,4-Anhydro-D-altritol with alkali gives 1,4-anhydro-D-altritol and 1,5-anhydro-L-glucitol by way of 2,3-anhydro-D-iditol (Scheme 10) (J.G. Buchanan and A.R. Edgar, Carbohydrate Res., 1969, 10, 295).

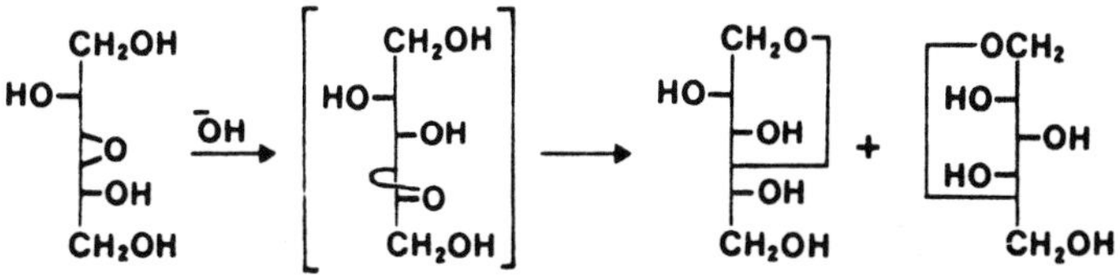

SCHEME 10

4. Aminodeoxyalditols

The sorbistins are a new class of aminoglycoside antibiotics which consist of 1,4-diamino-1,4-dideoxy-D-glucitol substituted with a 4-amino-4-deoxy-α-D-glucopyranosyl derivative at $O_{(3)}$ (T. Ogawa, K. Katano and M. Matsui, Tetrahedron, 1980, 36, 2727).

1-Amino-1-deoxyalditols can be prepared by hydrogenation of aldose oximes or 1-deoxy-1-nitroalditols, and, in these ways, the heptitol compounds with the D-*glycero*-D-*gulo*-, D-*glycero*-D-*galacto*- and D-*glycero*-L-*manno*-configurations have been prepared. Deamination results in mixtures of heptitols and anhydro-compounds and the second of these gives a crystalline ulose on enzymic oxidation with *Acetobacter suboxydans*. Presumably oxidation occurs at $C_{(6)}$ (H.J.F. Angus and N.K. Richtmyer, Carbohydrate Res., 1967, 4, 7).

5,6-Epimino-derivatives of L-iditol are obtained from D-glucitol 5,6-disulphonates by selective displacements of the primary ester groups with azide followed by reduction with lithium aluminium hydride. Nucleophilic ring opening occurs at the terminal position to give access to 5-amino-5-deoxy-L-iditol compounds (Scheme 11) (A.D. Barford and A.C. Richardson, *ibid.*, 1970, 14, 217). Related studies have been carried out

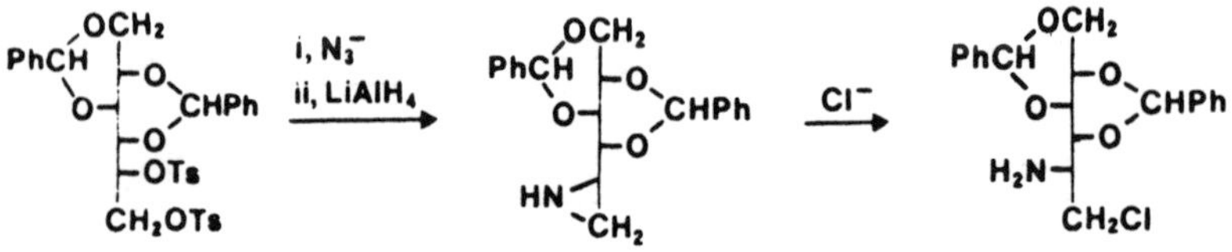

SCHEME 11

on 3,4-epimino-D-iditol derivatives (*idem.ibid.*, 1970, 14, 231).

Aminodeoxyalditols which have suitably sited good leaving groups readily form *N*-heterocyclic compounds, and 2-amino-2-deoxy-L-arabinitol 5-sulphonates react in this way despite the fact that arabinose derivatives are resistant to such ring closures (Scheme 12) (S.D. Gero *et al.*, *ibid.*, 1970, 15, 322; 1972, 24, 474).

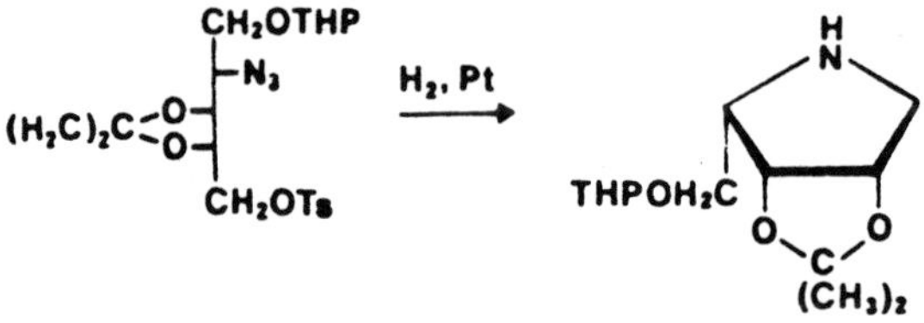

SCHEME 12

5. Thioalditols

1-Thio-L-xylitol is obtainable from 1,2-*O*-isopropylidene-5-*O*-tosyl-D-xylose by nucleophilic displacement with benzylthiolate followed by removal of the acetal group and then reduction. The compound oxidises to give a disulphide (1), and also affords a gold derivative (2) (P. Wirz, J. Staněk and E. Hardegger, Helv., 1971, 54, 2027).

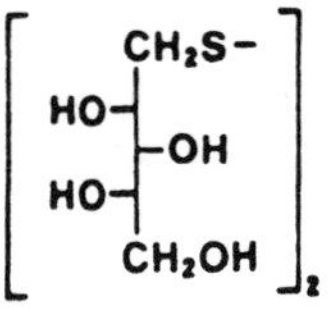

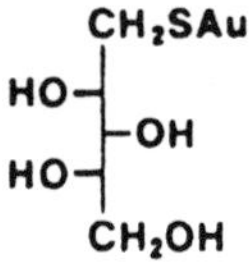

(1) (2)

The first tetrathiohexitol was made as outlined in Scheme 13 and is appreciably more soluble than normal hexitols, being crystallised from isopropyl ether (G.E. McCasland, A.B. Zanlungo and L.J. Durham, J.org.Chem., 1974, 39, 1462).

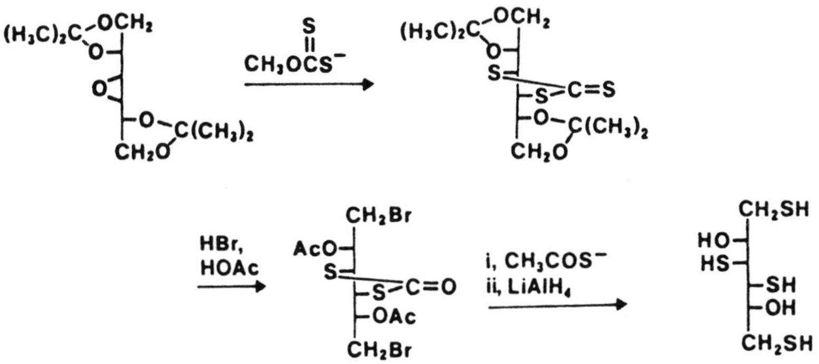

SCHEME 13

The isomeric D-*ido*-trithiocarbonate which is also obtained from the initial epoxide gives access to 1,3,4,6-tetrathio-D-iditol (*idem.ibid.*, 1976, 41, 1125).

1,2:5,6-Diepithio-D-mannitol and -glucitol, obtainable from diepoxides, are less active than the corresponding diepoxides as antitumour agents (J. Kuszmann and L. Vargha, Acta chim. Acad.Sci.Hung., 1970, 66, 307).

Pummerer rearrangement of the isomeric 1,4:3,6-di(thioanhydro)-D-iditol sulphoxides, induced by heating in acetic anhydride, did not give the 1,6-diacetoxy analogues as expected, but rather bicyclic thieno-[3,2-b]thiophene derivatives (J. Kuszmann, P. Sohár and G. Horváth (Tetrahedron, 1971, 27, 5055).

Acknowledgments

Dr. Richard Furneaux is thanked for reading the manuscript and for his helpful criticism. Ann Beattie's and Eric Stevens' excellent typing and drawing are also gratefully acknowledged.

Chapter 23

MONOSACCHARIDES AND THEIR DERIVATIVES

R.J. FERRIER

1. Monosaccharides

The following survey is based directly upon the very thorough treatment provided by Hough and Richardson in the Second Edition, but makes no effort to deal with the various aspects of the subject in parallel fashion. On the contrary, emphasis is placed on the growth areas, so that, for example, under "Nuclear magnetic resonance spectroscopy", scant attention is paid to proton spectroscopy, the basis of which was well laid by the previous authors, but ^{13}C spectroscopy is dealt with in some detail since its development has taken place largely within the intervening years.

A notable event occurred in the documentation of monosaccharide chemistry during the year of the publication of the Second Edition (1967) with the appearance of the first of the Chemical Society's "Specialist Periodical Reports - Carbohydrate Chemistry". This publication has been produced since then on an annual basis, each volume aiming to record and abstract every contribution made to the subject over a twelve month period.

No treatment will be given of the important branch of monosaccharide chemistry represented by the nucleosides, although it is recognised that this has led to the overlooking of much significant material.

During the period under review the chemistry of the sugars has been expanded very substantially and has become more integrated with general organic chemistry following the observations, for example, that sugar derivatives are unusually suited to study by ^{1}H N.M.R. spectroscopy and are consequently appropriate for use in general spectroscopy and in conformational analysis studies. Many non-carbohydrate natural products have now been synthesised in optically pure form consequent upon the recognition of the suitability of monosaccharides as chiral starting materials (B. Fraser-Reid and R.C. Anderson, Progress Chem.org.nat.Prods., 1980, 39, 1; S. Hanessian, Acc.chem.Res., 1979, 12, 159). Such has been the rate of progress and the volume of output in carbohydrate chemistry that this survey is necessarily highly selective, and in many ways it does not fully supplement the Second Edition. Nevertheless, it contains much new information and evidence of a rapidly expanding subject.

(a) Conformational analysis of the monosaccharides

(i) General

Rules for the description of the idealised regular forms of pyranoid rings [chairs (C), boats (B), skews (S) and half-chairs (H)] and of furanoid rings [envelopes (E) and twists (T)] are now in general use (J.C.P. Schwarz, Chem.Comm., 1973, 505; J.chem.Soc.Perkin I, 1973, 1913), and the method upon which they are based has replaced earlier conventions. Reference planes are selected which contain the maximum number of ring atoms (four for all the above forms except the twist conformations of furanoid rings), and the sides of these planes are distinguished according to whether the ring atom numbering appears clockwise (e.g. the "β-side" of the D-hexopyranoses) or anticlockwise. Out-of-plane locants which lie on the former side are then used as superscripts which precede the ring symbols; those which lie on the other side become subscripts which follow. Examples are shown below.

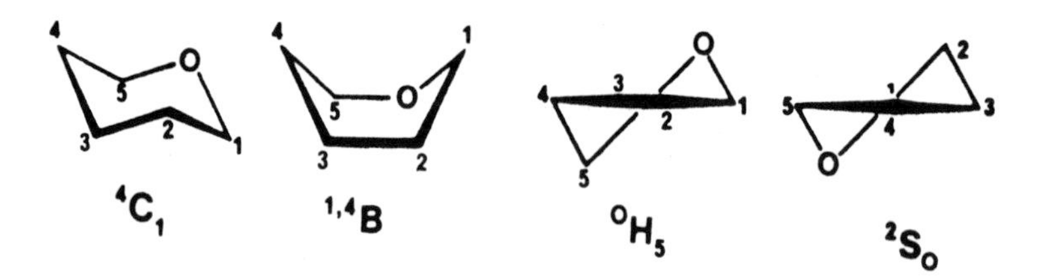

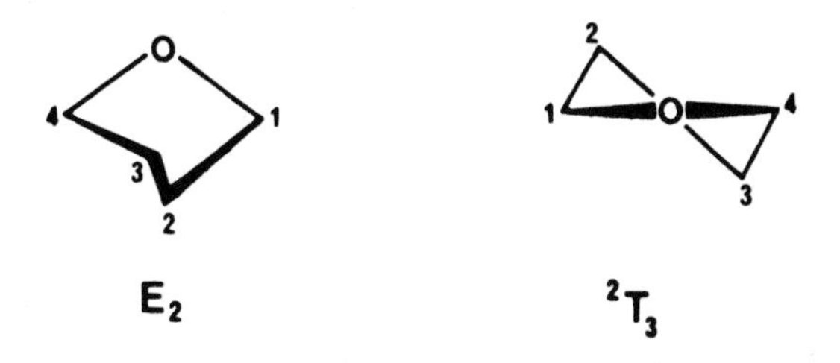

Except in the cases of rigid molecules, "preferred conformation" of a carbohydrate must now be taken to mean the time-averaged shape adopted by the molecules as they oscillate between different conformations. This follows from the observation (see below) that, at low temperatures, different conformations of pyranoid compounds can be "frozen out" and recognised by N.M.R. methods. When more than one conformation is energetically favoured relative to all others (for example the chair forms of pyranoid rings) the equilibrium between them can be considered, and N.M.R. data can be used to determine equilibrium constants, and, in principle, should be employed for this purpose rather than for the assignment of particular conformations. Nevertheless, such data often allow the characterisation of the geometric states in which compounds spend most time i.e. their energy minima. Particularly this is so for pyranoid compounds for which the energy differences between conformations is often large. In the cases of furanoid, and particularly acyclic, compounds, however, these energy differences are often small. Furthermore, the energy barriers between interconvertible conformations are small, and so they tend to be particularly flexible and conformationally non-specific. The concept of time-averaged, preferred conformations is however still valid, and these can often be determined.

Factors upon which relative stabilities of conformations depend can be steric or electronic in character, the most significant of the former kind in 6-membered cyclic compounds being the interactions which occur between *syn*-related axial substituent groups. On the other hand, the most significant electronic-based factor is the "anomeric effect" which relates to the instability caused by the presence of an equatorial electronegative group at the anomeric centre. In the cases of pyranoid compounds with axial substituents at the adjacent carbon atoms this factor is of particular significance (previously the Δ2-instability factor). Both types of influence apply in the cases of furanoid compounds; with acyclic derivatives the interactions between non-bonded groups are preeminently important.

It will be emphasised below that the quantitive assessment of all these factors is significantly subject to the detailed conditions of measurement; thus the anomeric effect varies with the nature of the anomeric substituent, the nature and stereochemistry of the ring substituents and, very significantly, with the solvent. Likewise, interactions between oxygen-bonded substituent groups are highly dependent upon the nature of the groups and again upon the solvent.

General aspects of conformational analysis have been covered in detail by J.F. Stoddart, "Stereochemistry of Carbohydrates" (Wiley-Interscience, New York, 1971) and P.L. Durette and D. Horton (Adv.carbohydrate Chem.Biochem., 1971, 26, 49). R.U. Lemieux (Pure appl.Chem., 1971, 25, 527; 27, 527) has reviewed subtler and newer aspects associated with solvation and the preferred orientation of substituent groups the investigations of which have largely been pursued in his laboratory.

(ii) Pyranoid rings

It is now well established that electronegative groups adjacent to ring oxygen atoms on saturated 6-membered rings show strong preference for the axial orientation, and that this effect is of dominant significance in determining both the position of the equilibrium between interconvertible anomers and the ring conformation adopted by compounds possessing this structural feature. Thus, for example, tri-*O*-acetyl-α-D-xylopyranosyl chloride (1) is more stable than the β-anomer, but both compounds exist in chloroform solution predominantly in the conformation with the chlorine axial (P.L. Durette and D. Horton, Carbohydrate Res., 1971, 18, 57). In the latter case the effect is strong enough to overcome the instabilities introduced into the molecule by interactions between the *syn*-axial substituents. Much debate and experimental and theoretical study have been devoted to the phenomenon (R.U. Lemieux and S. Koto, Tetrahedron, 1974, 30, 1933), and an early rationalisation attributed it to the greater

(1)

electrostatic repulsion occurring between an equatorial electronegative anomeric substituent (X_E) and the unshared electron pairs on the ring oxygen atom than is the case for an axial substituent (X_A).

$C_{(1)}$-$O_{(5)}$ projection of a pyranoid ring in the $^4C_1$ conformation

Considerable experimental evidence adds strength to this interpretation of the anomeric effect, but an alternative explanation suggests that axial anomers are stabilised by interaction between the ring oxygen lone pairs and antibonding σ^* orbitals of the $C_{(1)}$-X bonds. This would lead to lengthening of these bonds and shortening of $C_{(1)}$-$O_{(5)}$ bonds in axial anomers relative to equatorial anomers, and this has been observed in several X-ray diffraction analyses (G.A. Jeffrey, J.A. Pople and L. Radom, Carbohydrate Res., 1972, 25, 117).

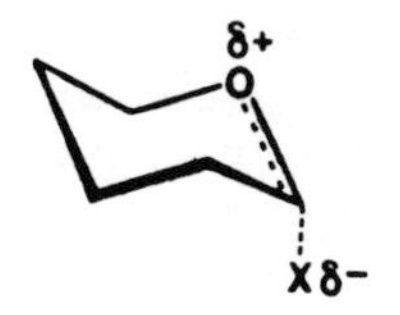

Groups attached to anomeric centres through electropositive atoms are subject to the "reverse anomeric effect" (R.U.

Lemieux, Pure appl.Chem., 1971, 25, 527; W.F. Bailey and E.L. Eliel, J.Amer.chem.Soc., 1974, 96, 1798) which preferentially stabilises configurations and conformations with equatorial anomeric substituents. Thus *N*-α-D-glucopyranosylimidazole tetra-acetate (2) adopts the $^{4}C_{1}$ chair conformation, but the *N*-methyl analogue (3), with the dipole of the $C_{(1)}$-N bond reversed, has small ring proton-proton coupling constants indicating that it prefers the inverted chair (Lemieux, *loc. cit.*)

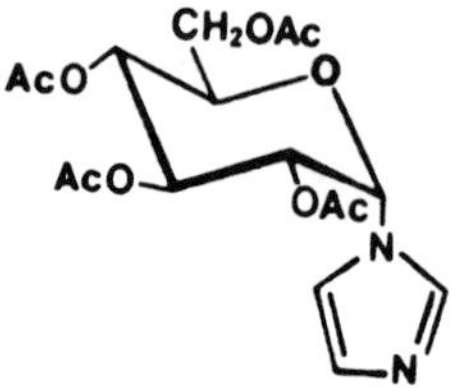

(2) (3)

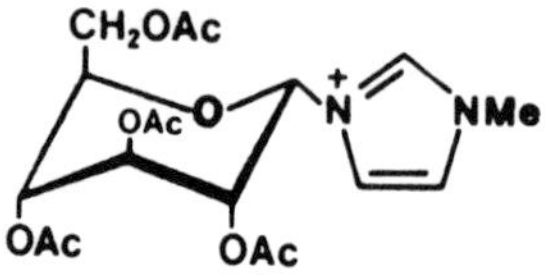

	(2)	(3)
$J_{2,3}$	10.4 Hz	6.5 Hz
$J_{3,4}$	9.2 Hz	∿5.5 Hz
$J_{4,5}$	9.6 Hz	∿6.5 Hz

That a small reverse anomeric effect will also apply in the case of the former compound is, however, suggested by the fact that 1-(tri-*O*-acetyl-α-D-xylopyranosyl)imidazole exists in the crystal, and to a considerable extent in solution, in the $^{1}C_{4}$ conformation (P. Luger, G. Kothe and H. Paulsen, Ber., 1974, 107, 2626).

In non-polar solvents 2-methoxytetrahydropyran exists predominantly in the chair conformation with the methoxy group axial consequent upon the dominant influence of the anomeric effect. This influence is reduced in solvents of high dielectric constant, as would be expected for a dipolar factor, and the compound exists as equal proportions of both chair forms. With methyl 3-deoxy-β-L-*erythro*-pentopyranoside (4), a compound which is similarly conformationally unstable, this effect is enhanced by the solvent dependence of the mutual interactions between the hydroxyl groups. In non-polar solvents intramolecular hydrogen bonding between them stabilises the $^{4}C_{1}$ chair form, whereas in polar, hydrogen bonding solvents, solvation not only removes this attractive interaction, but increases the electron densities on the oxygen atoms and thereby the repulsive interactions between them. The $^{1}C_{4}$ chair is consequently favoured. (R.U. Lemieux and A.A. Pavia, Canad.J.Chem., 1969, 47, 4441). On the other hand, the diacetate of the diol

(4)

favoured in non polar solvents   favoured in polar solvents

adopts the $^{4}C_{1}$ conformation in all solvents, whereas the analogous dimethyl ether exists mainly in the all-equatorial chair form regardless of solvent, and these significant observations indicate that the electron densities on the ring oxygen atoms determine the repulsive interaction between them. With the methoxy groups the densities and the consequent interactions are high; with the esters, the opposite is the case, and they assume the *syn*-diaxial relationship.

It must be concluded, therefore, that the conformational preference of an oxygen atom bonded to a six-membered ring which carries other such groups cannot be considered without reference to the nature of the substituent on the atom, and that electron-withdrawing substituents diminish substantially the propensities of the groups to be equatorial. Consistent with this, many compounds [for example (5)-(7)] adopt in solution conformations in which favourable anomeric effects outweigh the influences of several axial ester groups, and

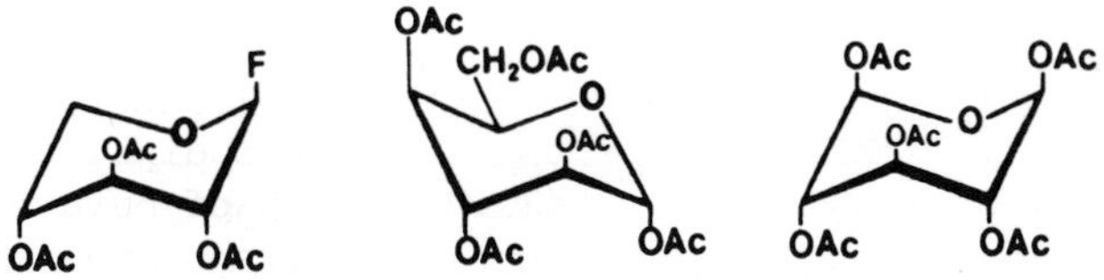

(5)   (6)   (7)

some such compounds, 1,2,3,4-tetra-*O*-benzoyl-β-D-xylopyranose, for example, (H. Paulsen et al., Carbohydrate Res., 1979, 68,

207) also crystallise in the "all-axial" form.

When the ^{1}H N.M.R. spectrum of tetra-*O*-acetyl-β-D-ribopyranose (8) in acetone-d_6 is measured at -84°C (220 MHz), the normal spectrum is resolved into those of the $^{1}C_4$ and $^{4}C_1$ chair conformations in the ratio 2:1 (P.L. Durette, D. Horton and N.S. Bhacca, Carbohydrate Res., 1969, 10, 565) from which it is generally concluded that most pyranoid compounds at room temperature are in conformational equilibrium. From the coalescence temperature it is calculated that the rate of ring inversion corresponds to a free energy of activation of *ca.* 42 kJ mole^{-1}, which is similar to the value for tetrahydropyran. Thus the effects of the ring substituents are not large.

Often, conformational equilibria are such that the spectra of minor components cannot be observed even at low temperatures, but compositions can be calculated from N.M.R. parameters. Coupling constants in time-averaged spectra when related to values for the discrete chair conformations provide this data, so that a $J_{1,2}$ value of 4.8 Hz for this ester measured at 20°C can be related to values of *ca.* 1 Hz and 8.0 Hz for the equatorial and axial-axial proton pairs of the

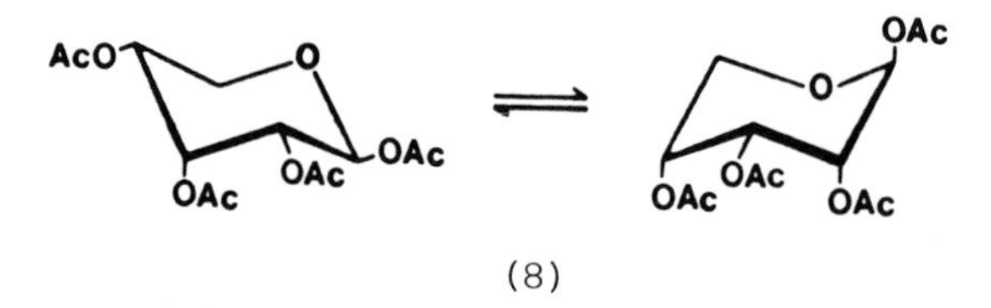

(8)

$^{1}C_4$ and $^{4}C_1$ chairs, respectively. These data give an equilibrium constant for 20°C of 1.2 in favour of the $^{1}C_4$ chair, whereas 2.0 is the constant for equilibrium at -84°C determined from relative intensities of resonances. Horton and his colleagues (P.L. Durette and D. Horton, Adv.carbohydrate Chem. Biochem., 1971, 26, 49) have carried out extensive studies of $^{4}C_1$, $^{1}C_4$ equilibria particularly of pentopyranose derivatives, (Table 1) from which, for example, the importance of the anomeric effect is apparent - particularly for compounds with strongly electronegative anomeric substituents.

TABLE 1

Percentages of $^4C_1$ conformations of D-pentopyranose derivatives in solution at 31°C

Configuration	Methyl tri-*O*-acetyl-pyranosides	Tri-*O*-acetyl-glycosyl acetates	Tri-*O*-acetyl-glycosyl chlorides
α-Ribo	65[a]	77[a]	-
β-Ribo	39[a]	43[a]	6[b]
α-Arabino	17[b]	21[b]	-
β-Arabino	3[a]	4[a]	2[b]
α-Xylo	>98[a]	>98[a]	>98[b]
β-Xylo	81[a]	72[a]	21[b]
α-Lyxo	83[a]	71[a]	91[b]
β-Lyxo	58[a]	39[a]	-

[a]Measured in $(CH_3)_2CO-d_6$; [b]measured in $CDCl_3$

(iii) Furanoid rings

Since the energy barriers between the envelope and twist conformations of tetrahydrofuranyl rings are small (*ca.* 15 kJ $mole^{-1}$) relative to the energy required for conformational inversion of pyranoid compounds, it follows that furanose derivatives in solution do not adopt specific conformations, but undergo rapid conformational change. N.M.R. evidence therefore provides information on time-averaged conformations which, nevertheless, represent energy minima. Vicinal ^1H coupling constants have been used for conformational analysis on several occasions, thermodynamic procedures have been applied with the methyl pentofuranosides (C.T. Bishop and F.P. Cooper, Canad.J.Chem., 1963, 41, 2743), but ^{13}C N.M.R. spectroscopic methods have proved to be most valuable. N. Cyr and A.S. Perlin, (Canad.J.Chem., 1979, 57, 2504), using

$^1J_{C(1),H(1)}$, $^2J_{C(1),H(2)}$, $^3J_{C(1),H(3)}$, $^3J_{C(1),H(4)}$,

$^{2}J_{C(2),H(1)}$, $^{2}J_{C(2),H(3)}$ and $^{3}J_{C(2),H(4)}$ values,

have concluded that the following are the conformations preferred by the methyl D-aldopentofuranosides in aqueous solution: α-riboside, E_1; β-riboside, ^{1}E; α-arabinoside, E_0 or ^{4}E; β-arabinoside, E_3 or ^{4}E; α-xyloside, ^{2}E; β-xyloside, E_2 or ^{3}E; α-lyxoside, E_4, and these conclusions are in many instances similar to others drawn for other furanoside derivatives and measured under different conditions. With small energy differences between conformations, this need not have been so. These results also show, in particular, that anomeric bonds tend to be *quasi*-axial (as would be anticipated through the application of the anomeric effect which would, however, be somewhat modified by the geometric constraints of the 5-membered ring), and that the largest ring substituent ($C_{(5)}$) favours the *quasi*-equatorial orientation in which it experiences minimal steric interference. The puckerings of the rings which these factors induce also relieve strong eclipsing interactions between adjacent *cis*-related substituents. Methyl α-D-xylofuranoside and methyl α-D-lyxofuranoside represent compounds whose preferred conformations illustrate the influence of the anomeric effect, and the $C_{(5)}$ factor, respectively, whereas methyl β-D-xylofuranoside in the E_2 and ^{3}E conformations incorporates each feature.

O OH MeO HOH₂C OH

α-D-*xylo* (^{2}E)

O OMe OH HOH₂C OH

α-D-*lyxo* (E_4)

OMe O OH HOH₂C OH

β-D-*xylo* (E_2)

OMe O OH HOH₂C OH

β-D-*xylo* (^{3}E)

Many X-ray diffraction analyses have been carried out on furanosyl derivatives - especially on β-D-ribofuranosyl nucleosides which have been shown normally to adopt, in the crystal, conformations close to ^{2}E or ^{3}E. (P. Murray-Rust and S. Motherwell, Acta Cryst., 1978, B34, 2534).

The destabilising factors within furanose rings are discussed further in the later section on "equilibration of free sugars in solution" (p. 32).

(iv) Acyclic compounds

Rotations about single bonds of acyclic carbohydrates in solutions can occur readily at room temperature, and their consequent flexibility is usually taken to account for the low optical activities observed for such compounds. Nevertheless, the measurement of N.M.R. vicinal proton coupling constants often provides good means of determining favoured, time-averaged conformations. Results indicate that these have the carbon atoms in planar, zig-zag orientations, so that, for example, 2,3,4,5-tetra-*O*-acetyl-*aldehydo*-D-arabinose (9) is determined to adopt the illustrated form with the *syn*- and

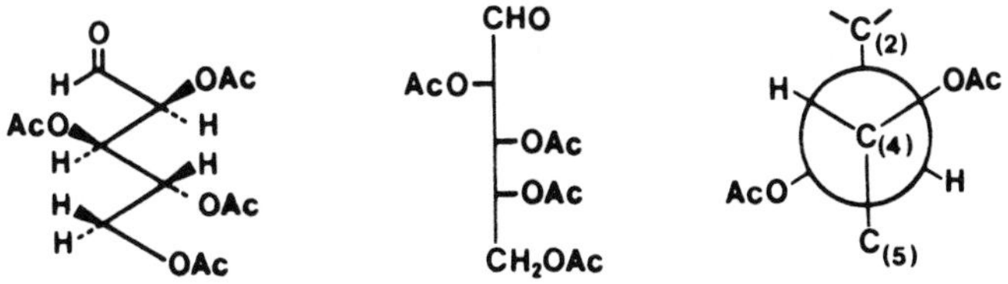

(9)

anti-relationships between the ester groups at $C_{(2)}$, $C_{(3)}$ and $C_{(3)}$, $C_{(4)}$, respectively. They thus have the opposite steric relationships to those revealed by the Fischer projection formula which represents a very high energy conformation with the carbon atoms in a smooth arc. From the $C_{(4)}$-$C_{(3)}$ bond Newman projection of the zig-zag conformation it can be seen that the dihedral angle between the hydrogen atoms at these positions is 180° which is consistent with the observed coupling constant of 8.6 Hz (D. Horton and J.D. Wander, Carbohydrate Res., 1970, 15, 271). Vicinal coupling constants are therefore highly suited to this type of analysis; $J_{2,3}$ is 2.2 Hz, indicative of a *gauche* related hydrogen pair.

In compounds, however, in which there are *cis*-oxygen substituents on 1,3-related carbon atoms of the sugar (in either the Fischer or zig-zag projections), the substituents interact in the same way as do *syn*-diaxial substituents on saturated six-membered rings, and the molecules are distorted in consequence. Of the aldopentose derivatives, *ribo*- and *xylo*-

compounds have this structural feature and show this effect, so that 2,3,4,5-tetra-*O*-acetyl-*aldehydo*-D-ribose (10) avoids the destabilising interaction between the ester groups at $C_{(2)}$ and $C_{(4)}$ by undergoing a distortion by rotation around the $C_{(2)}$-$C_{(3)}$ bond and adopting a "sickle" conformation. This generalisation has been found to extend over a wide range of

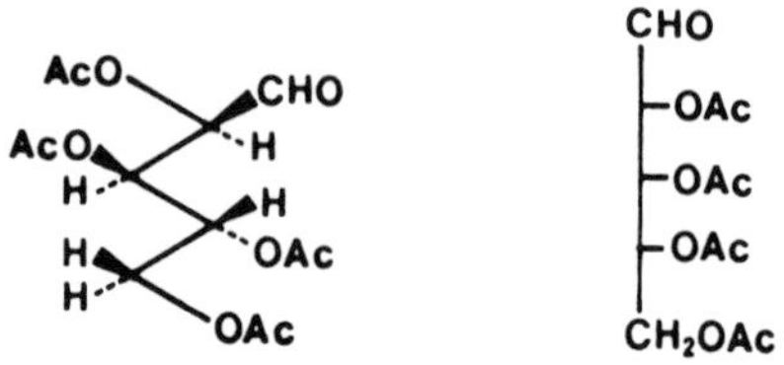

(10)

acyclic derivatives and to apply also in the hexose series in which only *manno*- and *galacto*-isomers are devoid of the distorting 1,3-*cis*-interactions, and only these adopt planar zig-zag conformations.

Likewise, these generalisations apply to acyclic compounds in the crystalline state, so that, for example, D-mannitol and galactitol have symmetrical, planar carbon skeletons whereas all of the other hexitols adopt distorted shapes (G.A. Jeffrey and H.S. Kim, Carbohydrate Res., 1970, 14, 207).

(v) The orientations of substituent groups on exocyclic atoms

Since substituents bonded through oxygen atoms to carbohydrates or their derivatives can adopt different orientations with respect to the sugar frameworks, full conformational analysis of a molecule requires the determination of these orientations as well as of the shapes of the central frameworks. Like basic molecular conformations, group orientations are dependent upon both electrostatic and steric interactions, but are also influenced strongly by other intramolecular and solvation-associated factors.

Extension of the electrostatic arguments upon which the anomeric effect has been based has led to the postulation of the "exo-anomeric effect" which influences the preferred orientation of an aglycone with respect to a pyranoside ring, i.e. the relative energies of the rotamers formed by rotation about the $C_{(1)}$-$O_{(1)}$ bond (R.U. Lemieux and S. Koto, Tetrahedron, 1974, 30, 1933). In the case of pyranose compounds with axial anomeric substituents the rotamer with the aglycone at position *a* avoids direct orbital interaction between the unshared electrons of $O_{(1)}$ and the equatorial pair on the ring oxygen atom, and is thereby favoured. With anomers having equatorially oriented substituents at $C_{(1)}$, related inter-

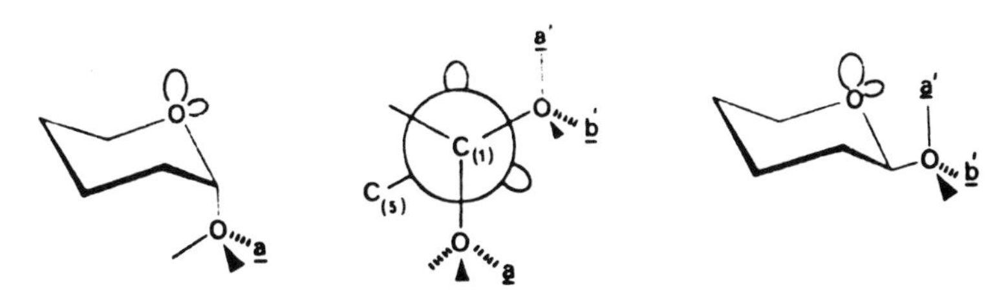

actions cannot be avoided - and this contributed to their relative instability (anomeric effect) - and the aglycone preferentially adopts sites *a*' or *b*' (although in the former position it interacts with any axial substituent at $C_{(2)}$. This factor is of significance in determining the conformations and energies of glycosides including di- and higher saccharides (R.U. Lemieux, Ann. New York Acad. Sci., 1973, 222, 915; S. Pérez and R.H. Marchessault, Carbohydrate Res., 1978, 65, 114); the relative orientation of the glucose units of methyl β-cellobioside has been studied by measurement of ^{13}C-O-C-^{1}H coupling across the glycosidic bond (A.S. Perlin *et al.*, Canad.J.Chem., 1978, 56, 3109).

The question of the orientation of $O_{(6)}$ with respect to $H_{(5)}$ of hexopyranose derivatives has been investigated by ^{1}H N.M.R. and polarimetric methods, and for D-*gluco*-compounds in aqueous solution the rotamer with the bonds $C_{(6)}$-$O_{(6)}$ and $C_{(4)}$-$C_{(5)}$ in the antiparallel orientation is preferred (A.De Bruyn and M. Anteunis, Carbohydrate Res., 1976, 47, 311; R.U. Lemieux and J.T. Brewer in "Carbohydrates in Solution", Adv. Chem. Ser., 1973, 117, 121). In a related study of acetylated derivatives in benzene solution, a similar conclusion was drawn (D. Gagnaire, D. Horton and F.R. Taravel, Carbohydrate Res., 1973, 27, 363).

Measurements of $^{3}J_{H,OH}$ values for solutions in DMSO of methyl 4,6-*O*-benzylidene-α- and β-D-glucopyranoside and their mono-substituted derivatives indicate that only in the cases of 3-*O*-substituted derivatives of the α-compound is the hydroxyl group *anti*- to the $C_{(2)}$-H bond (Y. Kondo and K. Kitamura, Can.J.Chem., 1977, 55, 141).

(vi) Equilibration of free sugars in solution

Improving ^1H N.M.R. methods have allowed S.J. Angyal and V.A. Pickles (Aust.J.Chem., 1972, 25, 1695) to detect and determine quantitatively the cyclic modifications of all the aldopentoses and -hexoses in equilibria in aqueous solution (Table 2).

TABLE 2

Equilibrium compositions of the aldo-pentoses and -hexoses in D_2O at 31°C and the calculated preferred conformations of the pyranoses

Sugar	α-Pyranose	β-Pyranose	α-Furanose	β-Furanose
Ribose	21.5 ($^4C_1$, $^1C_4$)	58.5 ($^4C_1$, $^1C_4$)	6.5	13.5
Arabinose	60 ($^1C_4$)	35.5 ($^4C_1$, $^1C_4$)	2.5	2
Xylose	36.5 ($^4C_1$)	63 ($^4C_1$)	– <1 –	
Lyxose	70 ($^4C_1$, $^1C_4$)	28 ($^4C_1$)	1.5	0.5
Allose[a]	15.5 ($^4C_1$)	76 ($^4C_1$)	3	5.5
Altrose	30 ($^4C_1$, $^1C_4$)	41 ($^4C_1$)	18	11
Glucose	38 ($^4C_1$)	62 ($^4C_1$)	0	0[c]
Mannose[b]	65.5 ($^4C_1$)	34.5 ($^4C_1$)	0[c]	0[c]
Gulose[b]	16 ($^4C_1$)	78 ($^4C_1$)	– 6 –	
Idose	38.5 ($^4C_1$, $^1C_4$)	36 ($^4C_1$)	11.5	14
Galactose	39 ($^4C_1$)	64 ($^4C_1$)	3	4
Talose[b]	37 ($^4C_1$)	32 ($^4C_1$)	17	14

[a]At 24°C; [b]at 44°C; [c]small proportions of α-D-mannofuranose (0.6%), β-D-mannofuranose (0.3%) and β-D-glucofuranose (0.13%) have been detected by ^{13}C N.M.R. methods (A. Allerhand *et al.*, J.Amer.chem.Soc., 1977, 99, 5450; Carbohydrate Res., 1977, 56, 173).

Angyal (Angew.Chem.internat.Edn., 1969, 8, 157) has independently refined his earlier empirical methods for calculating the free energies of all of the pyranose aldoses in aqueous solution in each chair conformation using destabilising factors arising from the following interactions: O_a/H_a (1.9 kJ $mole^{-1}$), C_a/H_a (3.8), O_a/O_a (6.3), O_a/C_a (10.5), O_e/O_a (vicinal) (1.5), O_e/O_e (vicinal) (1.5), C_e/O_e (vicinal) (1.9), C_e/O_a (vicinal) (1.9), C_a/O_e (vicinal) (1.9). Together with these, either of two terms for the anomeric effect is selected according to whether $O_{(2)}$ is axial (4.2 kJ $mole^{-1}$) or equatorial (2.3). His method gives excellent agreement with results obtained from ^{1}H N.M.R. studies when used to determine the favoured chair conformations of specific anomers, and also reasonable concurrence when used to calculate the equilibrium α:β pyranose ratios (c.f. Table 2).

The use by V.S.R. Rao *et al.* (Carbohydrate Res., 1971, 17, 341; 1972, 22, 413) of MO-LCAO *ab initio* methods leads to a further set of data on the free energies and preferred ring shapes of the aldopyranoses. The α:β-pyranose ratios which were calculated were somewhat less close to measured values than were Angyal's, but both the empirical method and the fundamental approach showed consistencies in their deviations. Related studies using Monte Carlo techniques (D.A. Rees and P.J.C. Smith, J.chem.Soc.Perkin Trans. II, 1975, 830; L.G. Dunfield and S.G. Whittington, *ibid*, 1977, 654) have produced both calculated energies and entropies for the aldopyranoses, and hence further theoretical values for α,β-pyranose ratios in aqueous solution. In the latter study ring distortions caused by steric compressions were taken into account, but the results were in agreement with experimental findings only after corrections were applied for specific solvations of equatorial hydroxyl groups which are considered to fit well into the solvent structure.

It has proved less easy to carry out related calculations for the furanoses (S.J. Angyal and V.A. Pickles, Austral.J. Chem., 1972, 25, 1711) partly because energy barriers between conformations and energy differences between them may be small, but the results given in Table 2 and other studies by Angyal lead to some generalisations. While the main destabilising interactions in furanoid compounds are between 1,2-*cis*-related substituents, a particularly significant one appears to occur when $O_{(3)}$ and the substituent at $C_{(4)}$ are *cis*-oriented. All aldoses with this relationship (xylose, lyxose, glucose, mannose, gulose, idose) have low furanose contents - except idose which might be expected to be exceptional since it has the most unstable pyranose forms. The influence of this factor was thought to have been dramatically illustrated by 3-deoxy-D-*ribo*-hexose (3-deoxy-D-glucose) (11) which has *ca.* 20% furanose forms while the parent sugar has tiny amounts.

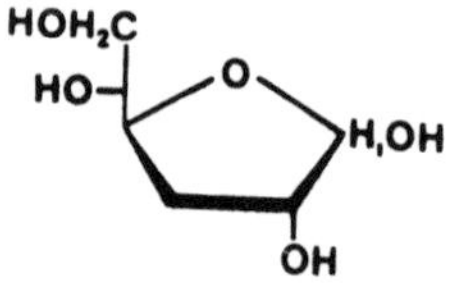

HOH₂C
HO
O
H,OH
OH

(11)

However, 3-*O*-methyl-D-glucose equilibrates to give a mixture containing 14% of the β-furanose which suggests that important factors other than steric (conceivably solvation effects involving the hydroxyl group at $C_{(3)}$) cause destabilisation of the glucofuranoses relative to the pyranoses, (P.J. Garegg, B. Lindberg and C.-G. Swahn, Acta chem.Scand., 1975, B29, 631).

As is to be expected, furanoses with the *cis*,*cis*-relationship of $O_{(2)}$, $O_{(3)}$ and $C_{(4)}$ (lyxose, gulose and mannose) are relatively unfavoured, but, on the other hand, arabinose and galactose with the *trans*, *trans* orientation of these atoms do not have particularly high concentrations which suggests that 1,3-interactions between groups brought together by the preferred ring conformations can also contribute to significant destabilisation.

In addition to the furanoses and pyranoses, aldehydic forms of free sugars (and their hydrates) are present in equilibrated mixtures, but the proportions are too small for detection by normal N.M.R. methods. Initially polarography indicated that the proportion of the *aldehydo*-form of D-glucose in aqueous solution was 0.003%, and this figure has been slightly amended to 0.002% by measurements of the circular dichroism exhibited at 290 nm ($n \rightarrow \pi^*$ transition of the carbonyl group), together with estimated values for $\Delta\varepsilon$ of the free carbonyl group (L.D. Hayward and S.J. Angyal, Carbohydrate Res., 1977, 53, 13). Largely, the proportions of the carbonyl forms of the aldohexoses vary with the energies of their pyranose rings, and glucose, with the most stable ring, has least *aldehydo*-form, while idose, the least stable pyranose, shows reducing properties with Schiff reagent, and contains 0.2% *aldehydo*-form. With aldopentoses, the most stable (xylose) and the least stable (ribose), have proportions (0.02 and 0.05%, respectively) which are higher than for the corresponding aldohexoses which illustrates that rings are more favoured in the latter series. With sugars which cannot form pyranose rings, equilibration necessarily occurs between the furanoses and the acyclic forms, and the latter are therefore more prominent in such cases. With 5,6-di-*O*-methyl-D-glucose, 1% *aldehydo*-form is present at equilibrium as determined by the circular dichroism method.

Little direct information is available on the aldehydrols of higher sugars, but because of the nature of the hydration of aldehydes they can be assumed to be present in higher proportions than the aldehydes. In the case of the aldotetroses *ca.* 1% of *aldehydo*-forms are present together with *ca.* 12% aldehydrols (S.J. Angyal and R.G. Wheen, Austral.J.Chem., 1980, 33, 1001).

Because of the absence of anomeric protons in the cyclic forms of the ketoses, ^{1}H N.M.R. spectroscopy does not provide information which allows ready identification of the individual isomers, although the chemical shifts and coupling constants for the major (β-pyranose) modification of D-fructose have been determined from the 300 MHz spectrum of the equilibrated sugar. Fortunately, ^{13}C N.M.R. spectroscopy serves ideally for the study of these compounds, particularly since the proton-decoupled hemiacetal carbon resonances are substantially deshielded relative to the other resonances, all of which have been assigned for the hexuloses (sometimes by use of specifically deuterated compounds; S.J. Angyal and G.S. Bethell, Austral.J.Chem., 1976, 29, 1249). It was thus possible to measure the proportions of the isomeric forms (Table 3) and correlate the determined ratios of the pyranose modifications with those calculated using Angyal's instability factors (above). For these compounds, the correlations are not good, which was interpreted as indicating that the instability factors associated with the geminally disubstituted anomeric centres are quantitatively different from the values empirically used for the aldoses. The hexuloses have been the subject of study of several groups (L. Que and G.R. Gray, Biochemistry, 1974, 13, 146; P.C.M.H. du Penhoat and A.S. Perlin, Carbohydrate Res., 1974, 36, 111).

TABLE 3

EQUILIBRIUM compositions of the hexuloses in D_2O at 27°C

Sugar	α-Pyranose	β-Pyranose	α-Furanose	β-Furanose
Psicose	22	24	39	15
Fructose	trace	75	4	21
Sorbose	98	0	2	0
Tagatose	79	16	1	4

When Angyal *et al.* (Austral.J.Chem., 1976, 29, 1239) extended their ^{13}C N.M.R. investigation to the 1-deoxyhexuloses in aqueous solution, they found substantial proportions of the acyclic *keto*-modifications and also significant alterations in the ratios of the cyclic isomers. Not only, therefore, does the CH_3CO- group form hemiacetals less readily than does the $CH_2OH.CO$- group, but the methyl and hydroxymethyl groups differ significantly in their interaction energies. Consistent with this, the circular dichroism method (Hayward and Angyal *loc.cit.*) detected the *keto*-forms of the 1-deoxyhexuloses in significant proportions (14-5%). Also 0.2-0.7% of the acyclic carbonyl forms of the hexuloses themselves were detected showing that these sugars ring close less readily than do the aldohexoses. Acyclic ketoses are believed not to hydrate in aqueous solution.

A detailed historical survey of the understanding of the composition of free sugars in solution has been presented. (Angyal, J.Carbohydr.Nucleosides Nucleotides, 1979, 6, 15).

(vii) Mutarotation and factors affecting equilibria

Mutarotation, the process by which free sugars equilibrate in solution, has been reviewed comprehensively (W. Pigman and H.S. Isbell, Adv.carbohydrate Chem., 1968, 23, 11; Adv.carbohydrate Chem.Biochem., 1969, 24, 13), and the mechanisms of both simple and complex processes have been analysed (B. Capon, Chem.Rev., 1969, 69, 407). The subject is extremely complicated, and although the rate determining step - the ring opening of the sugar to the *aldehydo* form - has been identified, details of the process vary with very many parameters. Rates of mutarotation are dependent upon such factors as temperature, solvent, substitution effects, intramolecular catalysis in the cases of sugars containing reactive functional groups (B. Capon and R.B. Walker, J.chem.Soc.Perkin II, 1974, 1600), and, of course, on the presence of acid or basic catalysts, as well as of various metal ions which also can apply catalytic influences.

The final position of equilibrium is also very sensitive, and the relative energies of the contributing species mentioned above are very dependent upon temperature, variation in the substitution patterns of the sugars, solvent and the presence of ions which have selective associating characteristics. As the substituent (R) at $C_{(1)}$ increases in size from H to CH_3 to CH_2OH to $CHOH.CH_2OH$ in the D-ribose series (12), the α-furanose contents at equilibrium in aqueous solu-

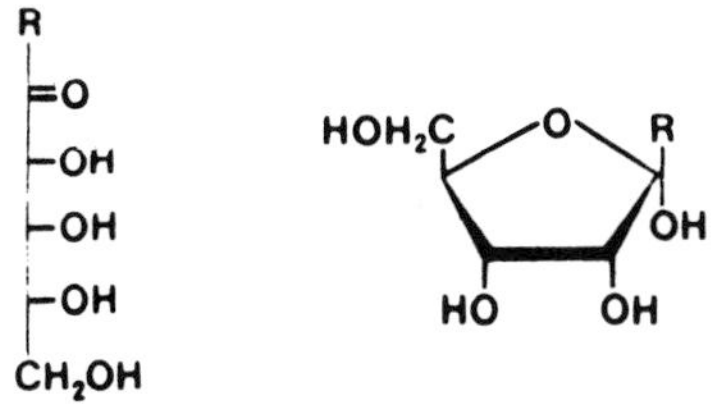

(12)

tion increase from 6.5 to 21 to 39 to 54% (S.J. Angyal *et al.*, Austral.J.Chem., 1976, 29, 1239); the striking example of 3-*O*-methyl-D-glucose has already been mentioned (p.34), and other methylated compounds to show high furanose proportions are 2,3-di-*O*-methyl-D-arabinose and 2,3-di-*O*-methyl-D-altrose. In particular this is so in non-aqueous solvents, these two sugars containing 65 and 80% furanose forms, respectively, in dimethyl sulphoxide (W. Mackie and A.S. Perlin, Canad.J.Chem., 1966, 44, 539, 2039).

Disturbance of equilibria by non-paramagnetic metal ions can be studied by N.M.R. methods, and so it has been shown that, with D-allose in aqueous solution, complexing occurs at the axial, equatorial, axial $C_{(1)}$-$C_{(3)}$ triol of the α-pyranose form, and in 0.85 M calcium chloride the β-pyranose:α-pyranose ratio is 1.5, whereas it is 5.6 in water. Similarly, there is preferential complexing with the $C_{(1)}$-$C_{(3)}$ *cis-*,*cis-*

triol of the α-furanose, and the β-furanose:α-furanose ratio changes from 1.6 to 0.84 in the salt solution (S.J. Angyal, Tetrahedron, 1974, 30, 1695). Selective complexing of this type involving various ions has been studied in this way, and La^{3+} has been found to be particularly effective and, like calcium, to give 1:1 complexes with contiguous *cis,cis*-triols of sugars (R.E. Lenkinski and J. Reuben, J.Amer.chem.Soc., 1976, 98, 3089; S.J. Angyal, Austral.J.Chem., 1972, 25, 1957).

Simple ions (e.g. Na^+) also form complexes (Angyal, *loc.cit.*) and this has wide biological significance. Studies of the complexes formed in pyridine solution have been carried out using ^{23}Na N.M.R. linewidths, and they have both solvent molecules and two (or three) sugar oxygen atoms arranged around the sodium ions (C. Detellier, J. Grandjean and P. Laszlo, J. Amer.chem.Soc., 1976, 98, 3375; Helv., 1977, 60, 259). Electrophoresis provides a way of studying the net interaction between sugars and ionic species, and all free sugars complex with calcium ions to some degree. Those, e.g. allose, which have forms which give relatively stable complexes show the effect of impeded mutarotation; they give complexes of specific isomeric forms which migrate rapidly, but their overall mobility is non-specific and they therefore appear as streaks after development (S.J. Angyal and J.A. Mills, Austral.J.Chem., 1979, 32, 1993).

Reduction in temperature retards mutarotation, and at low temperatures isomeric forms of free sugars have been separated on thin layer chromatography plates (G. Avigad and S. Bauer, Carbohydrate Res., 1967, 5, 417) and on anion exchange columns (O. Ramnäs and O. Samuelson, Acta chem.Scand., 1974, B28, 955).

(b) Identification and structural determination of monosaccharide derivatives by physical methods

(i) Crystal structure analysis

An entirely new status has been acquired by diffraction analysis following the development of computers in the 1950's and automatic diffractometers in the following decade, and X-ray analysis, in particular, is no longer the preserve of the specialist interested in solving geometric problems. Rather it is joining the spectroscopic methods as a generally applicable and powerful structural tool, and compared with the 6 carbohydrate analyses reported in 1967 almost 100 are now being published annually. Increasingly, results are appearing as integral parts of chemical (rather than crystallographic) papers. Neutron diffraction analysis has also increased in significance, particularly as more detailed information has been sought on the nature of the association between the hydroxyl groups of carbohydrate derivatives in the crystal lattice (G.A. Jeffrey, M.E. Gress and S. Takagi, J.Amer.chem. Soc., 1977, 99, 609; G.A. Jeffrey and S. Takagi, Acc.chem. Res., 1978, 11, 264). This is because hydrogen bonding leads to reduction in the electron densities about the hydrogen nuclei which makes them difficult to detect by X-ray studies; since hydrogen atoms have comparable neutron scattering power to carbon and oxygen atoms, they are located with similar precision by neutron diffraction methods.

Crystal structure data has been reviewed as follows: 1964-1970 literature, G. Strahs, Adv.carbohydrate Chem.Biochem., 1970, 25, 53; 1970-1972 literature, G.A. Jeffrey and M. Sundaralingham, *ibid.* 1974, 30, 445; 1973 literature, *idem. ibid.*, 1975, 31, 347; 1974 literature, *idem. ibid.*, 1976, 32, 353; 1975 literature, *idem. ibid.*, 1977, 34, 345.

X-Ray crystallography is now used almost routinely to solve the occasional structural problem which arises in the course of synthetic and natural product work, and which is not amenable to N.M.R. analysis. For example, with acyclic compounds it is often necessary to know the relative configuration before proton coupling analysis can be used to give the molecular conformation (and *vice versa*), and it was therefore not possible, by N.M.R. methods alone, to determine the stereochemistry of 4,5,6-tri-*O*-benzoyl-2,3-di-S-ethyl-2,3-dithio-D-allose diethyldithioacetal (13), which is the product of ethanethiolysis of 3,5,6-tri-*O*-benzoyl-1,2,*O*-isopropylidene-

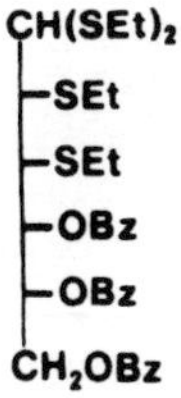

(13) 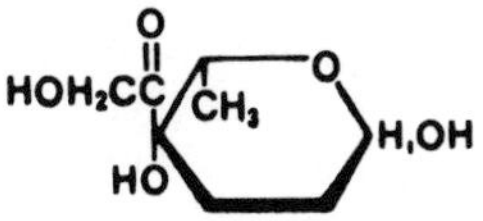 (14)

α-D-glucofuranose (W.T. Robinson *et al.*, Acta Cryst., 1977, B33, 2888). Similarly, pillarose (14), the constituent branched-chain sugar of the antibiotic pillaromycin A, had to be characterised by the X-ray diffraction method (J.Clardy, *et al.*, J.Amer.chem.Soc., 1975, 97, 6250).

The diffraction method is unique in its ability to reveal structural information on compounds in the crystalline form. Points of major structural significance are revealed; for example, D-*manno*-3-heptulose crystallises in the β-pyranose modification (T. Taga and K. Osaki, Tetrahedron Lett., 1969, 4433) whereas the D-*altro*-isomer (coriose) exists as the α-furanose (T. Taga, K. Osaki and T. Okuda, Acta Cryst., 1970, B26, 991). Points of lesser structural significance, which nevertheless assist with the understanding of the properties of compounds, are also determinable. Thus, 1,6-anhydro-β-D-gluco-pyranose (15) and methyl 3,6-anhydro-α-D-glucopyranoside (16) differ somewhat in properties which depend upon the relationships of the hydroxyl groups at $C_{(2)}$ and $C_{(4)}$ (e.g. reactions with diol complexing agents). This is unexpected since both pyranose rings are held in the $^1C_4$ conformation, but the anhydro-bridges cause different distortions in these conformations, crystallographic methods showing that the resulting $O_{(2)}$-$O_{(4)}$ bond distances are 3.30 A° and 2.76 A°, respectively (B. Lindberg, B. Lindberg and S. Svensson, Acta chem. Scand., 1973, 27, 373).

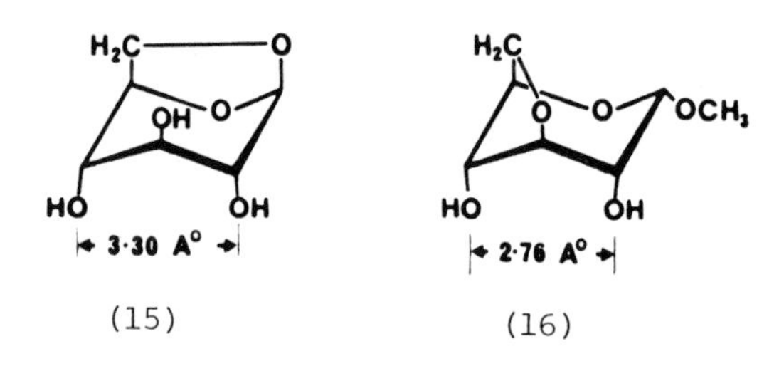

(15) (16)

Properties which belong specifically to the crystal can also be accounted for: decyl α-D-glucopyranoside shows liquid crystalline characteristics prior to melting (and also has micelle-forming properties) which are accounted for by the finding that the carbohydrate and alkyl parts of the molecules associate in the crystal to form an alternating polar and non-polar bilayer system with extended alkyl chains packing between layers of hydrogen bonded glucosyl moieties

(P.C. Moews and J.R. Knox, J.Amer.chem.Soc., 1976, 98, 6628).

It has been pointed out above, in the consideration of the anomeric effect (p.22), that its influence is reflected in the lengths of the C-0 bonds in the neighbourhood of the anomeric centre - for example, axial exocyclic bonds are specifically lengthened. Neutron diffraction studies have now led to the conclusion that anomeric hydroxyl groups are relatively strong hydrogen bond donors and corresponding weak acceptors (Jeffrey and Takagi *loc.cit.*). A further generalisation observed by Jeffrey in his analysis of the different types of hydrogen bond which occur in crystals of sugar derivatives is that "co-operative hydrogen bonding" or "hydrogen bond conjugation" occurs when possible, and leads to stabilisation. Such an effect arises when a hydroxyl group accepts a hydrogen bond which makes its own hydrogen atom a stronger hydrogen bond former.

It is now clear (G.A. Jeffrey in "Carbohydrates in Solution", Adv.Chem.Ser., 1973, 117, 177) that hydrogen bonding in the crystal is predominantly intermolecular, and of only minor influence in determining conformation. The same generalisation applies to other lattice forces, and it is therefore intramolecular stereochemical and electrostatic factors which largely determine the conformation of compounds (even flexible species such as acyclic derivatives) in the crystal as well as in solution. An elegant demonstration of this comes from the analysis of three polymorphs of D-mannitol all of which have a planar zig-zag carbon chain but differ in the association patterns of the hydroxyl groups (G.A. Jeffrey *et al.*, Acta Cryst., 1968, B24, 442, 1449). However, there are exceptions; one dimorph of potassium D-gluconate exists with the gluconate ion in the expected sickle conformation while the other has a straight chain (Jeffrey *et al.*, Acta Cryst., 1974, B30, 1421). There is also evidence that intramolecular factors, rather than lattice forces, govern the orientation of the primary hydroxyl group of hexopyranose derivatives (Jeffrey, in "Carbohydrates in Solution", Adv.Chem.Ser., 1973, 117:177), and the orientation of the $O_{(1)}$-aglycone bond of pyranosides (S. Pérez and R.H. Marchessault, Carbohydrate Res., 1978, 65, 114).

(ii) Infrared and Raman spectroscopy

An important review has appeared relating to all aspects of infrared spectroscopy of carbohydrates. The absorptions of relevant functional groups are considered, applications to conformational analysis are reported, and the use of the technique in analytical work is discussed [R.S. Tipson, Natl.Bur. Stand.(U.S.), Monogr. 110, 1968]. A more recent review covers practical aspects and the interpretation of spectra and contains much data on absorption bands (R.S. Tipson and F.S. Packer, in "The Carbohydrates" eds. W. Pigman and D. Horton, Second Edition, Vol. 1B, Academic Press, New York, 1980, p. 1394).

Laser-Raman studies have been begun; advantages over infrared are that solid samples can be used directly, and because of weak interference from OH absorptions aqueous solutions can be examined, and Raman absorptions are often sharper than infrared and give more precise information. The method has been used to study the anomers of glucose in aqueous solution, there being, e.g., 910 and 893 cm^{-1} absorptions characteristic of β-glucose and 860 and 842 cm^{-1} bands for the α-compound, and these have been used to show that a mutarotated mixture contained 62% β- and 38% α-D-glucopyranose. Other related assignments have been made in the part of the spectrum below 1700 cm^{-1} (M. Mathlouthi and D.V. Luu, Carbohydrate Res., 1980, 81, 203). Similar studies of D-fructose have led to 874 and 826 cm^{-1} bands being assigned to C-C stretching modes for furanoid and pyranoid rings, and to the conclusion that mutarotated mixtures contained 41 and 59% of these two types of isomers (*idem.ibid.* 1980, 78, 225). Related studies on sucrose (*idem.ibid.* 1980, 81, 203) and other glycosidic compounds (P.D. Vasko, J. Blackwell and J.L. Koenig, *ibid.*, 1971, 19, 297; H. Susi and J.S. Ard, *ibid.*, 1974, 37, 351; A.T. Tu, J. Lee and Y.C. Lee, *ibid.*, 1978, 67, 295) have been reported. Likewise, 2-acetamido-2-deoxy-D-glucose and D-glucuronic acid have been examined (C.Y. She, N.D. Dinh and A.T. Tu, Biochem. biophys.Acta, 1974, 372, 345), and hydrogen bonding work on methyl 3,6-dideoxy-β-D-*ribo*-hexopyranoside has been described (H.J. Bernstein, *et al.*, Canad.J.Chem., 1979, 57, 2640).

Theoretical studies of α- and β-D-glucopyranose using normal co-ordinate analysis have given results for calculated infrared and Raman bands in reasonably good agreement with observed data (P.D. Vasko, J. Blackwell and J.L. Koenig, Carbohydrate Res., 1972, 23, 407; J.J. Cael, J.L. Koenig and J. Blackwell, *ibid.*, 1974, 32, 79).

Reduction in the temperature of measurement affords considerable improvement in measured infrared band definition and can also cause the appearance of new bands and slight movements to higher or lower frequencies in a manner which indicates that the temperature dependence is related to hydrogen bonding (A.J. Mitchell, Austral.J.Chem., 1970, 23, 833).

Infrared methods have been advocated on several occasions for the distinction between axial and equatorial functional groups on pyranoid rings, but N.M.R. methods are much superior for this purpose, and a study of sulphonate esters of carbohydrates has shown that a previously recommended infrared method is unreliable (R.C. Chalk, *et al.*, Carbohydrate Res., 1978, 61, 549).

Intramolecular hydrogen bonding studies on a range of methyl 4,6-*O*-benzylidine-D-hexopyranosides and their monosubstituted derivatives has revealed a linear relationship between the lengths of the hydrogen bonds and the reduction in O-H stretching frequencies (H. Hönig, and H. Weidmann, *ibid.*, 1979, 73, 260).

(iii) Mass spectrometry

The basic studies having been completed on the fragmentation paths followed under electron impact by the main classes of carbohydrate derivatives which can be studied by mass spectrometry, the technique is now of major significance. This is particularly so when it is coupled with gas-liquid chromatography and applied to the investigation of complex mixtures. Such mixtures often result during work on the elucidation of structures of polysaccharides in which research the G.L.C.-M.S. procedures have become indispensible. Two comprehensive articles have reviewed electron impact mass spectrometry applied to the most important categories of cyclic and acyclic sugar derivatives: S. Hanessian (Methods biochem.Anal., 1971, 19, 105) extended his survey to include various nucleoside and antibiotic compounds, while J. Lönngren and S. Svensson (Adv.carbohydrate Chem.Biochem., 1974, 29, 41) also discussed inositols, *C*-glycosides and oligosaccharides. A briefer survey in a reference text covers basic studies of carbohydrate derivatives including deoxy- and amino-sugar and uronic acid compounds (T. Radford and D.C. Dejongh in "Biochemical Applications of Mass Spectrometry" ed. G.R. Waller, Wiley-Interscience, New York, 1972, P.313). See also D.C. Dejongh in "The Carbohydrates" ed. W. Pigman and D. Horton, Second Ed.,

Vol. 1B, Academic Press, New York, 1980, p. 1327.

Careful study of the breakdown of permethylated hexopyranoses involving the use of deuterium labelling methods has led to a fuller understanding of the most important pathways followed, and this serves as a basis for characterising the cleavages undergone by a range of pyranoid compounds (usually derivatised as methyl or trimethylsilyl ethers or acetyl or trifluoroacetyl esters). The codings of the modes of degradation and of the ions formed which were initially used by Kochetkov and Chizhov (Adv.carbohydrate Chem., 1966, 21, 39) have gained acceptance (Scheme 1).

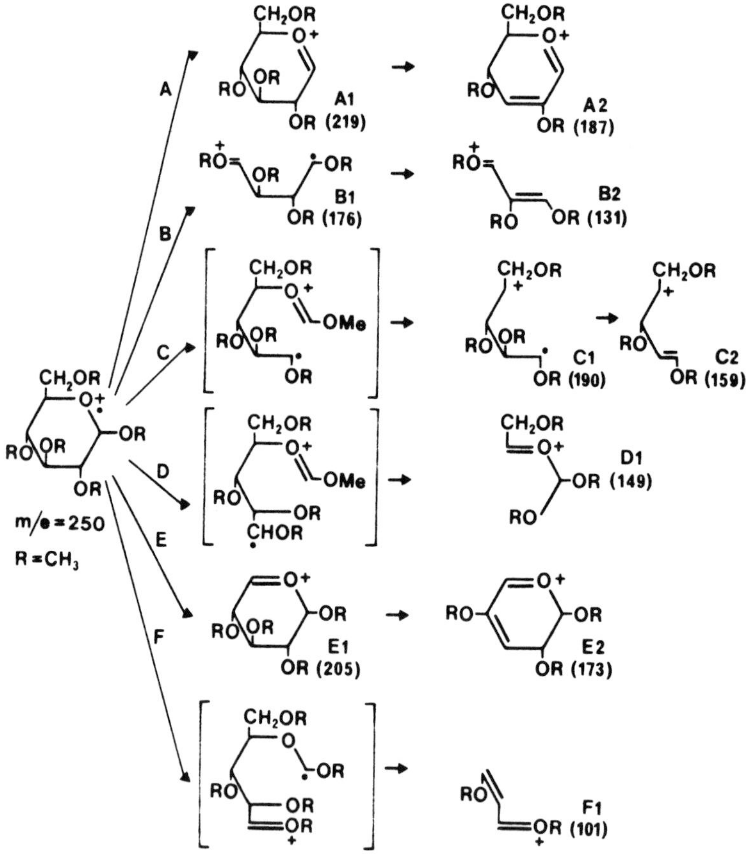

SCHEME 1

Variation in the substituents of the ring can be determined by comparing the ions formed with those of the above pathways and this leads to an important structural tool, for example, for the characterisation of partially methylated compounds. If, however, a major change is effected in a pyranoid ring - as with the introduction of a double bond - new pathways may dominate, and these too can be used in structural analysis. A characteristic fragmentation route for unsaturated pyranoid compounds involves retrodienic cleavage with the loss, for example, of the $C_{(5)}$-$O_{(5)}$ unit for 2,3-unsaturated compounds and $O_{(5)}$-$C_{(1)}$ for hex-3-enopyranose derivatives (O.S. Chizhov *et al.*, Carbohydrate Res., 1970, 13, 269) (Scheme 2).

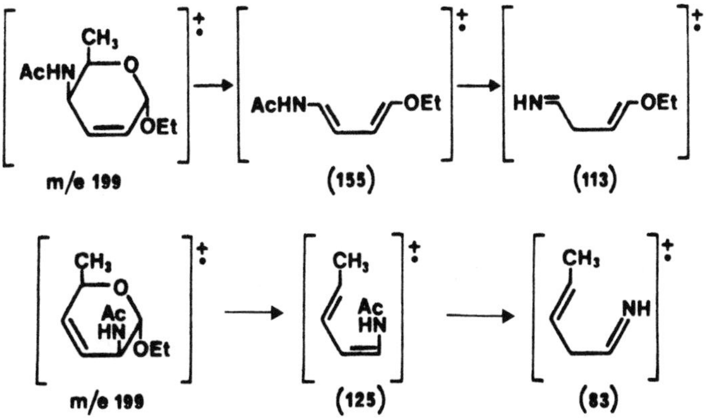

SCHEME 2

A set of paths has been elucidated for the fragmentation of the molecular ion of methyl tetra-*O*-methyl-α-D-glucofuranoside with the aid of deuteriomethyl labelling (E.G. de Jong *et al.*, Carbohydrate Res., 1978, 60, 229); some of the significant initial fragmentations are shown in Scheme 3. Particularly characteristic of the spectrum of this furanosyl derivative is the presence of the ion with m/e 161 and the absence of the ion with m/e 176 (B_1 of pyranoid isomer). These observations are of value in determining glycoside ring size.

The mass spectrometry of acyclic aldose derivatives has received considerable attention particularly because such compounds are frequently prepared to permit the G.L.C. analysis of free sugars and remove the isomer problem associated

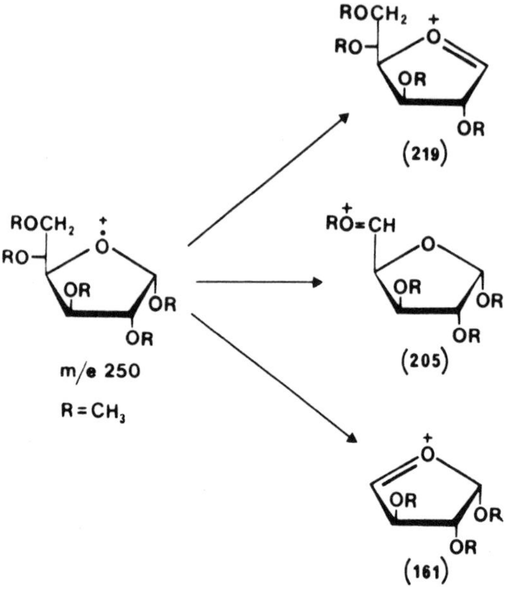

SCHEME 3

with having a chiral anomeric centre in cyclic compounds. Alditols are frequently analysed as their acetates, trifluoroacetates, methyl ethers or trimethylsilyl ethers and aldononitrile and aldose oxime derivatives have likewise been investigated. Primary fragmentation occurs by cleavages of the C-C bonds of the molecular ions, and this facilitates the identification and location of any groups present in place of hydroxyl groups in the initial sugar. Deoxy or partially methylated sugars are thus susceptible to analysis in this way.

Deuterium labelling of methyl groups has played an important role in the elucidation of the detailed fragmentations of methyl ethers and acetyl esters, and some studies have involved deuterium substitution of sugar hydrogen atoms, but analyses using this isotope can be complicated by intramolecular migration processes. ^{13}C is less vulnerable, and fragmentations of 1,2:5,6-di-*O*-isopropylidene-α-D-glucofuranose (17) have been more fully interpreted by use of samples independently labelled at $C_{(1)}$ and $C_{(6)}$.

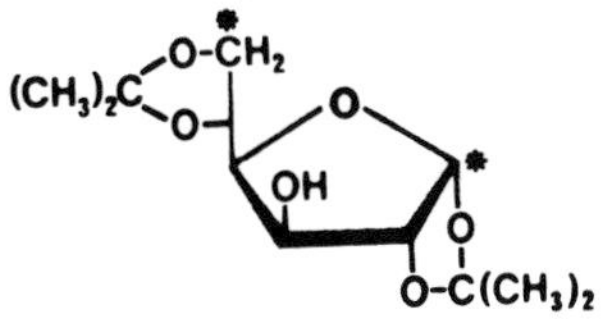

$(CH_3)_2C$ O-CH₂ ...

(17)

Related studies on the ketone derived from this alcohol have used ^{18}O and ^{2}H labelling (A. Glangetas, A. Buchs and J.M.J. Tronchet, Org.mass Spectrometry, 1977, 12, 402), and nitrogen-containing fragments derived from peracetylated aldononitriles have been identified by use of ^{15}N-labelled hydroxylamine in the original syntheses (F.R. Seymour, E.C.M. Chen and S.H. Bishop, Carbohydrate Res., 1979, 73, 19).

Normally for electron impact mass spectrometry thermally unstable and non-volatile compounds like carbohydrates must be derivatised, but a method involving the rapid heating of samples adsorbed on a rhenium wire affords significant ions from polyhydroxy compounds. Temperatures of the wire are taken to >1200°C in 0.1-0.2 sec, and the MH^+ ion for sucrose (m/e 343), fragment ions formed by successive loss of 3 water molecules, and an ion with m/e 163 derived by inter-unit bond cleavage are all observed (W.R. Anderson, W. Frick and G.D. Daves, J.Amer.chem.Soc., 1978, 100, 1974).

More commonly, non-derivatised compounds have been examined by the Field Desorption technique (H.D. Beckey and H.-R. Schulten, Angew.Chem.internat.Edn., 1975, 14, 403) in which the sample is deposited on a tungsten anode covered with micro-needles and inserted into the ion source at 10^{-6} Torr. The anode is then taken to 3-10 kV and a potential difference of 10-12 kV is generated between this and the cathode set a few mm. distant. Electrons are lost by tunnelling, and the cations formed require only the field desorption energy rather than the usual sublimation energy to escape. Most often MH^+ ions and only few fragment ions are observed, and the method therefore complements the electron impact technique especially for larger molecules. For example, a tetrasaccharide antibiotic gives an MH^+ ion at m/e 615, whereas no ion above m/e 447 is seen in the electron impact spectrum run at 18 eV. High resolution field desorption mass spectrometry has provided both molecular weights and molecular formulae for the sodium salts of sugar phosphates (H.D. Beckey *et al.*, Chem. Comm., 1973, 616). Attachment of alkali metal ions derived from tetraphenylborates to non-volatile compounds under field desorption conditions also permits ionisation and affords $(M + metal)^+$ species (H.J. Veith, Tetrahedron, 1977, 33, 2825).

It has often been accepted that electron impact mass spectra are insensitive to configurational changes and that the method cannot be utilised for the distinction between epimers. J.F.G. Vliegenthart *et al.* (J.Amer.chem.Soc., 1972, 94, 2542) have however shown that patterns of intensities of sets of peaks are characteristic of specific stereoisomers, and have suggested that computer matching offers a method for specific characterisation. Reduction in the energy of the ionising beam has been shown to enhance the differences between the spectra derived from epimeric uronic acid derivatives, D-*gluco*- and D-*galacto*-compounds being distinguishable when examined at 12 eV. (V. Kováčik, V. Mihálov and P. Kováč, Carbohydrate Res., 1977, 54, 23).

It appears, however, that the milder idirect chemical ionisation method has more potential in this regard. Usually derivatised carbohydrates are examined by this procedure, but this need not be so, D-glucose giving mainly an $(M + \overset{+}{N}H_4)$ peak at m/e 198 when examined using ammonia as reagent gas. Other substituted and unsubstituted compounds likewise give $(M + \overset{+}{N}H_4)$ ions primarily, so the method is ideally suited to molecular weight determination (A.M. Hogg and T.L.Nagabhushan, Tetrahedron Lett., 1972, 4827) and has been applied, sometimes with isobutane as alternative reagent gas, in studies of, for example, aldononitrile acetates (F.R. Seymour, E.C.M. Chen and S.H. Bishop, *loc.cit.*; B.W. Li, T.W. Cochran and J.R. Vercellotti, Carbohydrate Res., 1977, 59, 567), permethylated glycosylalditols (O.S.Chizhov *et al.*, J.org.Chem., 1976, 41, 3425) and peracetylated oligosaccharides up to pentasaccharides (*idem.*, *ibid.*, 1974, 39, 451). M. McNeil and P.Albersheim (Carbohydrate Res., 1977, 56, 239) have highlighted the usefulness of the chemical ionisation method for distinguishing between epimers, isobutane ionised 1,4,5-tri-*O*-acetyl-2,3,6-tri-*O*-methyl-D-glucitol and -D-galactitol giving (M + 1), (M + 1 -32) and (M + 1 -60) ions in markedly and reproducibly different ratios. In related manner, the isomeric trimethylsilyl ethers of D-glucopyranose, D-galactopyranose and D-mannopyranose give simple but different spectra when ionised with ammonia. The β-anomers give $(M + \overset{+}{N}H_4)$ ions which are 3-12 times more intense than those derived from the α-compounds, and the intensities of fragment ions vary profoundly with the sugars. Even the base peaks vary, and have m/e 198, 468 and 204 for both epimers of the three hexoses, respectively. Ion intensities also permit the distinction between furanose and pyranose isomers (T. Murata and S. Takahashi, *ibid.*, 1978, 62, 1).

Negative ion mass spectrometry also affords a method of determining molecule weight of carbohydrates including underivatised oligosaccharides (A.K. Ganguly *et al.*, Chem. Comm., 1979, 148). Bombardment with negative ions produced from methane with electrons of 500 eV energy give intense M^- ions with a few fragment ions. With chloride ions produced from CF_2Cl_2 (Freon 12), $(M + Cl^-)$ ions are observed for hydroxylated compounds, but poor spectra are obtained for methylated derivatives. Fragment ions are seen in the spectra of oligosaccharides which permit the sequencing of the sugar units. Negative ions have been produced from fully methylated compounds by the dissociative capture of electrons of 1-10 eV energy, fragments formed from the molecular ion by loss of methoxyl and methanol being observed (L.A. Baltina *et al.*, Zhur.obshch.Khim., 1977, 47, 2379).

(iv) Nuclear magnetic resonance spectroscopy

(1) General. The period under review has seen such expansion in the use of ^{1}H N.M.R. spectroscopy that the technique is now clearly the most widely used in structural monosaccharide chemistry. Of particular significance has been the introduction of Fourier transform procedures and of superconducting solenoid instruments with vastly increased resolving power, so that 220, 270 and 300 MHz spectral data are frequently described, and reports of work carried out at 500 MHz are beginning to appear. Perhaps even more important has been the introduction of ^{13}C N.M.R. spectroscopy and the development of instrumentation to permit the technique to be used in an ideally complementary fashion to ^{1}H spectroscopy.

(2) ^{1}H N.M.R. spectroscopy. G. Kotowycz and R.U. Lemieux (Chem. Rev., 1973, 73, 669) and T.D. Inch (Ann.Reports N.M.R. Spectroscopy, 1969, 2, 35; 1972, 5A, 305) have reviewed ^{1}H N.M.R. spectroscopy in carbohydrate chemistry paying particular attention to applications, while B. Coxon (Adv.carbo-

hydrate Chem.Biochem., 1972, 27, 7) devoted more attention to the experimental methods, and L.D. Hall (*ibid.* 1974, 29, 11) considered in detail the chemical devices (e.g. the use of solvent shifts or paramagnetic shift reagents) and physical methods (e.g. the application of spin-decoupling methods or high resolution spectrometers) that can be adopted to solve the problems associated with spectra which contain overlapping resonances. Hall has also reviewed general aspects (in "The Carbohydrates", ed. W. Pigman and D. Horton, Second Edition, Vol. 1B, Academic Press, New York, p.1300).

While the above physical procedures are based simply on the principles underlying the experiment, the chemical approaches are dependent on empirical observation and are sometimes rather difficult to rationalise. For example, when acetylated sugars are examined in chloroform and then in benzene the acetyl resonances suffer an up-field shift of *ca.* 0.4 p.p.m., while some ring protons are shielded and others are deshielded (M.H. Freemantle and W.G. Overend, J.chem.Soc.(B), 1969, 547). Sometimes the effects of the interactions between benzene and carbonyl groups of carbohydrate derivatives can be dramatic, and the relative positions of resonances of the protons of the methylene group adjacent to the carbonyl group in methyl 4,6-*O*-benzylidene-2-deoxy-α-D-*erythro*-hexopyranosid-3-ulose reverse in the two solvents. In chloroform the axial proton resonates downfield from the equatorial atom, and in benzene the opposite is the case (R.F. Butterworth, P.M. Collins and W.G. Overend, Chem.Comm., 1969, 378).

The use of paramagnetic, chelated lanthanide complexes to induce chemical shift changes, following the discovery of the effect in 1969, has represented a major development in ^{1}H N.M.R. spectroscopy. Fast equilibrium is established between the reagents and the complexing sites on the carbohydrates - usually hydroxyl or substituted hydroxyl groups - and the spectra observed are therefore the weighted average spectra of the complexed and uncomplexed compound. Since the shift effects are often very substantial, it is therefore necessary to use only small proportions of reagents, and often experiments involve the addition of successive small amounts so that altered spectra may be related to the spectra of the uncomplexed compound. This procedure also allows recognition of whether complexing causes conformation alteration or line broadening, and frequently neither effect is significant when europium(III) or praseodymium(III) complexes are used. These therefore are usually the favoured lanthanide elements and are used in nonpolar solvent as diketone complexes - often the lanthanide tris(dipivalomethanato) [Ln(dpm)$_3$] or tris(1,1,1, 2,2,3,3-hepta-fluoro-7,7-dimethyl-4,6-octanedionato) [Ln(fod)$_3$] complexes, the latter having the advantage of increased solubility. Generally europium complexes cause downfield shifts while praseodymium analogues have the opposite effect. Uncomplexed lanthanides can also cause dramatic changes in spectra, and S.J. Angyal and D. Greeves (Austral. J. Chem., 1976, 29, 1223) have compared the chemical shifts (sometimes >20 p.p.m.) induced by the additions of thirteen lanthanide salts to aqueous solutions of *epi*-inositol.

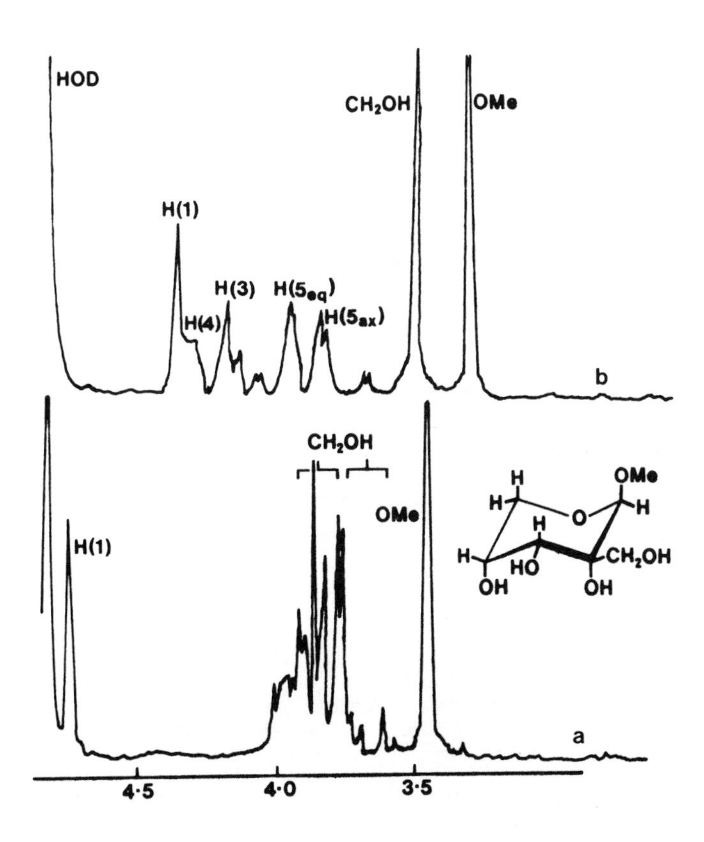

Fig. 1 (a) 100 MHz spectrum of methyl β-D-hamamelopyranoside in D_2O; (b) with 0.22 equivalents of $PrCl_3$ added (p.p.m. from $Me_3SiCH_2CH_2CH_2SO_3Na$)

The former author found that the addition of 0.22 equivalents of praseodymium chloride to an aqueous solution of methyl β-D-hamamelopyranoside resolves a complex 100 MHz spectrum (Fig. 1), so that every signal can be recognised, the coupling constants determined and the ring conformation identified (Carbohydrate Res., 1973, 26, 271). It can be seen that most resonances are moved to lower fields, but that some protons are shielded.

Effects are largest in the neighbourhood of the sites of complexing [but see Hall (*loc.cit.*)]. For hydroxylic compounds in nonaqueous solvents association involves the hydroxyl groups, and the configurations at tertiary alcohol centres of branched-chain sugars can, in consequence, be determined by use of these reagents. In Fig. 2 the chemical shifts of some of the protons of methyl 4,6-*O*-benzylidene-2-*O*-toluene-*p*-sulphonyl-α-D-glucopyranoside (a) and the *allo*-epimer (b) are plotted against equivalents of $Eu(fod)_3$ added to deuteriochloroform solutions, and the effects can be seen to be greatest on the position of resonance of the hydroxyl protons.

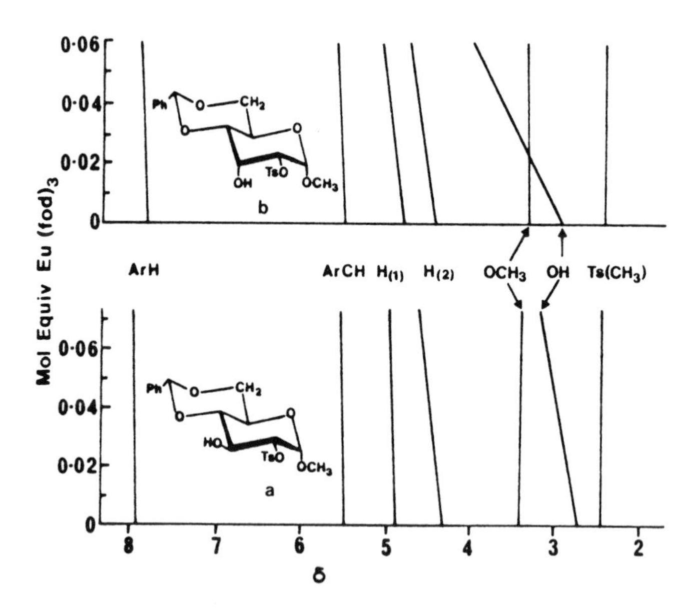

Fig. 2. Effects of adding $Eu(fod)_3$ on ^1H resonances of epimers.

Derivatives of these compounds with tertiary centres at $C_{(3)}$ give linear plots under these conditions with slopes which correlate with one of the two sets of reference lines (J.J. Nieuwenhuis and J.H. Jordaan, Carbohydrate Res., 1976, 51, 207).

The reagents also complex with non-hydroxylic compounds, the resonances for the methyl groups of fully methylated sugars being better resolved by their use. Again $Pr(fod)_3$ and $Eu(fod)_3$ cause shielding and deshielding, respectively, and differential effects on different groups indicate the relative propensities of different sites to bind the reagents (A.M. Stephen, *et al.*, *ibid.*, 1975, 39, 136; 1976, 49, 13). In similar fashion, the acetyl proton resonances of hexose per-acetates are affected differently, with the signals of the primary groups undergoing most change. *N*-Acetyl resonances undergo much larger shifts so that they can be specifically identified (R.F. Butterworth, A.G. Pernet and S. Hanessian, Canad.J.Chem., 1971, 49, 981).

(3) ^{13}*C N.M.R. spectroscopy.* Unlike ^{12}C, ^{13}C has nuclear spin (I = ½) and undergoes resonance in a magnetic field when suitably irradiated. However, its natural abundance is only 1.1%, and its gyromagnetic ratio is relatively small, which makes it inherently several thousand times less sensitive in the N.M.R. experiment than is the proton. This factor, together with the fact that ^{13}C couples significantly to both bonded protons and protons on adjacent carbon atoms, necessitated new technicological developments before, by the late 1960's, ^{13}C spectroscopy could be used comparably with ^{1}H spectroscopy. Frequently, the two techniques are now applied to complement each other successfully.

Introduction of wide band proton decoupling, resulting in the collapse of ^{13}C-^{1}H spin multiplets, leads to signal enhancement, not just because of the generation of singlets, but also because of Nuclear Overhauser Enhancement (NOE) which follows from a redistribution of the populations of ^{13}C nuclei in the upper and lower energy states consequent upon the disturbance of the Boltzmann distribution of adjacent protons.

The other major development was the introduction of Fourier Transform (F.T.) N.M.R. in which a short high energy pulse of radiofrequency is used to excite all the ^{13}C nuclei in the sample. (In the continuous wave experiment each nucleus is brought into resonance in turn either in a "field-sweep" or "frequency-sweep" experiment). The resulting free induction decay signal contains information relatable to the precessional frequencies of all of the resonating nuclei, and automated Fourier Transform procedures extract this data to provide chemical shifts and signal shapes. A sensitivity gain of >10 is acquired relative to a continuous wave experiment, and repetitive pulsing (often many thousands of times), and time averaging processes are used to further increase sensitivity. These procedures clearly had to await appropriate computational facilities, and are now used routinely to provide natural abundance ^{13}C N.M.R. spectra on samples of *ca.* 10-100 mg. A.S. Perlin (in International Review of Science, Organic Chemistry, Series Two, ed. G.O. Aspinall, Vol.7, Butterworths, London, 1976, p.1) has reviewed ^{13}C N.M.R. spectroscopy applied to both simple and complex carbohydrates.

Figure 3 illustrates the finding that carbon resonances of carbohydrate derivatives occur roughly within the range 0-200 p.p.m. downfield of TMS (while proton signals lie approximately within 10 p.p.m. of the resonance of this reference substance). It also shows the similarity between the relative chemical shifts of many carbon atoms and their associated protons. Because of the good dispersion of the carbon signals and also their sharpness, excellent resolution can be achieved without recourse to high field technology, and measurements are frequently made using 18.5 or 23.5 kG magnetic fields in which ^{13}C nuclei resonate at 20 or 25 MHz, respectively. In these fields protons resonate at 80 or 100 MHz.

While the resolution of signals is seldom a problem with carbohydrate derivatives, their assignment when several carbon atoms have similar chemical shifts - as is often the case with, for example, $C_{(2)}$, $C_{(3)}$ and $C_{(4)}$ of similarly substituted pyranose derivatives - can be difficult. Methods based on analogy with related substances are fraught with danger because of far-reaching stereochemical or substitution influences, and unambiguous assignments are best made by use of compounds specifically labelled with ^{13}C or ^{2}H (see below). Other methods can, however, be of appreciable value: *(a)* Single frequency off resonance decoupling (SFORD) results in the loss of all ^{13}C-^{1}H coupling except one-bond coupling, and causes carbon atoms with n bonded hydrogen atoms to resonate as $(n + 1)$ multiplets. Signals for $C_{(6)}$ of hexopyranose derivatives and the anomeric carbon atoms of cyclic ketose deri-

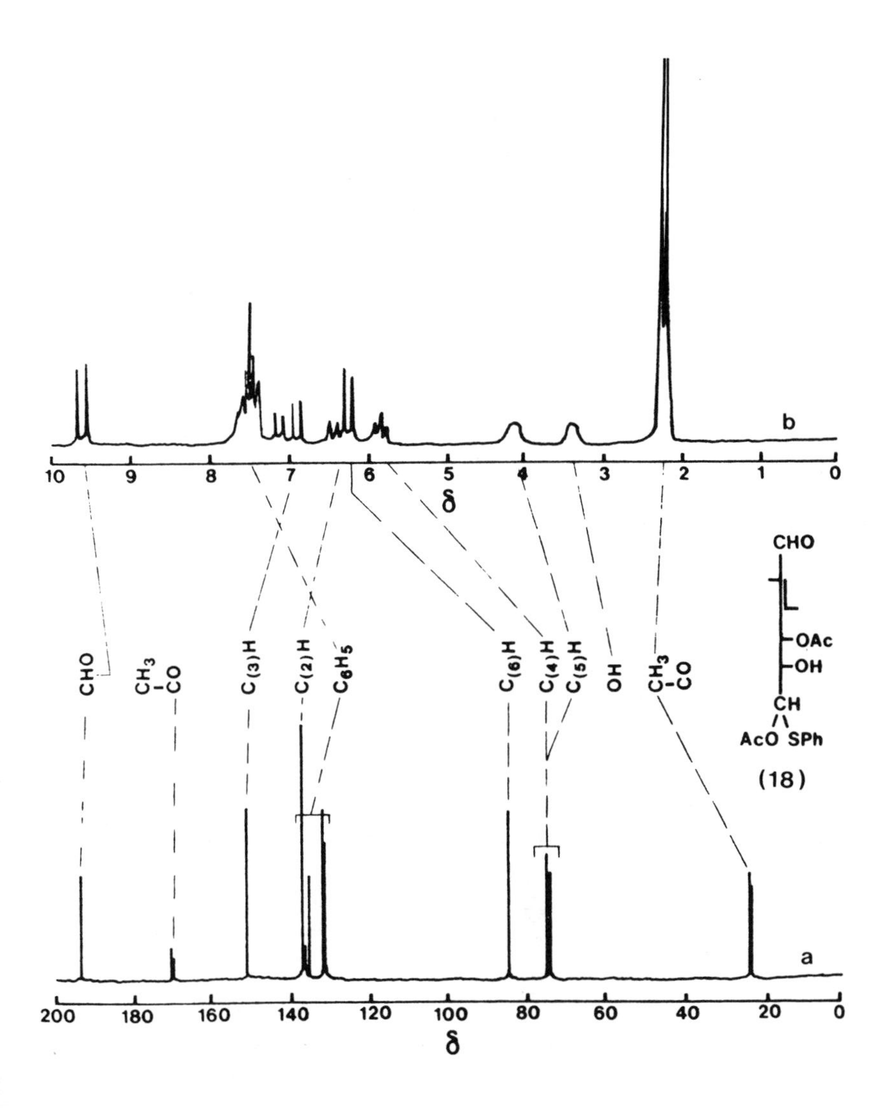

Fig. 3. ^{13}C (a) and ^{1}H (b) N.M.R. spectra of (E)-4,6-di-*O*-acetyl-2,3-dideoxy-6-phenylthio-D-*erythro*-hex-2-enose (18). Measurements were made in $CDCl_3$ solutions at 20 MHz for carbons and 80 MHz for protons.

vatives therefore give characteristic triplets and singlets, respectively. *(b)* A fully interpreted proton spectrum permits specific proton decoupling and consequent collapse of coupled carbon resonances and therefore their assignment. *(c)* $^{13}C-^{1}H$ and $^{13}C-^{13}C$ (labelled samples only) coupling constants are quantitatively dependent on both structural and stereochemical relationships (see below), and for compounds of known structure can be used for spectral interpretation. *(d)* While in ^{1}H N.M.R. spectra peak areas correlate with numbers of protons from which the peaks are derived, in ^{13}C spectroscopy peak areas and heights are highly dependent upon specific relaxation characteristics and Overhauser enhancement factors. These, in turn, depend upon the detailed parameters used in the experiment, and it is common to find resonances of unequal intensities for different carbon atoms of a sugar derivative. This is especially so for nuclei bearing different numbers of hydrogen atoms (which facilitate spin-lattice relaxation), but observations of this kind can be used for signal assignments only with due regard for the experimental methods used.

Initial reports of the ^{13}C N.M.R. spectra of free sugars appeared around 1970, mutarotated mixtures giving data for the main isomers (L.D. Hall and L.F. Johnson, Chem.Comm., 1969, 509; A.S. Perlin, B. Casu and H.J. Koch, Tetrahedron Letters, 1969, 2921; Canad.J.Chem., 1970, 48, 2596; D.E. Dorman and J.D. Roberts, J.Amer.chem.Soc., 1970, 92, 1355). Normally these are the pyranoses, but the spectrum of D-allose, for example, shows obvious resonances for both furanoses and pyranoses, and C. Williams and A. Allerhand (Carbohydrate Res., 1977, 56, 173), using a large (20 mm) probe, later detected the β-furanose of D-glucose present in aqueous solution to the extent of only 0.14 ± 0.02% (characterised by resonances at δ 103.8 [$C_{(1)}$], 82.1 [$C_{(2)}$] and 81.8 [$C_{(4)}$]. The method has been used with [1-^{13}C]-enriched compounds to show that the tetroses contain *ca.* 12% of the aldehydrol forms in dilute aqueous solution, while the remainders are in the forms of the furanoses (erythrose 25% α, 63% β; threose 51% α, 37% β). (A.S. Serianni, E.L. Clark and R. Barker, Carbohydrate Res., 1979, 72, 79. *c.f.* S.J. Angyal and R.G. Wheen, Austral.J.Chem., 1980, 33, 1001).

Assignments of the $C_{(1)}$ and $C_{(6)}$ resonances of the aldohexoses are readily made by correlating them with those of the bonded protons either by specific decoupling methods or, for the latter, by application of the SFORD technique. Methods involving isotopic labelling are, however, sometimes required for unambiguous assignment of the $C_{(2)}$-$C_{(5)}$ resonances. Deuterium has the effect on occasion of causing the disappearance of the ^{13}C resonance of the bonded carbon atom (quadrupole broadening and decreased NOE), or sometimes causing it to appear as a triplet (spin coupling; I for deuterium = 1); H.J. Koch and A.S. Perlin (Carbohydrate Res., 1970, 15, 403) using D-[3-^{2}H]glucose resolved an ambiguity concerning the assignment of the resonances for $C_{(2)}$ and $C_{(3)}$ of the sugar. As well as having these effects on the bonded carbon atoms (α-effect), deuterium causes slight shielding to adjacent atoms (β-effect; *ca.* 0.1 p.p.m.), and this also can be used to assist with assignments, a $C_{(5)}$ resonance being, for example, recognisable by deuteration at $C_{(6)}$ of a hexose (P.A.J. Gorin and M.Mazurek, Canad.J.Chem., 1974, 52, 458; 1975, 53, 1212). Otherwise, ^{13}C labelling can be used, and in the spectra of [1-^{13}C] enriched aldoses the signals for $C_{(2)}$ appear as doublets (J *ca.* 46 Hz) in consequence of large one-bond coupling with $C_{(1)}$. Similarly, the signals for $C_{(3)}$ can be assigned because $J_{C(1),C(3)}$ is observed (4 Hz) only for β-pyranoses, and likewise $C_{(5)}$ is recognisable since $J_{C(1),C(5)}$ is 2 Hz only for α-anomers (R. Barker *et al.*, J.Amer.chem.Soc., 1976, 98, 5807). Full assignments for the carbon atoms of some common hexopyranoses are given in Table 4.

TABLE 4

Chemical shifts of the carbon atoms of D-glucose, D-mannose and D-galactose (in H_2O with TMS as external reference)

Sugar	$C_{(1)}$	$C_{(2)}$	$C_{(3)}$	$C_{(4)}$	$C_{(5)}$	$C_{(6)}$
α-D-glucopyranose	93.6	73.2	74.5	71.4	73.0	62.3
β-D-glucopyranose	97.4	75.9	77.5	71.3	77.4	62.5
α-D-mannopyranose	95.5	72.3	71.9	68.5	73.9	62.6
β-D-mannopyranose	95.2	72.8	74.8	68.3	77.6	62.6
α-D-galactopyranose	93.8	70.0	70.8	70.9	72.0	62.8
β-D-galactopyranose	98.0	73.6	74.4	70.4	76.6	62.6

The anomeric carbon atoms, with two bonded oxygen atoms, are deshielded by 20-30 p.p.m. relative to the other secondary carbon atoms which, in turn, resonate 6-15 p.p.m. downfield from the exocyclic atoms. Inversion of an equatorial hydroxyl group has a shielding effect not only on the bonded atom but on all but the most distant atoms. Thus, $C_{(1)}$, $C_{(2)}$, $C_{(3)}$ and $C_{(5)}$ of α-D-glucopyranose are all significantly shielded (mean value 3.5 p.p.m.) relative to these atoms of the all-equatorial β-anomer. α- And β-D-galactopyranose, with axial hydroxyl groups at $C_{(4)}$, show analogous shielding effects, but there is surprisingly little change in the $C_{(4)}$ chemical shifts with inversion, and $C_{(1)}$, in these cases, undergoes least change amongst the other ring atoms. The anomeric D-mannopyranoses behave in almost analogous fashion at $C_{(2)}$-$C_{(5)}$ (the $C_{(5)}$ atoms undergoing least shift as is consistent with their being "*para*" to the inverted centre), but there is an anomaly at $C_{(1)}$ for the α-isomer. Inversion at $C_{(2)}$ of α-D-glucopyranose therefore does not give rise to the expected shielding of the anomeric carbon atom, but rather to a deshielding of *ca.* 2 p.p.m. This is consistent with an electronic influence arising from the operation of the enhanced anomeric effect in this specific situation (p.33), and the slight deshielding of $C_{(5)}$ is also consistent with this effect. However, a *gauche* interaction is removed by this change, so the phenomenon may have, at least partly, a steric basis. (A.S. Perlin, B. Casu and H.J. Koch *loc.cit.*).

Since effects on the protons bonded to the carbon atoms undergoing these changes operate in the opposite directions (i.e. they are deshielded if the attached carbon atoms are shielded), it has been concluded (Koch and Perlin, *loc.cit.*) that introduction of axial groups leads to alteration in the polarisation of the neighbouring H-C-O units consequent upon the resultant steric compression, and thus to the observed shifts. However, there may be factors other than those based on steric strain behind these observations since some saturated six-membered ring compounds with *syn*-diaxial oxygen atoms do not show the expected enhancements.

The effects of acetylating the hydroxyl groups of free sugars is remarkably small, the spectra of the anomeric penta-*O*-acetyl-D-glucopyranoses being very similar to those of the unsubstituted compounds. Although each carbon atom is apparently shielded by *ca.* 3 p.p.m. with respect to the anomers of D-glucose (Table 4) (D.E. Dorman and J.D. Roberts, J.Amer.

chem.Soc., 1971, 93, 4463; M.R. Vignon and J.A. Vottero, Tetrahedron Lett., 1976, 2445), this could result from solvent change; acetylation of simple alcohols results in deshielding by about this amount. Methylation of the anomeric hydroxyl groups, on the other hand, causes deshielding of $C_{(1)}$ by *ca.* 7 p.p.m. irrespective of configuration, and similar substitution at other ring oxygen atoms results in deshielding by 9-10 p.p.m. of the α-carbon atoms, and usually shielding by 0.5-1.5 p.p.m. of the β-atoms (H. Sugiyama, *et al.*, J.chem. Soc. Perkin I, 1973, 2425). This latter factor can, however, be as high as 4.5 p.p.m. when a hydroxyl group on the β-carbon atom is axial (W. Voelter *et al.*, Tetrahedron, 1973, 29, 3845).

Methods involving the use of labelled methyl groups (^{13}C and ^{2}H) and of specific proton decoupling have been applied to assign fully the sugar and methyl carbon resonances of methyl tetra-*O*-methyl-α- and β-D-glucopyranoside (J. Haverkamp, J.P.C.M. van Dongen and J.F.G. Vliegenthart, Tetrahedron, 1973, 29, 3431). Again the substitution effects are smaller at the anomeric centres (deshielding of 5.4 and 7.6 p.p.m. for the α- and β-compounds with respect to the free sugars; acetonitrile solutions) than at $C_{(2)}$, $C_{(3)}$, $C_{(4)}$ and $C_{(6)}$ (deshielding of 8.7-10.1 p.p.m.). $C_{(5)}$, on the other hand, is shielded by 2 p.p.m. in both the α- and β-series, whereas this atom is shielded by 1.2 and 1.5 p.p.m., respectively, when α- or β-D-glucose is specifically methylated at $C_{(4)}$ or $C_{(6)}$. The methyl carbon atoms for methyl 2,3,4,6-tetra-*O*-methyl-α- and β-D-glucopyranoside resonate as follows (δ values in acetonitrile):

$C_{(1)\alpha}$, 55.3; $C_{(1)\beta}$, 57.0; $C_{(2)\alpha}$, 58.4; $C_{(2)\beta}$, 60.4;

$C_{(3)\alpha}$, 60.7; $C_{(3)\beta}$, 60.7; $C_{(4)\alpha}$, 60.5; $C_{(4)\beta}$, 60.5;

$C_{(6)\alpha}$, 59.2; $C_{(6)\beta}$, 59.3.

This data can be used to characterise partially methylated glucoses following their conversion into fully substituted compounds by use of labelled methylating agents.

The ring carbon atoms of the methyl pentofuranosides resonate at somewhat lower fields than do the corresponding atoms of the pyranosides (Table 5), and *O*-methylation causes similar effects in both series, *i.e.* α-carbon atoms are deshielded by 8-11 p.p.m., and adjacent atoms are shielded by *ca.* 1-3 p.p.m. (P.A.J. Gorin and M. Mazurek, Carbohydrate Res.,

TABLE 5

Chemical shifts of the carbon atoms of the methyl D-pentofuranosides (in D_2O with TMS as external reference)

Methyl D-glycoside	$C_{(1)}$	$C_{(2)}$	$C_{(3)}$	$C_{(4)}$	$C_{(5)}$
α-ribofuranoside	104.2	72.1	70.8	85.5	62.6
β-ribofuranoside	109.0	75.3	71.9	83.9	63.9
α-arabinofuranoside	109.3	81.9	77.5	84.9	62.4
β-arabinofuranoside	103.2	77.5	75.7	83.1	64.2
α-xylofuranoside	103.0	77.7	76.0	79.3	61.5
β-xylofuranoside	109.6	80.9	76.0	83.5	62.1
α-lyxofuranoside	109.1	77.0	72.0	81.3	61.2
β-lyxofuranoside	103.2	72.9	70.7	81.9	62.4

1976, 48, 171). Generally *cis*-interactions between vicinal substituents lead to significant shieldings (A.S. Perlin *et al.*, Canad.J.Chem., 1975, 53, 1424). ^{13}C N.M.R. spectroscopy has proved invaluable for the investigation of the ketoses in solution since, by this technique, but not by ^{1}H spectroscopy, the specifically deshielded resonances associated with the anomeric centre can be recognised (see p.35).

The chloroacetyl group deshields α-carbon atoms by 2 p.p.m. more than does the acetyl group, and the trichloroacetyl and methanesulphonyl groups, likewise, are more deshielding by about 5 and 4-10 p.p.m., respectively (Vignon and Vottero, *loc.cit.*). Benzoates have only slight effect, penta-*O*-benzoyl-β-D-glucose giving in chloroform solution a set of ring carbon resonances which are all upfield from those of β-D-glucopyranose measured in water (Table 4) by 1-5 p.p.m. (R.J. Ferrier and P.C. Tyler, J.chem.Soc.Perkin I, 1980, 1528). Phosphate and sulphate esters deshield α-carbon atoms by 2-3 p.p.m. and 6-10 p.p.m., respectively.

Isopropylidene and benzylidene acetals cause deshielding of α-carbon atoms by 5-10 p.p.m. Approximate chemical shifts of other carbon atoms frequently found in carbohydrates are (δ values): ester carbonyl carbon, 165-170; amide carbonyl carbon, 175-177; ketonic carbonyl carbon, 190-200; deoxy groups, 35-40; vinylic carbons, *ca.* 100 and 145 for β- and α-atoms, respectively, of vinyl ethers (e.g. glycals), and 125-130 for olefinic atoms of such unsaturated compounds as alkyl 2,3-dideoxyhex-2-eno-pyranosides (A.I.R. Burfitt, R.D. Guthrie and R.W. Irvine, Austral.J.Chem., 1977, 30, 1037). Common substituent group atoms resonate as follows (δ values): isopropylidene, 26-30 and 128-135; benzylidene, 99-102 (acetal) and 126-136; acetoxy, 19-20 and 165-170; acetylamino, 22-24 and 175-177; methoxyl, 55-60.

Examination of the couplings undergone between pairs of ^{13}C nuclei and between ^{13}C and protons has produced generalisations of value for the assignments of resonances and for structural and conformational analysis. Because of the low natural abundance of ^{13}C, C,C couplings cannot be measured without the use of enrichment techniques, but with [1-^{13}C]-enriched sugars Barker *et al. (loc.cit.)* have determined such couplings over one, two and three bonds for pyranoid sugars (p.57).

From the $^{1}J_{^{13}C,^{1}H}$ values given in Table 6 it follows that coupling between anomeric carbon atoms and bonded protons is very dependent on orientation, coupling constants being 170 Hz when the proton is equatorial and 11 Hz less when axial. While a similar trend is apparent at other ring positions the effect is not uniformally so marked, and the coupling constants are somewhat smaller. By use of this parameter operating at the anomeric centre, conformational equilibria can be determined, and in this way it has been shown, in agreement with data in Table 2, that, while α-D-arabinopyranose adopts the $^{1}C_{4}$ conformation in solution, the β-anomer has a less marked tendency to assume this chair form (K. Bock and C. Pedersen, Acta chem. Scand., 1975, B29, 258). The values of these coupling constants are, however, dependent to some degree on the electronegativities of the anomeric substituents, glycosyl acetates, chlorides and fluorides having $^{1}J_{C(1),H}$ values of 177, 184 and 186 Hz when $H_{(1)}$ is equatorial and again about 11 Hz less than these values when the proton is axial (K. Boch, I.Lundt and C. Pedersen, Tetrahedron Letters, 1973, 1037).

Very much smaller coupling occurs between ^{13}C and ^{1}H nuclei over longer ranges, but correlations have been observed which are of value for making assignments or for stereochemical analysis. $^{2}J_{C,H}$ Coupling of *ca.* 5 Hz is found when there is a

TABLE 6

$^1J_{C,H}$ values (Hz) for methyl D-xylopyranosides and their 2,4-phenylboronates

Compound	$^1J_{C,H(1)}$	$^1J_{C,H(2)}$	$^1J_{C,H(3)}$	$^1J_{C,H(4)}$	$^1J_{C,H(5)}$
A	170	146	145	146	145, 148
B	159	154	148	151	145
C	159	144	144	147	142, 150
D	171	156	154	155	145

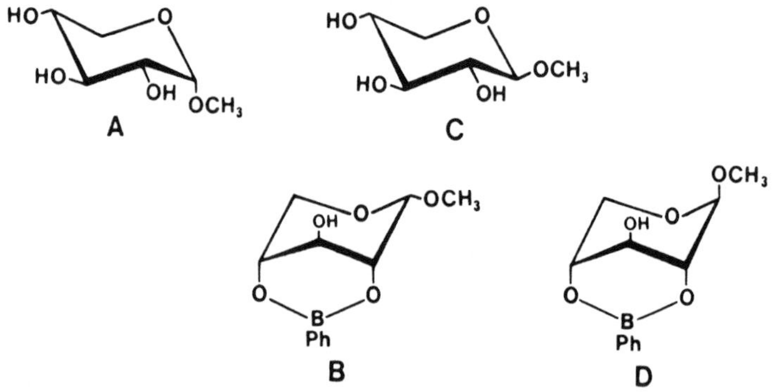

gauche relationship between the ^{13}C-O bond and the neighbouring C-H bond, but when this relationship is *anti* (i.e. when these bonds are both axial on a pyranoid ring) there is no coupling. An exception occurs at the anomeric centre, however, because of the presence of the second electronegative atom and no coupling is found between $C_{(2)}$ and $H_{(1)}$ for β-glucose or β-galactose derivatives (R.G.S. Ritchie, N. Cyr and A.S.Perlin, Canad.J.Chem., 1976, 54, 2301; K. Bock and C. Pedersen, Acta chem.Scand., 1977, B31, 354). For vicinally related ^{13}C and ^{1}H atoms a Karplus-like relationship is observed with $^3J_{C,H}$ = *ca.* 5 Hz occurring when the respective atoms are *anti*-related (e.g. $C_{(3)}$ and $H_{(1)}$ of α-D-glucopyranose). No substantial coupling is seen for *gauche*-related atoms (R.U. Lemieux, T.L. Nagabhushan and B. Paul, Canad.J.Chem., 1974, 50, 773). These data can be used in the assessment of the steric relationship between the aglycone and glycosyl components of glycosides and disaccharides (i.e. the rotamer state about the bond indicated) (R.U. Lemieux and S. Koto, Tetrahedron, 1974, 30, 1933; G. Excoffier, D.Y. Gagnaire and F.R. Taravel, Carbohydrate Res., 1977, 56, 229). They also are applicable to the determination

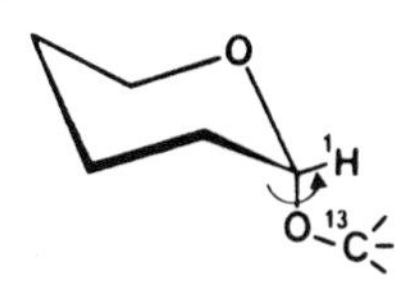

of the stereochemistry at branch points of branched-chain sugars.

(4) Heteronuclear N.M.R. studies. Considerable work has been carried out on fluorinated carbohydrates by use of ^{19}F spectroscopy, and the angular dependences of $J_{F,H}$ and $J_{F,F}$ (L.D.Hall *et al.*, Canad.J.Chem., 1971, 49, 236) and $^{3}J_{F,C}$ and $^{4}J_{F,C}$ values (V. Wray, J.chem.Soc.Perkin II, 1976, 1598) have been established.

References to studies into the N.M.R. spectroscopy of other isotopes incorporated in, or associated with, carbohydrates are:

^{2}H (J.R. Campbell, L.D. Hall and P.R. Steiner, Canad.J.Chem., 1972, 59, 504) - deuterium chemical shifts measured by internuclear double resonance (INDOR) spectroscopy;

^{11}B (A.M. Yurkevich, *et al.*, Zh.obshch.Khim., 1974, 44, 2327) - used to distinguish between 5- and 6-membered cyclic boronate esters;

^{15}N (R.E. Botto and J.D. Roberts, J.org.Chem., 1977, 42, 2247) - applied to 2-amino-2-deoxyhexose derivatives;

^{23}Na (C. Detellier, J. Grandjean and P. Laszlo, J.Amer.chem. Soc., 1976, 98, 3375) - line width studies of sodium-sugar interactions;

^{29}Si (A.H. Haines, R.K. Harris and R.C. Rao, Org.Magn.Resonance, 1977, 9, 432) - studies of trimethylsilylated glycosides;

^{31}P (R.D. Lapper, H.H. Mantsch and I.C.P. Smith, J.Amer.chem.

Soc., 1973, 95, 2878) - ^{1}H-^{13}P and ^{13}C-^{31}P couplings in cyclic nucleotides;

^{117}Sn and ^{119}Sn (L.D. Hall, P.R. Steiner and D.C. Miller, Canad.J.Chem., 1979, 57, 38) - ^{13}C-Sn couplings in C-stannane derivatives;

^{199}Hg (V.G. Gibb and L.D. Hall, Carbohydrate Res., 1977, 55, 239) - ^{1}H-^{199}Hg and ^{13}C-^{199}Hg couplings in carbon-mercury bonded compounds.

(v) Optical rotatory power, optical rotatory dispersion and circular dichroism

Very simple and elegant N.M.R. experiments allowed Lemieux *et al.* (Canad.J.Chem., 1969, 47, 4427) to conclude that when the optical rotation of a pyranoid compound alters with change of solvent this follows from a conformational change rather than, for example, from differences in molecular solvation. This led him, with J.C. Martin, (Carbohydrate Res., 1970, 13, 139) to revise Whiffen's empirical method for calculating the optical rotations ($[\alpha]_D$) of pyranoid compounds in specific chair conformations. Reciprocally, measured rotational values can be used to determine conformations or conformational equilibria despite the authors' emphasis on the imprecise nature of such empirical approaches.

Their method both extends and simplifies the procedures employed by Whiffen (Second Edition), by using four rotational parameters based on pairwise interactions between vicinal, *gauche*-related oxygen and carbon atoms. The approach ignores related interactions involving hydrogen atoms, but distinguishes between carbon atoms according to whether they are bonded to ring carbon or oxygen atoms in the projections. The parameters with their assigned notations and rotational contribution values are shown below.

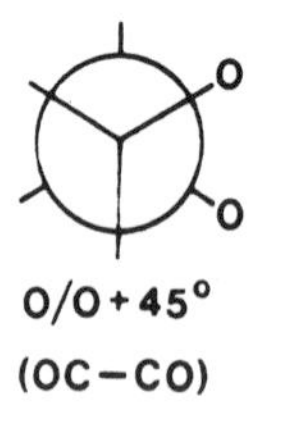

O/O +45°
(OC–CO)

O/C +10°
(OC–CC)

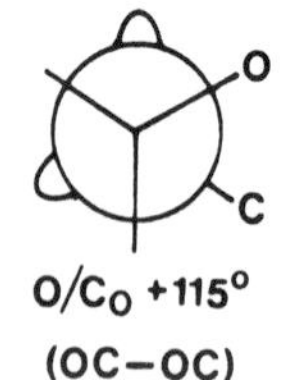

O/C$_O$ +115°
(OC–OC)

C/C$_O$ +60°
(CC–OC)

The analyses take into account contributions from rotamers about the $C_{(5)}$-$C_{(6)}$ bond of hexopyranoid compounds and the $C_{(1)}$-$O_{(1)}$ bond of glycosides, and can provide evidence on preferred rotamers and their solvent dependencies (R.U.Lemieux, Pure appl.Chem., 1971, 27, 527; R.U.Lemieux and S.Koto, Tetrahedron, 1974, 30, 1933).

Following several O.R.D. studies of sugars and simple derivatives (R.J. Ferrier in "The Carbohydrates", Vol. 1B, ed. D. Horton and W. Pigman, Academic Press, New York, 1980, p.1354), attention has turned to C.D. work because, for complex systems, simpler and more interpretable data are obtained. Techniques which allow study in the vacuum ultraviolet have led to the finding that free sugars show Cotton effects near 170 nm, but the signs do not correlate with anomeric configuration. α- And β-D-glucopyranose and α-D-galactopyranose all give positive effects, while β-D-galactopyranose affords a negative curve (R.G. Nelson and W.C. Johnson, J.Amer.chem. Soc., 1976, 98, 4290). Methyl glycopyranosides show bands near 185 nm (associated with the ring oxygen chromophore), 175 nm (aglycone) and below 165 nm, and the signs of the second and third of these correlate with the anomeric configurations (*idem. ibid.*, p.4296). Johnson (Carbohydrate Res., 1977, 58, 9) then compared the effects given by related pairs of pyranoid compounds and compiled a set of "fragment" C.D. spectra for pyranoid derivatives from which spectra of specific compounds can be predicted.

Several chiroptical methods are now available for the determination of the absolute configurations of *gauche*-related vicinal diols following the observation of N. Harada and K. Nakanishi (J.Amer.chem.Soc., 1969, 91, 3989) that the signs of Cotton effects near 233 nm ($\pi \rightarrow \pi^*$ intramolecular charge transfer transitions) observed for the derived dibenzoates are positive when their absolute orientation is as shown (*idem.*, Acc.chem.Res., 1972, 5, 257).

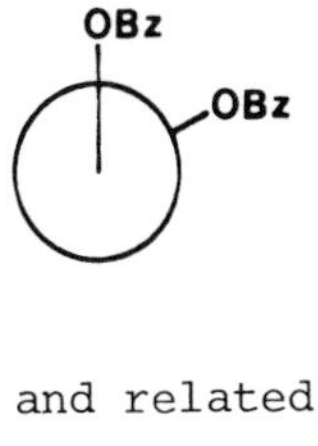

OBz
OBz

Complexes of such diols and related α-hydroxyamines with

cuprammonium ions (R.D. Guthrie *et al.*, Tetrahedron, 1970, 26, 3653; W. Voelter, H. Bauer and G. Kuhfittig, Ber., 1974, 107, 3602), Ni(acetylacetonate)$_2$ (J.Dillon and K. Nakanishi, J.Amer. chem.Soc., 1975, 97, 5409) and Pr(dpm)$_3$ (*idem.ibid.*, 1975, 97, 5417) likewise give C.D. maxima the signs of which can be correlated with the absolute chirality of the carbohydrate components. It has been found, however, that inconsistencies can arise, and that methyl 4,6-*O*-ethylidene-α-D-glucopyranoside, complexed with Pr(dpm)$_3$, gives a spurious result (C.W. Lyons and D.R. Taylor, Chem.Comm., 1976, 647). Conversion of α-diols into cyclic thionocarbonates gives yet another method for determining absolute stereochemistries since the configurations of the diols correlate with the signs of Cotton effects associated with n → π* bands at *ca.* 316 nm (A.H.Haines and C.S.P. Jenkins, J.chem.Soc.(C), 1971, 1438). A specific method advocated for similar analysis of α-diamines depends upon examination of the C.D. spectra of bis-2,4-dinitrophenyl derivatives (M.Kawai, U. Nagai and T. Kobayashi, Tetrahedron Letters, 1974, 1881).

Otherwise, by an extension of a method used for determining the absolute configuration of secondary alcohols, configurational analysis can be carried out on α-diols. Partial reaction with 2-phenylbutanoic anhydride causes partial kinetic resolution of the anhydride, and the optical activity (single wavelength) of unreacted material correlates with the required configuration (D.P.G. Hamon, R.A. Massy-Westropp and T. Pipithakul, Austral.J.Chem., 1974, 27, 2199).

Several series of compounds are known which have a heterocyclic ring bonded to $C_{(1)}$ of alditol derivatives, and chiroptical studies have often been applied to relate the configurations of the chiral centres adjacent to the heterocycles with the signs of the measured optical activities or Cotton effects (H.S. El Khadem and Z.M. El-Shafei, Tetrahedron Letters, 1963, 1887). El Khadem has amplified these generalisations to take into account important dipolar factors associated with the groups attached to the chiral centres (Carbohydrate Res., 1977, 59, 11).

A symmetry rule has been enunciated which relates the signs of the n → π* C.D. bands at 275-290 nm (observable with very sensitive techniques) derived from the carbonyl forms of free sugars in aqueous solution with the configurations at the chiral centres adjacent to the carbonyl groups. A positive band reveals an (S) centre, and the rule also applies to acetylated carbonyl forms of the free sugars (L.D. Hayward and S.J. Angyal, Carbohydrate Res., 1977, 53, 13). Changes in the ellipticities in this wavelength range of free sugars on dissolution in water have been correlated with mutarotation rates (G.D. Maier, J.W. Kusiak and J.M. Bailey, *ibid.*, 1977, 53, 1).

Compounds with formally free carbonyl groups usually absorb in the 280-300 nm region, but this may be as high as 330 nm for strained cyclic derivatives. Ketonic derivatives of 1,6-anhydropyranoses show ellipticities which correlate with expectations based on application of the carbonyl octant rule (K. Heyns, J. Weyer and H. Paulsen, Ber., 1967, 100, 2317).

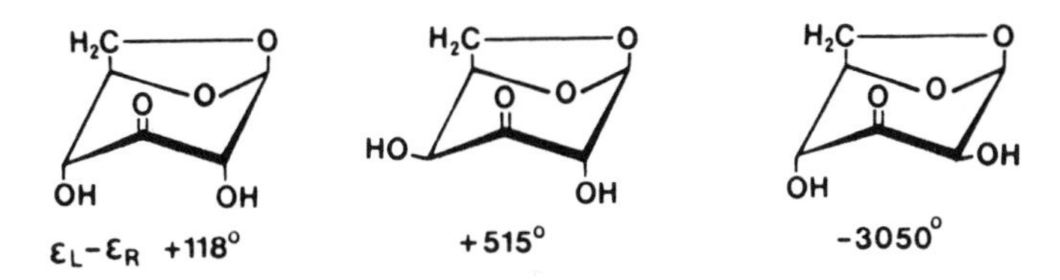

The signs of 240 nm C.D. bands derived from aldose ethylene dithioacetals or diethyl dithioacetals can be correlated with absolute configurations at $C_{(3)}$ (M.K. Hargreaves and D.L. Marshall, Carbohydrate Res., 1973, 29, 339). For 1-thioglycopyranosides, the signs and magnitudes of C.D. bands derived from $n \rightarrow \sigma^*$ transitions of the sulphur atoms can be correlated with the geometric relationships between these atoms and the oxygen atoms at $C_{(2)}$ and in the rings. α-D-Mannopyranosides show particularly strong positive dichroism, β-anomers show negative bands, and α- and β-D-glucopyranosides both exhibit intermediate positive effects, the former being stronger than the latter (E. Ohtaki and H. Meguro, Tetrahedron Letters, 1979, 3297).

Generalisations have also been noted regarding the signs of Cotton effects observed at 280-290 nm and the configurations at $C_{(4)}$ of 4-azido-4-deoxypentopyranose derivatives, those compounds with the *S* stereochemistry giving positive effects regardless of the conformations of the rings (T. Sticzay, P. Šipoš and Š. Bauer, Carbohydrate Res., 1969, 10, 469). H. Paulsen (Ber., 1968, 101, 1571) has extended this into a general octant rule for pyranoid carbohydrate azides.

(c) Separation, identification and estimation of monosaccharides by chromatography and related techniques

Perhaps the two main developments since Hough and Richardson dealt at some length with separatory methods in the Second Edition have been the coupling of mass spectrometry with gas-liquid chromatography and the introduction of high pressure liquid chromatography. These will be given brief treatment, and comment on established techniques will be minimal. A considerable body of literature on applications and adaptations of all methods has however built up since 1967, and readers are referred to the Chemical Society's Specialist Periodical Reports for correlated details.

The associated use of mass spectrometry and gas-liquid chromatography has revolutionised the analysis of mixtures of sugars and their derivatives. Applied to tiny samples of derivatised materials they provide information of both a qualitative and quantitative nature on the composition of such mixtures. G.G.S. Dutton and J. Lönngren and S. Swensson have commented on earlier aspects of this subject (Adv.Carbohydrate Chem.Biochem., 1974, 30, 9; 1974, 29, 41), and B. Lindberg *et al.* (Angew.Chem.internat.Ed., 1970, 9, 610) have discussed the application of the technique to the analysis of mixtures of methyl ethers produced during structural studies of polysaccharides.

To remove problems associated with the isomeric forms of cyclic free sugars, these are frequently reduced to the alditols prior to derivatisation for G.L.C., but during this process the terminal groups can become indistinguishable. This drawback can be overcome by converting the free sugars to oximes and thence, by use of acetic anhydride, into peracetylated aldononitriles which can be subjected to G.L.C.-M.S. analysis (F.R. Seymour, E.C.M. Chen and S.H. Bishop, Carbohydrate Res., 1979, 73, 19). Chemical ionisation M.S. gives molecular weights (and hence the number of aldehyde and hydroxyl groups) of the original sugars, and electron impact M.S. affords data on the nature and positions of groups other than the hydroxyl groups. Isopropylidene derivatives of isomeric sugars can differ structurally as well as configurationally, and can be subjected to the G.L.C.-M.S. procedure for identification purposes (S. Morgenlie, *ibid.*, 1975, 41, 285). Methaneboronic and butaneboronic esters can be used in analogous fashion after trimethylsilylation of unsubstituted hydroxyl groups for the characterisation of sugars and the determination of the nature of their interactions with the boronic acids (V.N. Reinhold, F. Wirtz-Peitz and K. Biemann, *ibid.*, 1974, 37, 203; P.J. Wood, I.R. Siddiqui and J. Weisz, *ibid.*, 1975, 42, 1).

High pressure (performance) liquid chromatography involves the use of columns of microparticulate adsorbents through which solutes are forced under high pressure. Because of the greatly increased adsorbent surface areas relative to those involved in normal column chromatograms, considerably enhanced separations are frequently achieved, and the method offers the further advantages of being rapid and applicable both analytically and preparatively. A considerable limitation is associated with the detection of components as they are eluted. Differential refractometers, polarimeters or ultraviolet spectrophotometers (preferably operating below 200 nm) can be used with carbohydrates, or compounds can be derivatised to afford readily detectable products e.g. *p*-nitrobenzoates (R. Schwarzenbach, J.Chromatog., 1977, 140, 304). Free sugars have been separated on cation resins (W. Funasaka *et al.*, J. Chromatog.Sci., 1974, 12, 517), anion resins (W. Voelter and H. Bauer, J.Chromatog., 1976, 126, 693) or silica (J.L.Rocca and A. Rouchouse, *ibid.*, 1976, 117, 216). A wide range of sugar derivatives have also been subjected to separation by the technique.

The following references contain compilations of data on separatory methods applied to sugars and their derivatives:
J.Chromatog., 1971, 62, 173 (previously published paper chromatographic data; see also other compilations in this Journal).
S.J. Angyal and J.A. Mills, Austral.J.Chem., 1979, 32, 1993 (sugars and derivatives separated electrophoretically as calcium ion complexes).
B. Pettersson and O. Theander, Acta chem.Scand., 1973, 27, 1900 (sugars and derivatives separated electrophoretically as anion complexes with sulphonated phenylboronic acid).
M. Lato *et al.*, J.Chromatog., 1968, 34, 26; 1969, 39, 407 (surveys of T.L.C. separations of sugars).
G.G.S. Dutton, Adv.Carbohydrate Chem.Biochem., 1973, 28, 11; 1974, 30, 9 (reviews of G.L.C. of carbohydrate derivatives).
S.C. Churms, *ibid.*, 1970, 25, 13 (review of gel chromatography of sugars).
R.J. Ferrier, *ibid.*, 1978, 35, 31 (review of separations of sugars as boronates or boronic acid complexes; paper, column,

G.-L. chromatography, electrophoresis).
P.Jandera and J. Churáček, J.Chromatog., 1974, 98, 55 (survey of separations of carbohydrates on anion-exchange resins).
K. Capek and J. Stanek, J.Chromatog.Libr., 1975, 3, 465 (review of column chromatographic separations of carbohydrates).

(d) Synthesis of monosaccharides

In the area of total synthesis of monosaccharides considerable attention has been devoted to the formose reaction and the subject has been reviewed (T. Mizuno and A.H. Weiss, Adv. Carbohydrate Chem.Biochem., 1974, 29, 173), but the most striking advance has been in the synthesis of specific sugars and their derivatives from non-carbohydrate materials.

One approach has elaborated DL-ribose from a furan derivative (O. Achmatowicz and G. Grynkiewicz, Carbohydrate Res., 1977, 54, 193) (Scheme 4).

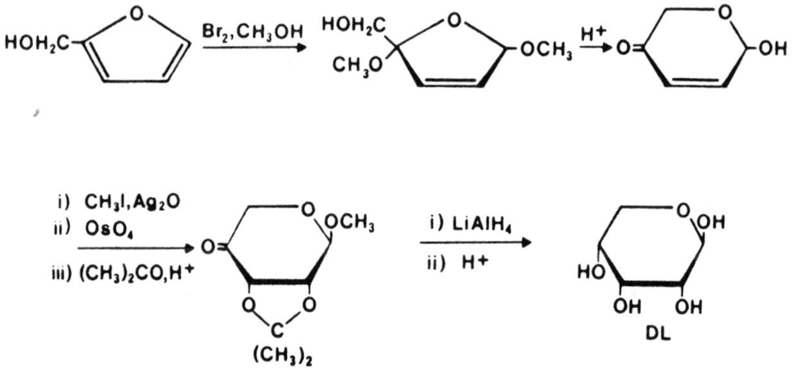

SCHEME 4

Applied to the resolved (S)-form of a 6-carbon furan derivative, this approach has also afforded a synthesis of methyl α-D-glucopyranoside (O. Achmatowicz and R. Bielski, *ibid.*, 1977, 55, 165) (Scheme 5).

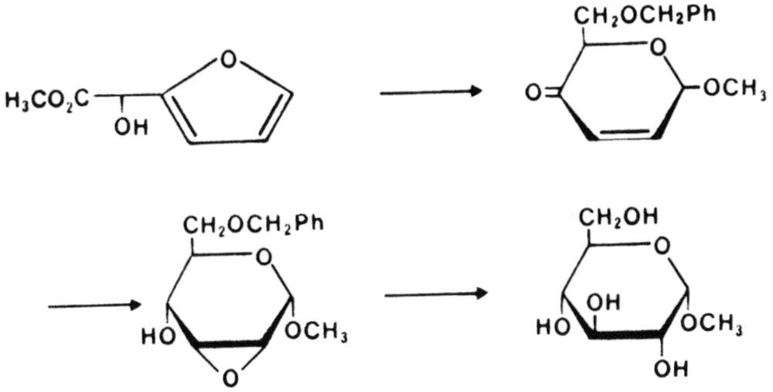

SCHEME 5

Dihydropyran rings can be made by use of 4+2 cycloaddition reactions of the Diels Alder type, with a carbonyl group in either the diene or the dienophile, and the products have been converted to free sugars in the racemic form. 2-Deoxy-DL-*erythro*-pentose has thus been made from butadiene (V.B. Mochalin and A.N. Kornilov, Zhur.obshch.Khim., 1973, 43, 218) (Scheme 6).

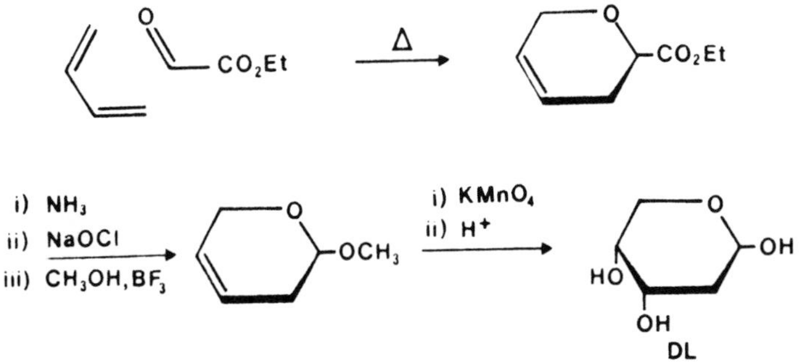

SCHEME 6

In related fashion, acrolein dimerises to give an aldehydic dihydropyran derivative from which DL-glucose has been synthesised as outlined in Scheme 7 (U.P. Singh and R.K. Brown, Canad.J.Chem., 1971, 49, 3342).

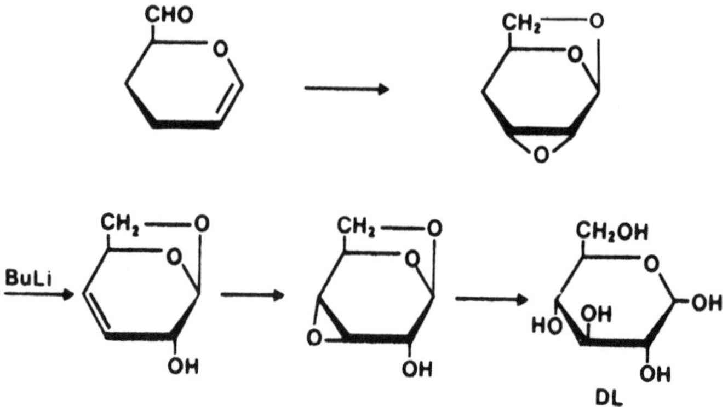

SCHEME 7

An entirely different approach to total synthesis involves the free radical polymerisation of vinylene carbonate in halogenated solvents. Amongst the products formed in methylene dibromide are dibromides (18) and (19) comprising two monomers and one solvent molecule from which 5-bromo-5-deoxy-DL-lyxose and -xylose are obtained (T. Tamura, T. Kunieda and T.Takizawa, J.org.Chem., 1974, 39, 38). Trimeric analogues offer a route to heptose derivatives (Y. Nii, J. Kunieda and T. Takizawa, Tetrahedron Letters, 1976, 2323).

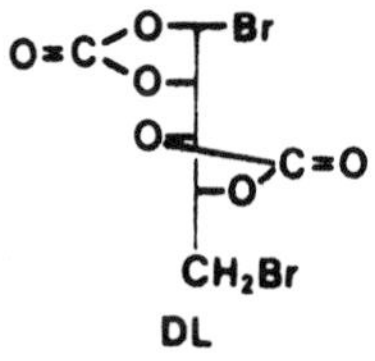

(18) 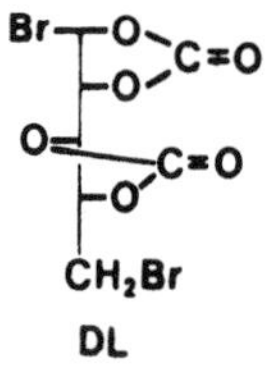 (19)

Very many syntheses which involve extension of the chain of sugar derivatives have been reported, the most useful approach being based on application of the Wittig reaction. The example outlined in Scheme 8 - one of the most complex so far achieved (J.A. Secrist and S.-R. Wu, J.org.Chem., 1979, 44, 1434) - illustrates the scope of this approach by elaborating the framework of an undecose.

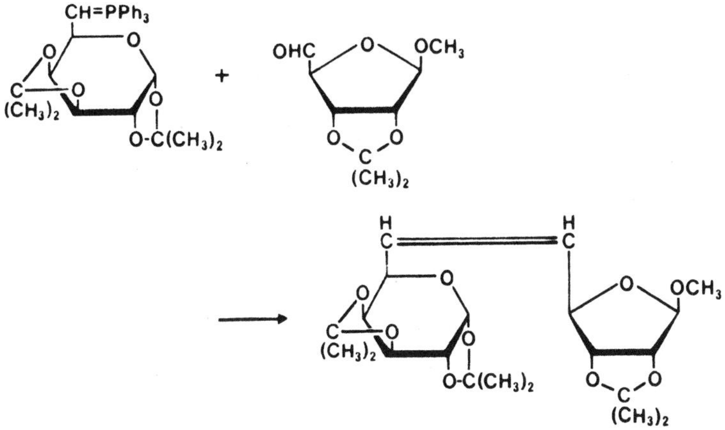

SCHEME 8

Developments in the reverse process, i.e. the descent of the sugar series, include a photochemical modification of the Wohl degradation by which D-lyxose can be produced from the oxime of D-galactose (W.W. Binkley and R.W. Binkley, Tetrahedron Letters, 1970, 3439). Otherwise, this degradation can be effected by treating 2,3,4,5,6-penta-*O*-acetyl-D-galactose phenylhydrazone with oxygen to give a $C_{(1)}$ hydroperoxide which, with dilute sodium methoxide, also gives D-lyxose (M. Schulz and L. Somogyi, Angew.Chem.internat.Edn., 1967, 6, 168). Related $C_{(1)}$-$C_{(2)}$ bond cleaving reactions can be effected on peroxyesters of aldonic acid derivatives (M. Schulz and P. Berlin, *ibid.*, 1967, 6, 950) and t-butyl peroxyglycosides (M. Schulz, H.-F. Boeden and P. Berlin, Ann., 1967, 703, 190). Applied in the ketose series this last method leads to loss of $C_{(1)}$ and $C_{(2)}$, L-threose and D-erythrose being obtainable from L-sorbose and D-fructose, respectively (M. Schulz *et al.*, Z.Chem., 1967, 7, 13). An ingeneous synthesis of L-glucose is based on periodate cleavage of the $C_{(6)}$-$C_{(7)}$ bond of D-*glycero*-D-*gulo*-heptonolactone obtained by hydrogen cyanide addition to D-glucose (W. Sowa, Canad.J.Chem., 1969, 47, 3931).

Many examples have been reported of configurational changes to covert one sugar into another. Epimerisation occurs with aldoses when they are heated in aqueous solution with molybdic acid, compounds with *trans*-2,3-diols (Fischer projection formulae) giving access to *cis*-related epimers. D-Talose is thus available, for example, directly from D-galactose (*ca.*

16%) (V. Bilik, W. Voelter and E. Bayer, Ann., 1974, 1162). Alternatively, furanose derivatives with the *cis*-configuration at $C_{(2)}$, $C_{(3)}$ on acetolysis with acetic acid, acetic anhydride in the presence of sulphuric acid undergo inversion at $C_{(2)}$. The reaction probably involves a 2,3-acetoxonium ion which then isomerises with configurational inversion at $C_{(2)}$. 1,2: 5,6-Di-*O*-isopropylidene-α-D-allose, in this way, gives D-altrose in 45% yield after deacetylation of the products (Scheme 9) (W. Sowa, Canad.J.Chem., 1971, 49, 3293; 1972, 50, 1092).

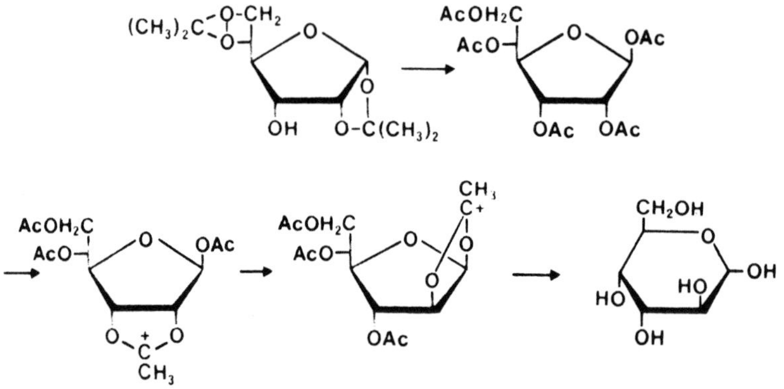

SCHEME 9

Related reactions occur when penta-*O*-acetyl-β-D-galactopyranose is treated with antimony pentachloride. In this case, an initial 1,2-acetoxonium ion partly isomerises to the 2,3-D-*talo*-ion by attack at $C_{(2)}$ of the carbonyl group of the ester at $C_{(3)}$. Hydrolysis of the mixed ionic products gives monohydroxy compounds which, after acetylation, afford a mixture from which penta-*O*-acetyl-α-D-talopyranose crystallises. By extension of this principle a most elegant route to D-idose from D-glucose has been devised, the α-D-*ido*-antimony hexachloride salt crystallising from the reaction mixture (Scheme 10) (H. Paulsen, Adv.carbohydrate Chem.Biochem., 1971, 26, 127).

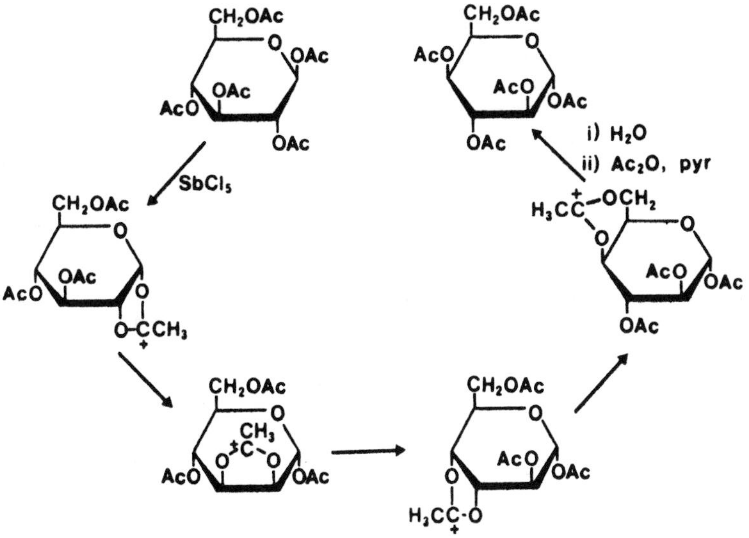

SCHEME 10

A new synthesis of ketoses in their acyclic, acetylated form is provided by the oxidation of acetylated monomethylene or monobenzylidine acetals of alditols. In this way, 1,3,4,6-tetra-*O*-acetyl-2,5-*O*-methylene-D-mannitol gives 1,3,4,6-tetra-*O*-acetyl-5-*O*-formyl-D-fructose in high yield (S.J. Angyal and K. James, Austral.J.Chem., 1971, 24, 1219).

Inversions at single asymmetric centres have been applied in the preparation of rare hexoses as indicated below:

D-Psicose; oxidation, reduction at $C_{(3)}$ of a D-fructose derivative (R.S. Tipson, R.F. Brady and B.F. West, Carbohydrate Res., 1971, 16, 383).

L-Psicose; oxidation, reduction at $C_{(4)}$ of an L-sorbose derivative (A. Armenakian, M. Mahmood and D. Murphy, J.chem. Soc.Perkin I, 1972, 63).

D. Tagatose; oxidation, reduction at $C_{(4)}$ of a D-fructose derivative (R.D. Guthrie *et al.*, Carbohydrate Res., 1971, 16, 474).

D-Gulose; oxidation at $C_{(3)}$ of a D-glucofuranose derivative, enolisation to $C_{(4)}$ and reduction of the derived enol acetate (W. Meyer zu Reckendorf, Ber., 1969, 102, 1071).

L-Idose; hydrolysis of 5,6-anhydro-1,2-*O*-isopropylidene-β-L-idofuranose prepared following reduction at $C_{(6)}$ of 1,2-*O*-

isopropylidene-5-*O*-toluene-*p*-sulphonyl-α-D-glucofuranurono-6,3-lactone (M. Blanc-Muesser and J. Defaye, Synthesis, 1977, 568; *c.f.* H. Weidmann *et al.*, Carbohydrate Res., 1980, 80, 45).

(e) Natural sources

Since Hough and Richardson's very full treatment of this topic, D-allose has been found to occur in the leaves of plants as substituted aryl glycosides (G.W. Perold, P. Beylis and A.S. Howard, J.chem.Soc.Perkin I, 1973, 643). D-Altrose has been reported as a component of a fungal glycoside (N. Cagnoli Bellavita *et al.*, Gazz.chim.Ital., 1969, 99, 1354), and L-gulose occurs in the bleomycin antibiotics (H. Umezawa *et al.*, J.Antibiot., 1977, 30, 388). It appears, therefore, that aldohexoses with all possible relative configurations have been identified in Nature except *ido*-compounds. It is noteworthy that, whereas the most abundant - glucose, mannose and galactose - all have relatively stable pyranose rings, idose has the least stable 6-membered ring.

Many new reports of the occurrence of hept-2-uloses and oct-2-uloses have appeared, but the majority of the isomers found were recorded in the Second Edition; the most commonly encountered of the latter is the D-*glycero*-D-*manno*-isomer. The first naturally occurring hept-3-ulose - the D-*altro*-compound (coriose) - has been described (T. Okuda and K. Konishi, Chem.Comm., 1968, 553).

(f) Isomerisation and degradation of monosaccharides in basic media

Although much evidence continues to support the belief that the base-catalysed isomerisation of free sugars takes place by way of 1,2-enolate anions, some data indicate that other factors and possibilities must be considered. To help account for discrepancies (for example in the differences in the positions of the *pseudo*-equilibria attained starting from epimeric aldoses) H.S. Isbell *et al.* (Carbohydrate Res., 1969, 9, 163) have postulated the significance of *pseudo*-acyclic species formed by momentary ring-opening which retain geometric features of their cyclic precursors. They then propose that enolisation leads to specific geometric isomers of the derived enediolates, α-D-glucopyranose and β-D-mannopyranose, for example, affording the same *cis*-species, and their anomers giving the isomeric *trans*-ion. With D-fructose, enolisation occurs specifically at $C_{(1)}$-$C_{(2)}$, and both ions can be produced following removal of the different hydrogen atoms at $C_{(1)}$. In this way glucose, mannose and fructose all afford two intermediate ions, but in different proportions. (Scheme 11). The subsequent fates of these species are controlled by

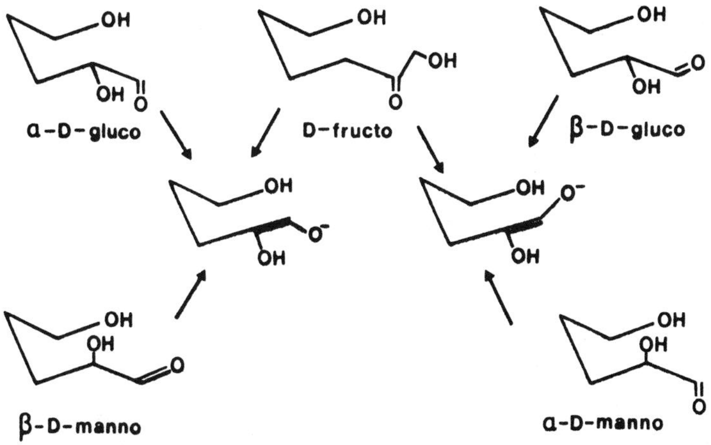

SCHEME 11

several competing forces which determine the natures and proportions of reaction products.

This enolisation mechanism for isomerisation requires that $H_{(2)}$ of an aldose is transferred to the solvent during the process, but it has been observed that D-arabinose, produced by treatment of D-[2-^{3}H]ribose with aqueous potassium hydroxide, contains appreciable label at $C_{(1)}$ indicating that reaction occurred to a significant degree by hydride (tritride) transfer and via the corresponding ketose (Scheme 12) (W.B. Gleason and R. Barker, Canad.J.Chem., 1971, 49, 1433)

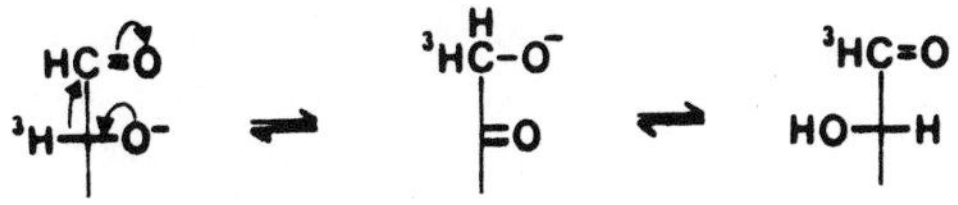

SCHEME 12

That this is not the general mode of reaction however, was established by Isbell (Advan.Chem.Ser., 1973, 117, 70) who found that D-fructose, produced by isomerisation of D-[2-^{3}H] glucose under the influence of aqueous calcium hydroxide, is unlabelled. However, the situation is further complicated by the opposing observation that enzymic isomerisation of D-[1-^{2}H] fructose 6-phosphate gives D-glucose 6-phosphate with a considerable proportion of label at $C_{(2)}$ (I.A. Rose and E.L. O'Connell, J.biol.Chem., 1961, 236, 3086).

For the purposes of preparing ketoses, pyridine is often used as isomerising base. In the case of the tetroses it avoids the aldol dimerisation caused by aqueous bases and, from D-erythrose and D-threose, gives access to D-*glycero*-tetrulose exclusively. This establishes that no enolisation occurs between $C_{(2)}$ and $C_{(3)}$ since this would result in racemisation. (K. Linek, M. Fedoroňko and H.S. Isbell, Carbohydrate Res., 1972, 21, 326). Again, however, the situation may be more complex than this, and all four D-aldopentoses, D-*erythro*- and D-*threo*-pent-2-ulose and D-*erythro*- and D-*threo*-pent-3-ulose are present in the products of pyridine-catalysed isomerisation of D-arabinose (M. Fedoroňko and K. Linek, Coll.Czech, chem.Comm., 1967, 32, 2177). Presumably, given enough time, enolisation will occur at all sites α- to carbonyl groups and lead to all possible products.

Aqueous solutions can be used for the production of some ketoses, and the proportions of products can be favourably affected by the addition of suitable complexing agents. Phenylboronic acid increases significantly the proportion of D-fructose available directly from D-glucose (up to 73%), and a technique involving the use of polymerised *p*-styrylboronic acid has been developed for the industrial production of the ketose (S.A. Barker *et al.*, Carbohydrate Res., 1973, 26, 41, 55). The reaction is not confined to the monosaccharide series, lactulose, for example, being obtainable in good yield by calcium hydroxide-isomerisation of lactose.

Oxidative reactions of free sugars which occur in alkaline solution in the presence of oxygen occur by way of enolate anions. These undergo both isomerisations and C-C bond cleavage reactions to give a range of carboxylic acids. D-Glucose and D-fructose afford mixed acids from D-arabinonic acid to formic acid; kinetic models have been developed for the processes (H.G.J. De Wilt *et al.*, *ibid.*, 1971, 19, 5; 1972, 23,

333, 343).

Added hydrogen peroxide causes more extensive degradation, aldohexoses giving formic acid exclusively while ketohexoses give one equivalent of glycolic acid as well as four of formic acid (H.S. Isbell *et al.*, *ibid.*, 1973, 26, 287; 28, 295).

In the presence of transition metals (e.g. platinum or rhodium on charcoal) sugars are oxidised efficiently in alkaline media to their aldonic acids (G. de Wit, *et al.*, Tetrahedron Letters, 1978, 1327).

M.S. Feather and J.F. Harris (Adv.carbohydrate Chem. Biochem., 1973, 28, 161) have summarised alkaline degradation reactions of carbohydrates which lead to both fragmentation products and saccharinic acids. That these reactions are extremely complex is further exemplified by the finding of cyclopentane derivatives amongst the products of alkaline degradation of several carbohydrates (R. Koetz and H. Neukom, Carbohydrate Res., 1975, 42, 365).

(g) The action of acids on monosaccharides

Reversion in 10% D-glucose solutions in the presence of dilute hydrochloric acid leads to the production of oligosaccharides up to heptaoses with bonding mainly 1 → 6 and 1 → 4 (A. Soler, M.C. Martinez Sanchez and P.S. Garcia Ruiz, An.Quim., 1976, 72, 910, 957, 961), but in concentrated sulphuric acid at low temperatures considerable proportions of non-dialysable products are formed. These non-specific materials comprise several sugars, each bearing 2-3 sulphate ester groups (K. Nagasawa and Y. Inoue, Carbohydrate Res., 1973, 28, 103). In anhydrous hydrofluoric acid at room temperature glucose also polymerises, but in this case the products contain 3,6-anhydro-D-glucose residues as well as glucose (U. Kraska and F. Micheel, *ibid.*, 1976, 49, 195).

Studies with labelled sugars have shown that acid-catalysed isomerisation occurs substantially by processes which involve hydride shift between $C_{(1)}$ and $C_{(2)}$. Thus D-[2-^{3}H]glucose gives D-fructose with label at $C_{(1)}$, and D-[1-^{3}H]fructose isomerisms to D-glucose and D-mannose with tritium at both $C_{(1)}$ and $C_{(2)}$. The reaction paths are therefore different from those usually followed during alkaline isomerisations, but similar to those used by enzymes (*c.f.* p.78) (D.W. Harris and

M.S. Feather, J.Amer.chem.Soc., 1975, 97, 178). D-Xylose and D-*threo*-pentulose interconvert in acid by a similar mechanism (S. Ramchander and M.S. Feather, Arch.Biochem.Biophys., 1977, 178, 576).

As well as taking part in the above reactions, aldohexoses form 1,6-anhydro-derivatives (intramolecular glycosides) in aqueous acid (M. Černý and J. Staněk, Adv.carbohydrate Chem. Biochem., 1977, 34, 23). The percentages of 1,6-anhydro-pyranoses found for each of the D-aldohexoses at equilibrium are as follows, and agree with expectations based on thermo-dynamic calculations: allose, 14.0; altrose, 65.5; glucose, 0.2; mannose, 0.8; gulose, 65.0; idose, 86.0; galactose, 0.8; talose, 2.8. Because of this, all preparations of these sugars which involve acid-hydrolysis (e.g. cleavage of a glycoside) result in equilibrium mixtures which can contain, for example in the case of idose, dominant proportions of anhydrides. Aldoheptoses equilibrate with appreciably more of the 1,6-anhydrides than do the homomorphous aldohexoses (allose, altrose and mannose analogues, for example, giving 54, 98 and 15%, respectively), suggesting that secondary hydroxyl groups at $C_{(6)}$ participate in anhydride formation more readily than do primary hydroxyl groups (S.J. Angyal and R.J. Beveridge, Carbohydrate Res., 1978, 65, 229).

While these 1,6-anhydropyranoses are the main anhydrides formed in aqueous solution, the furanose isomers are much more significant in aprotic solvents, and in some cases largely dominate the products. In the case of D-galactose, 87% of the 1,6-anhydrofuranose is produced on heating in DMF solution with toluene-*p*-sulphonic acid and with azeotropic removal of water, and 33% can be isolated by direct crystallisation (Scheme 13).

SCHEME 13

Similarly, allose, glucose, mannose and talose give 78, 35, 22 and 86% 1,6-anhydrofuranoses, respectively, under these conditions (S.J. Angyal and R.J. Beveridge, Austral.J.Chem., 1978, 31, 1151). In aqueous acid the first of these equilibrates with about 1% of this derivative together with the pyranose isomer (14%) (K. Heyns and P. Köll, Ber., 1972, 105, 2228).

Occasionally anhydro-derivatives of dimeric products of reversion reactions of sugars have been encountered: hexuloses are notable in affording a range of compounds of this type. These are tricyclic and usually contain a 1,4-dioxane ring formed by reciprocal condensations involving the hydroxyl groups at $C_{(1)}$, and $C_{(2)}$ of two sugar units (T. Fujiwara and K. Arai, Carbohydrate Res., 1979, 69, 97, 107). D-Ribofuranose, alternatively, can cyclise through positions 1 and 5 both intramolecularly and in dimeric fashion. Heating the sugar with benzaldehyde, zinc chloride and acetic acid at 80° gives a mixture of diastereomeric monomer (20) and dimer (21) acetalated products (T.B. Grindley and W.A. Szarek, *ibid.*, 1972, 25, 187). 6-Deoxy-D-allose, a homomorphous sugar, reacts simi-

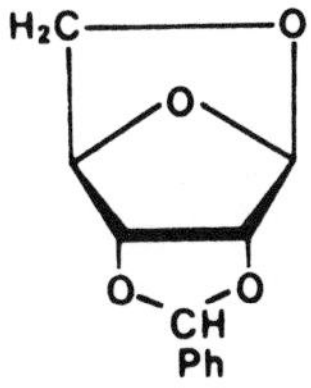

(20)

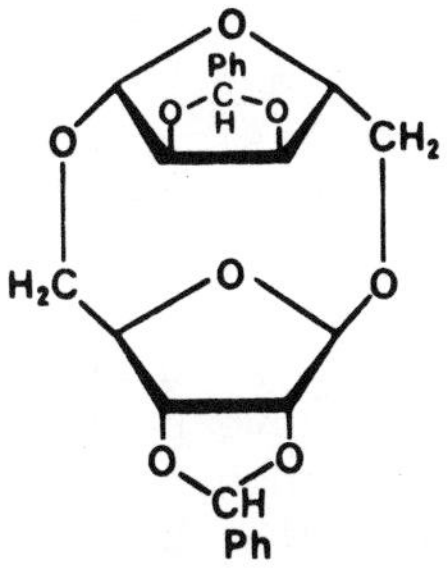

(21)

larly under these conditions (W.A. Szarek *et al.*, *ibid.*, 1974, 32, 279).

Since reversion can occur under the acidic conditions used for the hydrolysis of polysaccharides, it might be expected that derived reducing disaccharides with relatively stable glycosidic bonds would undergo intramolecular glycosidation reactions provided sterically suitable hydroxyl groups were available on the non-reducing moieties. Such is the case with 2-*O*-(α-D-galactopyranosyluronic acid)-L-rhamnose (Scheme 14) (T. Fujiwara and K. Arai, *loc.cit.*) and hydrolysis of poly-

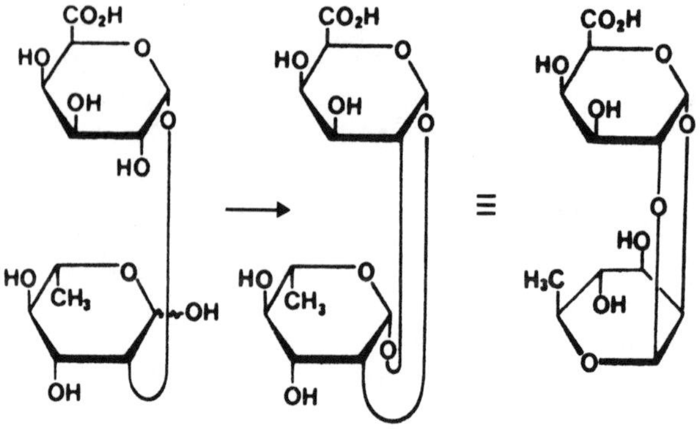

SCHEME 14

saccharides containing this aldobiouronic acid would be expected to afford some of this dimer. The situation is complicated by the possibility that intramolecular glycosylation may occur non-specifically, and analogues of the above anhydride comprising 4-*O*-methyl-α-D-glucopyranuronic acid and α- or β-D-xylopyranose and α-D-xylofuranose units have been isolated from the hydrolysis products of a wood polysaccharide (K. Larsson, Carbohydrate Res., 1975, 44, 199).

The dehydration reactions by which, under acidic conditions, sugars are converted into unsaturated products - notably furan derivatives - have been reviewed by M.S. Feather and J.F. Harris (Adv.carbohydrate Chem.Biochem., 1973, 28, 161), and their complexities have been well illustrated by T. Popoff and O. Theander (Acta chem.Scand., 1976, B30, 397). After heating solutions of D-glucose and D-fructose at pH 4.5 to high temperatures they identified four furans, seven substituted benzenes, an isobenzofuranone, a benzopyranone and a benzofuran amongst the products. A thorough study of the factors operating during the dehydration of D-fructose (B.F.M. Kuster *et al.*, Carbohydrate Res., 1977, 54, 159, 165, 177, 185) culminated in the development of a continuous process for the manufacture of 5-hydroxymethyl-2-furaldehyde in 65-85% yield (*idem.*, Stärke, 1977, 29, 99).

(h) The radiolysis of monosaccharides

The study of the radiation chemistry of sugars has been promoted by the desirability of knowing how biological systems are affected by radiation at the molecular level. ^{60}Co γ-radiolysis of water leads to the formation of hydroxyl radicals, solvated electrons and hydrogen atoms, and the effective concentration of hydroxyl radicals can be increased by saturating the solution with nitrous oxide.

$$H_2O \rightarrow {}^{\bullet}OH + e^-_{aq} + {}^{\bullet}H$$

$$e^-_{aq} + N_2O \rightarrow {}^{\bullet}OH + N_2 + {}^{-}OH$$

These react with sugars by hydrogen abstraction processes to afford carbon radicals the reactions of which depend upon whether or not molecular oxygen is available (Scheme 15).

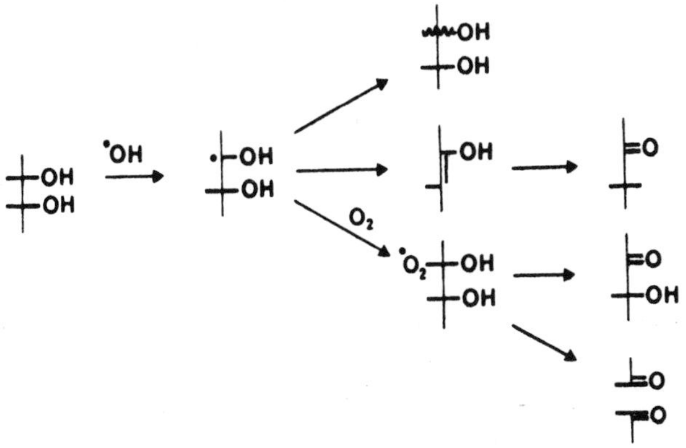

SCHEME 15

Products of epimerisation (N.K. Kochetkov *et al.*, Carbohydrate Res., 1973, 28, 86), elimination, oxidation and carbon-carbon bond cleavage are therefore found.

D-Glucose irradiated in aerated aqueous solution gives a wide range of products derived from different carbon radicals (S. Kawakishi, Y. Kito and M. Namiki, *ibid.*, 1975, 39, 263) (Scheme 16).

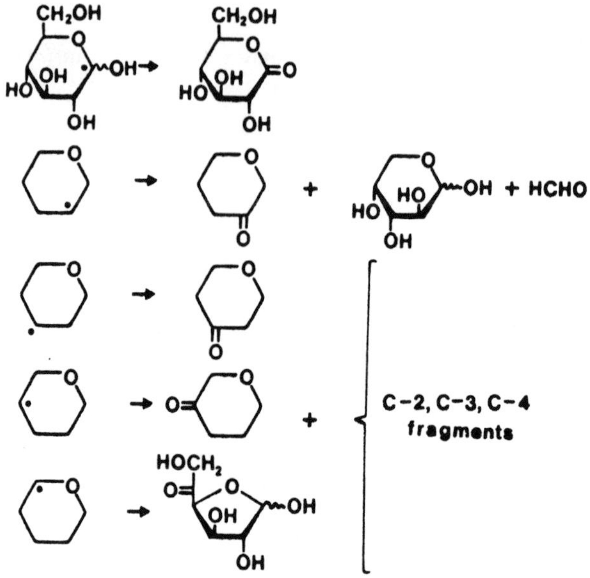

SCHEME 16

Products derived from the $C_{(6)}$ radical were detected in a related study in which relative rates of elimination of $HO_2^{\bullet}$ were measured (M.N. Schuchmann and C. von Sonntag, J.chem.Soc. Perkin II, 1977, 1958). In the absence of oxygen the main products are correspondingly deoxyhexuloses formed by dehydration steps (S. Kawakishi, Y. Kito and M. Namiki, Carbohydrate Res., 1973, 30, 220). For example, a 2-deoxyhexos-3-ulose and a 3-deoxyhexos-4-ulose are derived from the $C_{(3)}$ and $C_{(4)}$ radicals, respectively. Analogous results are obtainable from D-ribose, the nature of the products again being highly dependent on the availability of oxygen (C. von Sonntag and M. Dizdaroglu, *ibid.*, 1977, 58, 21). The radiation chemistry of carbohydrates has been surveyed in a book of that title (by N.K. Kochetkov, L.I. Kudrjashov and M.A. Chlenov, Pergamon Press, Oxford, 1979) and in a review article (C. von Sonntag, Adv.carbohydrate Chem.Biochem., 1980, 37, 7).

(i) Oxidation of monosaccharides

Main consideration will be given to the oxidation of hydroxylic carbohydrate derivatives to carbonyl-containing products which have become invaluable starting materials in synthesis. Considerable developments have, nevertheless, taken place in all other aspects of carbohydrate oxidation reactions.

(i) *Aldonic acids*

The mechanism of the oxidation of aldoses with bromine has been surveyed together with aspects of reactions of aldonolactones, aldonamides and aldose cyanohydrins (B. Capon, Chem. Rev., 1969, 69, 407). Oxidation of substituted aldoses occurs stoichiometrically with hypoiodite, and automated procedures have been developed to control the reaction and to identify the sites of substitution of the aldonic acid products (A.M.Y. Ko and P.J. Somers, Carbohydrate Res., 1974, 34, 57).

In alkaline aqueous solution, and in the presence of catalytic platinum or rhodium on charcoal, free sugars undergo dehydrogenations to give aldonic acids (G. de Wit *et al.*, Tetrahedron Letters, 1978, 1327), and substituted sugars with free anomeric centres can be similarly oxidised by use of silver carbonate on Celite in boiling benzene. However, in boiling methanol, specific C-C bond cleavage occurs and 2,3-*O*-isopropylidene-L-erythrose is, for example, produced from 3,4-*O*-isopropylidene-L-arabinose (Scheme 17) (S. Morgenlie,

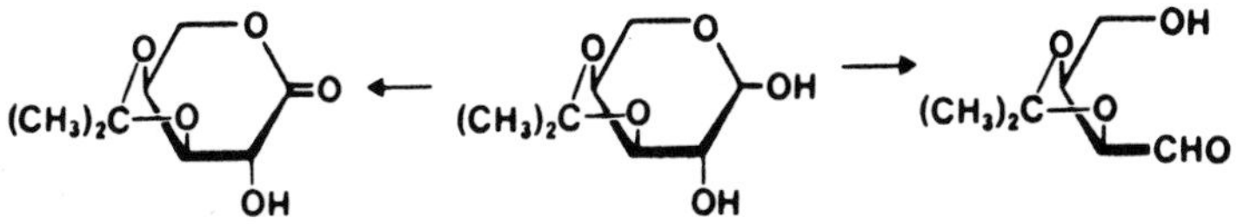

SCHEME 17

Acta chem.Scand., 1972, 26, 2518). In related fashion, with this reagent, 3-*O*-methyl-D-glucose gives 2-*O*-methyl-D-arabinose (*idem.ibid.*,1971, 25, 2773), and L-threose and D-erythrose are produced from L-sorbose and D-fructose, respectively (*idem.ibid.*, 1972, 26, 2146; 1973, 27, 1557).

An alternative approach to the synthesis of aldonic acid derivatives involves the ozonolysis of glycosides which is a reaction highly specific in two ways. Ortho-acid intermediates are formed exclusively from glycosides having equatorially oriented aglycones. These then decompose specifically to give acyclic esters rather than lactones (Scheme 18). For such

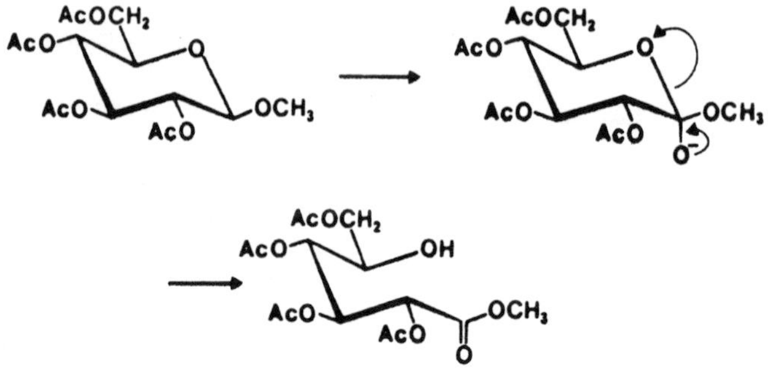

SCHEME 18

oxidation reactions to occur, each of the acetal oxygen atoms must have a lone pair orbital oriented antiperiplanar to the C-H bond; the reaction of the intermediates is likewise controlled by stereoelectronic factors (P. Deslongchamps, Tetrahedron, 1975, 31, 2463).

Spiro-orthoester derivatives of aldonolactones, analogues of those which occur in some antibiotics, can be prepared by treatment of suitably protected lactones with epoxides in the presence of boron trifluoride, or with diols and catalytic acid (Scheme 19) (J. Yoshimura and M. Tamaru, Carbohydrate Res., 1979, 72, C9).

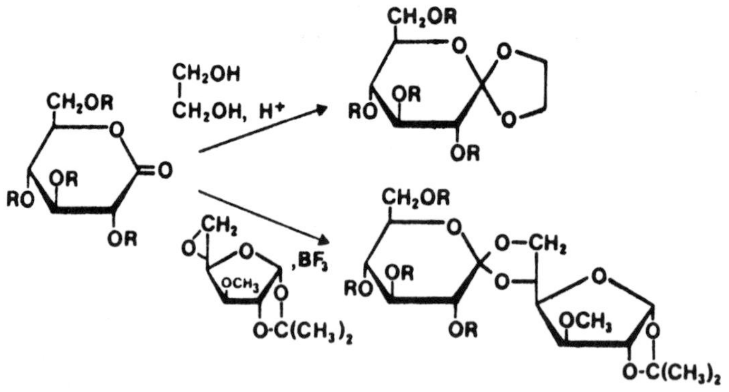

SCHEME 19

Aldonolactone derivatives also take part in the Reformatsky reaction (Scheme 20) (V.K. Srivastava and L.M. Lerner, J.org.

Chem., 1979, 44, 3368).

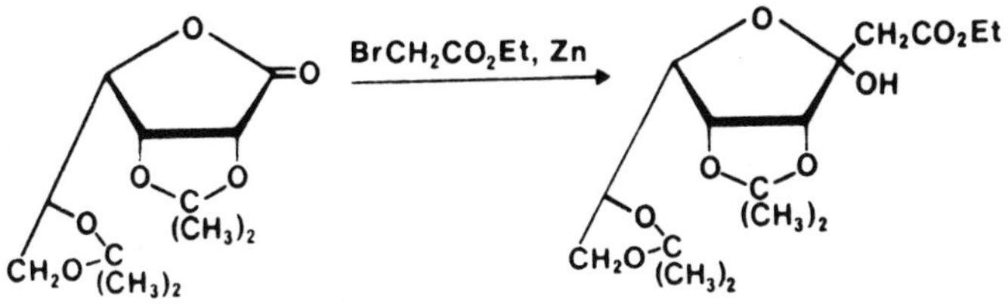

SCHEME 20

(ii) *Alduronic acids*

Considerable developments have occurred, and several aspects of the chemistry of alduronic acid derivatives have been reviewed by D. Keglević (Adv.carbohydrate Chem.Biochem., 1979, 36, 57). She covered glycosiduronic acids, dealing with synthetic approaches based on both glycosylation and oxidation methods, reduction and hydrolysis reactions, and 1-thioglycoside and glycosylamine analogues. K. Dax and H. Weidmann have summarised the reactions of D-glucofuranurono-6,3-lactone (*ibid.*, 1976, 33, 189), and J. Kiss has comprehensively treated β-elimination reactions of uronic acid derivatives (*ibid.*, 1974, 29, 229).

Studies of microbial metabolites have revealed a range of new naturally occurring uronic acid derivatives. Compounds (22) and (23) are, respectively, components of gougerotin (F.W. Lichtenthaler *et al.*, Tetrahedron Letters, 1975, 3527) and the polyoxins (K. Isono, K. Asahi and S. Suzuki, J.Amer. chem.Soc., 1969, 91, 7490), and compound (24) is octosyl acid A - an octuronic acid derivative - which is produced together with the polyoxins by a *Streptomyces* (K. Isono, P.F. Crain and J.A. McCloskey, *ibid.*, 1975, 97, 943).

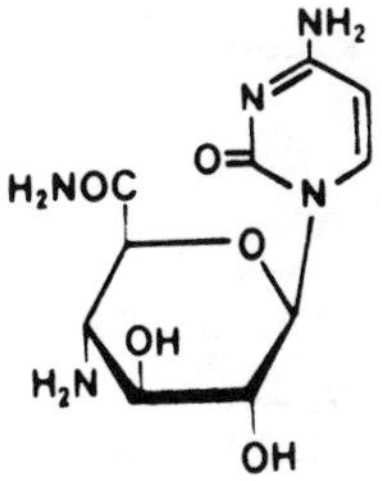

(22) 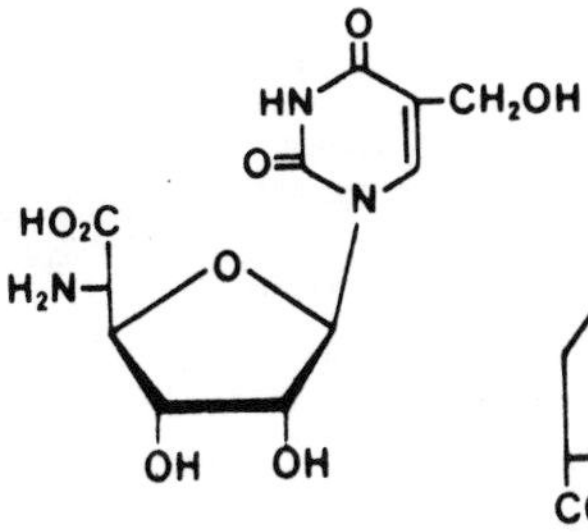 (23) 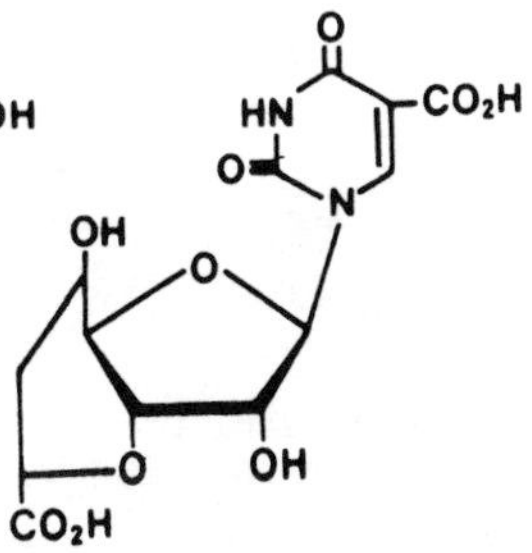 (24)

Oxidation of aldohexopyranosides with potassium ferrate followed by sodium hypoiodite affords alternative means of preparing hexopyranosiduronic acids (J.N. BeMiller, V.G.Kumari and S.D. Darling, Tetrahedron Letters, 1972, 4143); pento-furanosyl methyl uronates are obtained on oxidation of 1,2-*O*-isopropylidenehexofuranoses with silver carbonate on Celite in boiling methanol (Scheme 21) (S. Morgenlie, Carbohydrate Res., 1977, 59, 73).

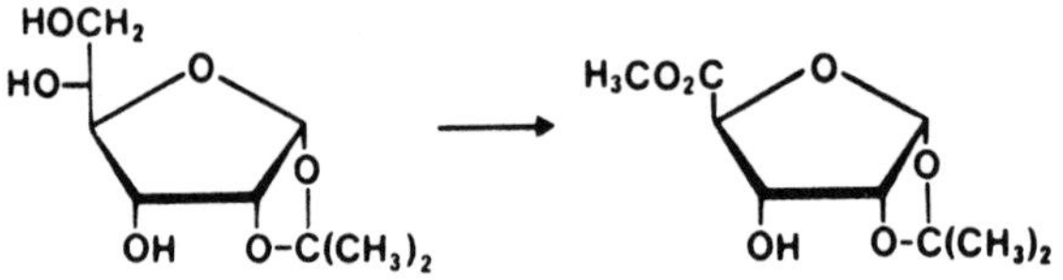

SCHEME 21

Alternatively, uronic acid derivatives can be prepared by reactions which extend carbon chains (Scheme 22) (N. Berg and O. Kjolber, *ibid.*, 1977, 57, 65; D. Horton and J.-H. Tsai, *ibid.*, 1977, 58, 89; O. Kjolberg and T.B. Sverreson, Acta chem.Scand., 1972, 26, 3245).

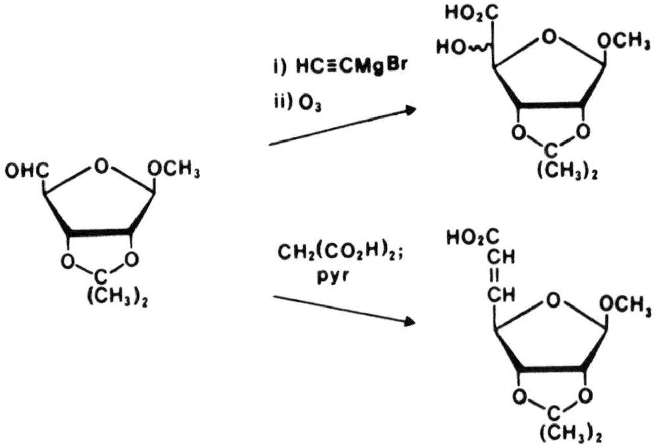

SCHEME 22

More thorough studies have revealed the complexities of the Lobry de Bruyn-van Ekenstein isomerisations which occur when glucuronic acids are heated in neutral solution (O. Samuelson *et al.*, *ibid.*, 1969, 23, 318; Carbohydrate Res., 1969, 11,

347, 1973; 31, 81). D-Galacturonic acid gives the corresponding ketose, D-*arabino*-hex-5-ulosonic acid, as main product, but all other isomers obtainable by rearrangement of the $C_{(1)}$-$C_{(3)}$ portion of the molecule are also formed (Scheme 23).

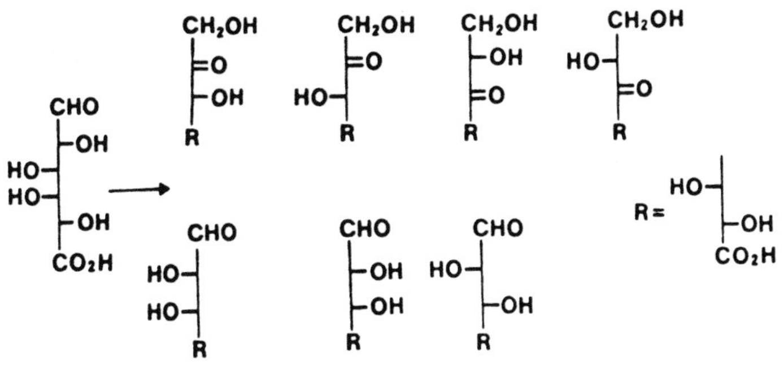

SCHEME 23

Complete oxidation of hexuronic acids with alkaline hydrogen peroxide occurs by two concurrent means which, respectively, give rise to five equivalents of formic acid and carbon dioxide and four of formic acid and one of oxalic acid (H.S. Isbell, H.L. Frush and Z. Orhanovic, *ibid.*, 1974, 36, 283). Heated in acidic solution these compounds give mainly furan derivatives and one equivalent of carbon dioxide which can be used for their determination. Penturonic acids behave in analogous fashion, but hepturonic acids release very much less carbon dioxide (M.A. Madson and M.S. Feather, *ibid.*, 1979, 70, 307).

Eliminative decarboxylation of methyl 2,3-di-*O*-benzyl-α-D-glucopyranosiduronic acid occurs on treatment with DMF dineopentylacetal in DMF (Scheme 24) (K.D. Philips, J. Žemlička and J.P. Horwitz, *ibid.*, 1973, 30, 281).

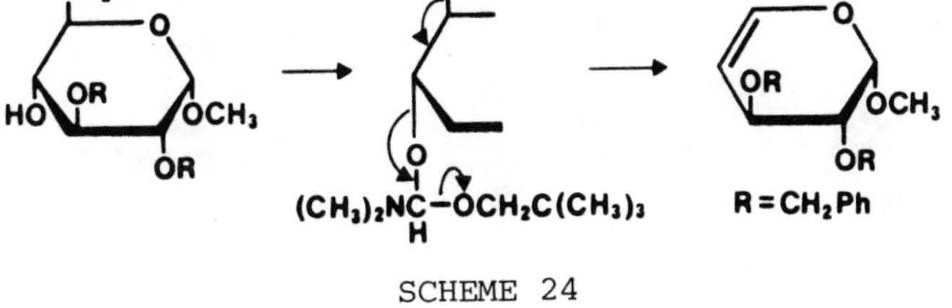

SCHEME 24

Treatment of related glycosiduronic acids with lead tetra-acetate in benzene also leads to decarboxylation and affords products with acetoxy groups replacing the carboxylic acid function. Base-catalysed deacylation then gives cyclic hemi-acetals which spontaneously ring open and dissociate by loss of the aglycone. The method, therefore, affords specific means of cleaving glycosidic bonds of hexuronic acid-containing materials (I. Kitagawa *et al.*, Tetrahedron Letters, 1976, 549). Analogous cleavage of derived uronate esters can be achieved photochemically (*idem.*, *ibid.*, 1973, 3997), a model experiment showing that the carbohydrate moiety undergoes complex changes (Scheme 25) (A.G.W. Bradbury and C. von Sonntag, Carbohydrate Res., 1978, 60, 183).

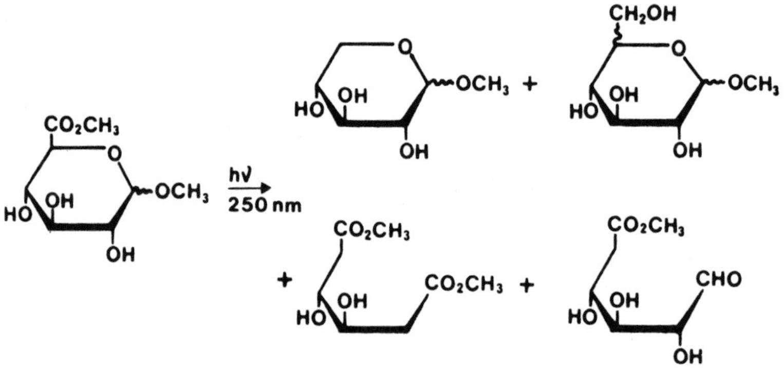

CO_2CH_3 ... OCH_3 $\xrightarrow[250\,nm]{h\nu}$ ... OCH_3 + CH_2OH ... OCH_3 + CO_2CH_3 ... CO_2CH_3 + CO_2CH_3 ... CHO

SCHEME 25

Amides of hexopyranosiduronic acids undergo the Hofmann degradation to give acid-labile glycosylamine derivatives, and as this process also affords specific means of cleaving associated glycosidic bonds, it can be applied to the selective degradation of uronic acid-containing polysaccharides (Scheme 26) (N.K. Kochetkov, O.S. Chizhov and A.F. Sviridov, *ibid.*, 1970, 14, 277).

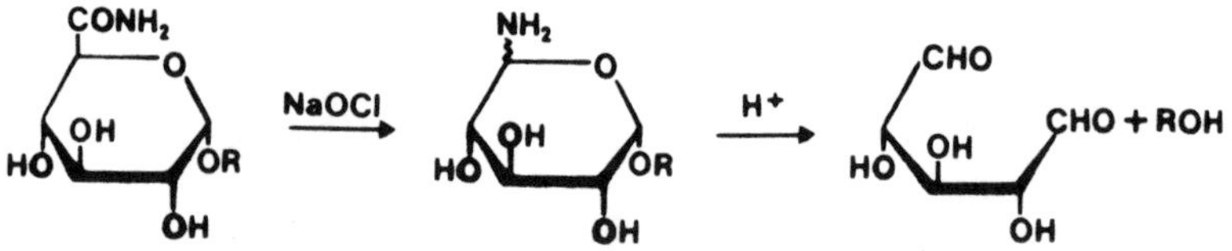

$CONH_2$ ... OR $\xrightarrow{NaOCl}$ NH_2 ... OR $\xrightarrow{H^+}$ CHO ... CHO + ROH

SCHEME 26

(iii) *Aldaric acids*

D-Galactaric acid has been isolated by aqueous extraction from a succulent (R. Kringstad and A. Nordal, Acta chem.Scand., 1973, 27, 1432), and D-allaric acid has been found as its 4-phosphate bonded glycosidically through $O_{(2)}$ to D-glucose as part of a nucleosidic bacterial toxin (L. Kalvoda, M. Pryštaš and F. Šorm, Tetrahedron Letters, 1973, 1873). Daucic acid (25), an anhydroheptaric acid derivative, occurs in a range of food plants including wheat and sugar beet (D.H.R. Barton *et al.*, J.chem.Soc.Perkin I, 1975, 2069).

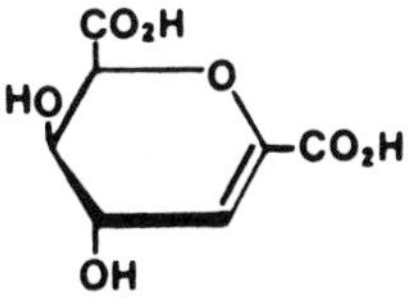

(25)

Oxidation of D-glucose or D-gluconic acid with oxygen over a platinum catalyst gives D-glucaric acid (J.M.H. Dirkx *et al.*, Carbohydrate Res., 1977, 59, 63), and an NAD-linked dehydrogenase has been isolated from a *Pseudomonas* which oxidises D-glucuronic acid and D-galacturonic acid to the corresponding aldaric acids (D.F. Bateman, T. Kosuge and W.W. Kilgore, Arch. Biochem.Biophys., 1970, 136, 97).

Oxidation of 2,6-anhydro-D-*glycero*-L-*manno*-heptonic acid with fuming nitric acid gives the analogous anhydroheptaric acid (26) from which several unsymmetrical derivatives are obtainable since its anhydride ring opens selectively following nucleophilic attack at $C_{(1)}$ (Scheme 27) (E.-F. Fuchs and J. Lehmann, Carbohydrate Res., 1975, 45, 135).

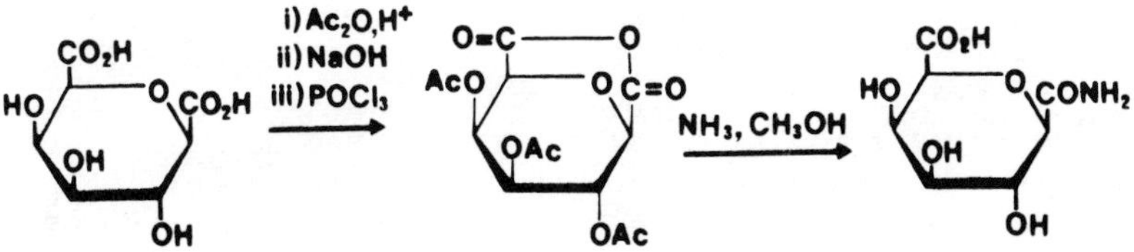

(26) SCHEME 27

Additions of hydrogen cyanide to 6-cyano-6-deoxy-D-glucose, followed by hydrolysis, gives 2-deoxy-L-*glycero*-L-*gulo*-octaric acid which is isolated as its 1,4:8,5-dilactone (27) (I. Dijong, *ibid.*, 1969, 11, 428).

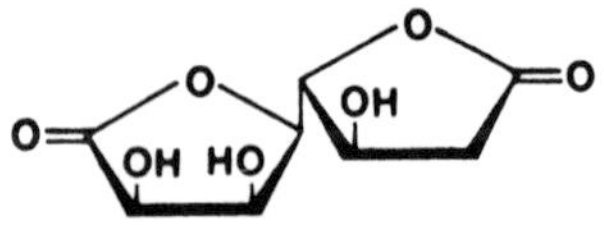

(27)

Oxidation of D-mannaro-1,4:6,3-dilactone with potassium permanganate gives the 2-keto-derivative and then the diketone (K. Heyns and A. Linkies, Ber., 1975, 108, 3633).

Tri-*O*-acetylxylaryl chloride, treated with diazomethane leads to the bisdiazoketone and this, with acetic acid containing copper(II) acetate, affords a cyclohexanone derivative by way, probably, of a dicarbene intermediate (Scheme 28) (D.E. Kiely *et al.*, Carbohydrate Res., 1972, 23, 155; Tetrahedron Letters, 1973, 4379).

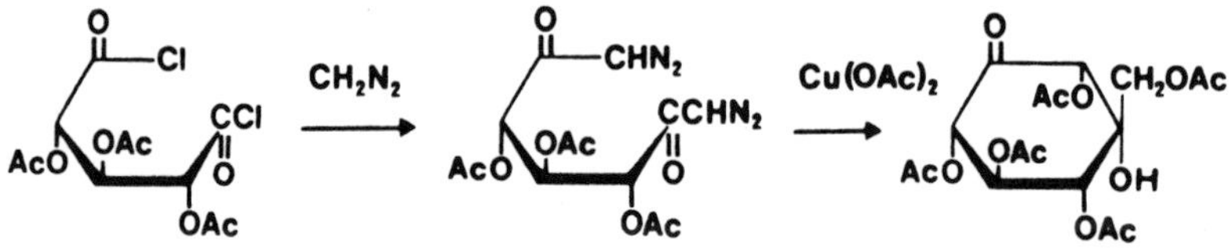

SCHEME 28

(iv) *Derivatives of dicarbonyl compounds*

Selective oxidations of hydroxyl groups of sugar derivatives such as glycosides represent common reactions providing products which are derivatives of dicarbonyl compounds. These exhibit a useful range of chemical reactivity and afford access to many modified carbohydrates such as deoxy-, enolic-, amino- and branched-chain compounds. The nature of the products depends upon the materials oxidised, and may be aldos-2-ulose (osone), other aldosulose, dialdose or diketose derivatives. R.F. Butterworth and S. Hanessian have reviewed the main oxidation methods (Synthesis, 1971, 70). (See also J.S. Brimacombe, Angew.Chem.internat.Edn., 1969, 8, 401 and G.H. Jones and J.G. Moffatt in "Methods in Carbohydrate Chemistry" Vol. 6, ed. R.L. Whistler and J.N. BeMiller, Academic Press, 1972, p.315).

(1) *DMSO - based oxidations*. (See W.W. Epstein and F.W. Sweat, Chem.Rev., 1967, 67, 247 for a comprehensive review). These reactions have become the most frequently adopted means of oxidising a carbohydrate alcohol to the corresponding carbonyl compound, and depend upon the initial interaction between DMSO and electrophilic agents to give sulphonium complexes with which alcohols react. The derived alkoxysulphonium species, under the influence of bases, then undergo deprotonation and loss of dimethyl sulphide to afford aldehydes or ketones.

The electrophilic reagents usually employed are acetic anhydride or dicyclohexylcarbodi-imide (DCC), but phosphorus pentoxide and sulphur trioxide-pyridine can be used amongst others. With acetic anhydride, the first intermediate leads to the carbonyl product only when nucleophilic attack by alcohol occurs at sulphur. However, attack at the acyl carbon atom can also occur to afford the product of acetylation, and attack at one of the carbon atoms bonded to sulphur in deprotonated species gives the analogous methyl-thiomethyl ether as a further reaction byproduct (Scheme 29).

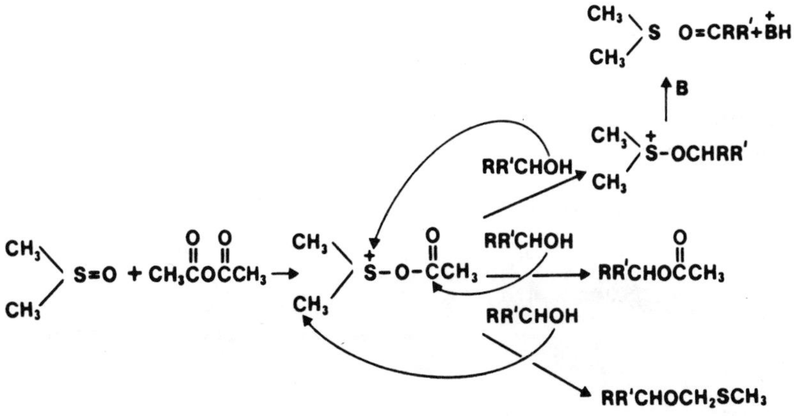

SCHEME 29

The products of substitution are particularly significant when acetic anhydride is used with primary alcohols, and the aldehydic oxidation products may be formed in only low yields. Alternatively, acetic anhydride is to be preferred to DCC for the oxidation of hindered secondary alcohols, and it has the added advantage of affording readily removed by-products. Small proportions of methylthiomethyl ethers may also be formed during DCC activated oxidations, but these can be minimised by use of pyridinium trifluoroacetate or phosphate as the proton sources which must be used with this system. Separation of dicyclohexylurea, the side-product of the oxidation reaction, from the carbonyl product can be troublesome; this can be overcome by use of alternative di-imides.

A wide range of monohydroxy compounds have been converted uneventfully to aldehyde or ketones by these procedures, but several concurrent reactions have been encountered.

Epimerisation (Scheme 30) (H. Kuzuhara *et al.*, Carbohydrate Res., 1972, 23, 217).

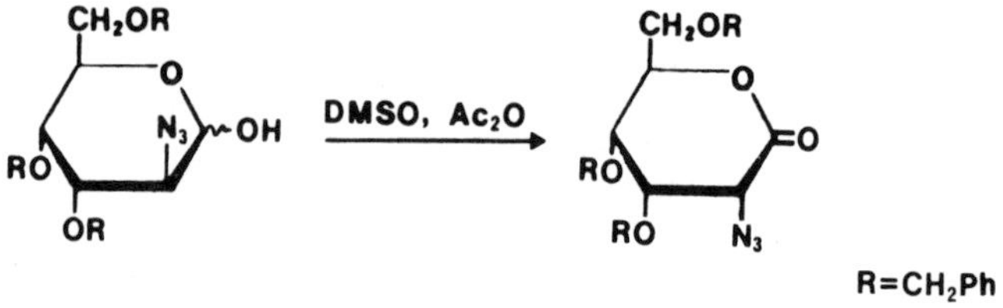

DMSO, Ac_2O

$R=CH_2Ph$

SCHEME 30

β-Elimination (Scheme 31) (D.M. Mackie and A.S. Perlin, *ibid.*, 1972, 24, 67).

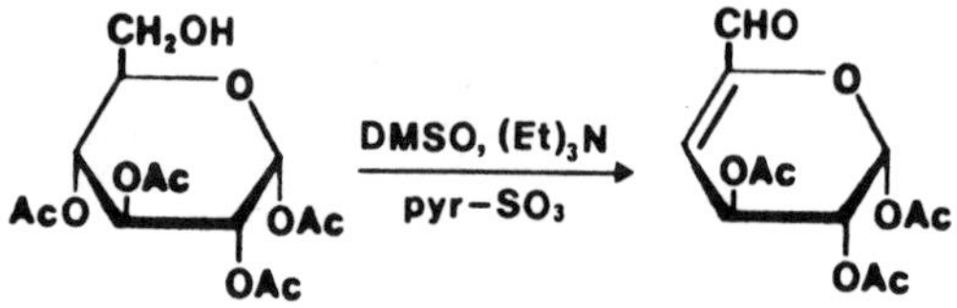

DMSO, $(Et)_3N$, pyr$-SO_3$

SCHEME 31

Enol acetate formation (Scheme 32) (S. Hanessian and A.P.A. Staub, Chem. and Ind., 1970, 1436).

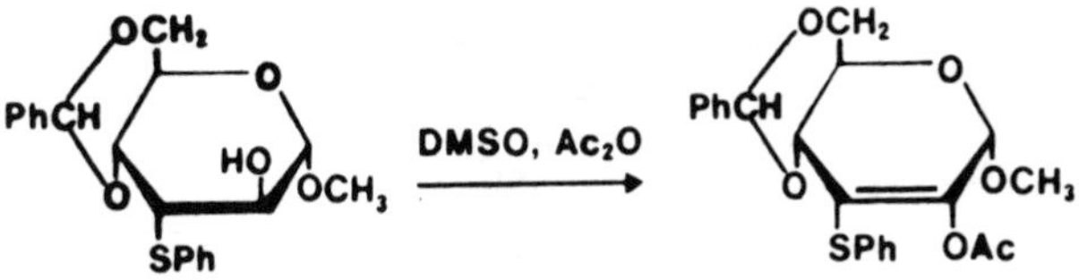

SCHEME 32

Mixed products (Scheme 33) (Y. Kondo and F. Takao, Canad.J. Chem., 1973, 51, 1476; J. Defaye and A. Gadelle, Carbohydrate Res., 1974, 35, 264).

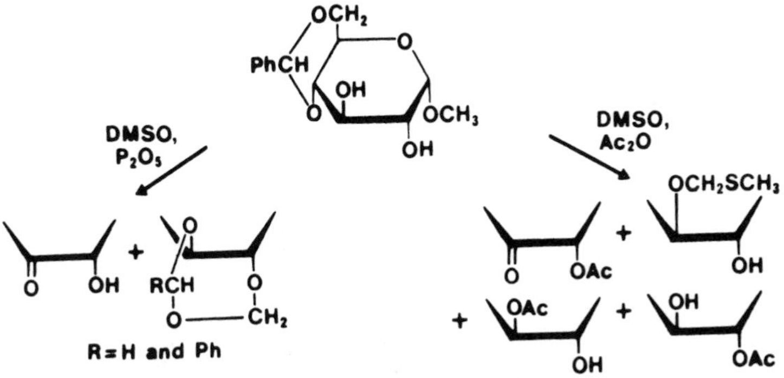

SCHEME 33

Selective oxidation (Scheme 34) (J. Defaye and A. Gadelle, *ibid.*, 1977, 56, 411).

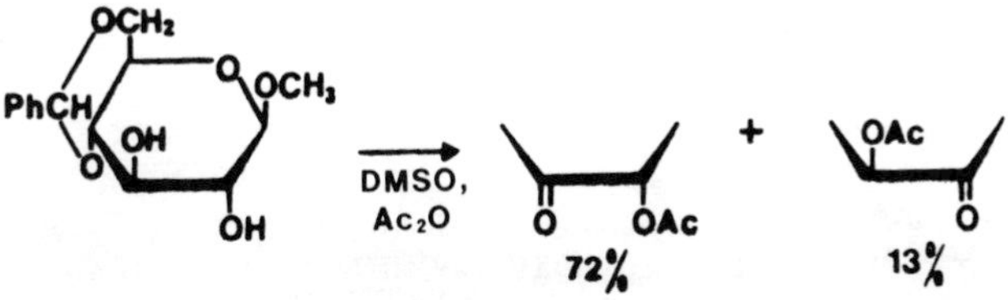

SCHEME 34

Complete and partial oxidation of a diol (Scheme 35) (C.-K. Lee, *ibid.*, 1975, 42, 354).

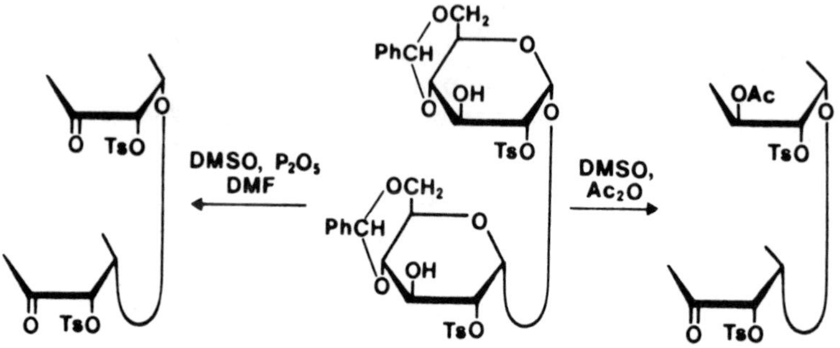

SCHEME 35

Treated with DMSO in the presence of boron trifluoride, epoxides give carbonyl products (Scheme 36) (G. Hanisch and G. Henseke, Ber., 1968, 101, 2074) in reactions which are similar to those involving the oxidation of compounds with good leaving groups such as sulphonyl esters; the latter (Kornblum reactions) have not been used extensively.

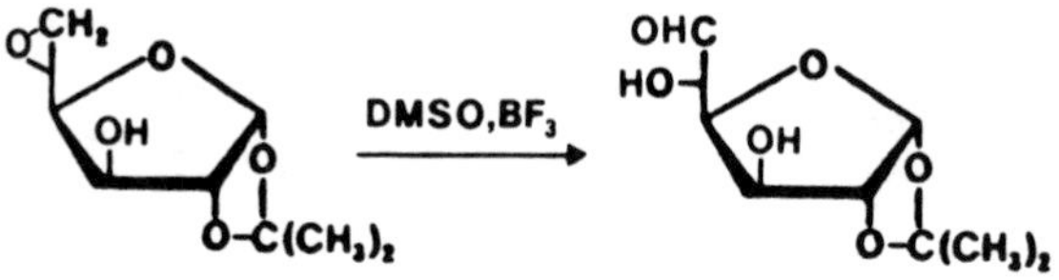

SCHEME 36

(2) *Ruthenium tetroxide oxidations*. If this reagent is prepared by oxidation of ruthenium dioxide with the periodate ion, it is important that the hydrated form of the starting dioxide be used (W.G. Overend *et al.*, Carbohydrate Res., 1968, 6, 431). If, alternatively, the catalytic oxidation method involving the use of the dioxide and the periodate ion is employed, the choice of sparingly soluble potassium periodate and control of the pH with potassium carbonate minimise over-oxidation of the carbonyl products and afford high yields of ketones from secondary monohydroxy compounds (B.T. Lawton, W.A. Szarek and J.K.N. Jones, *ibid.*, 1969, 10, 456). These reagents operate without causing epimerisation at sites α- to the carbonyl groups which is often observed with DMSO-based oxidations (P.M. Collins and B.R. Whitton, *ibid.*, 1974, 33, 25). They also permit the isolation of ketones with ester

groups in the β-position (I. Lundt and C. Pedersen, *ibid.*, 1974, 35, 187).

Over-oxidation analogous to the Baeyer-Villiger conversion of cyclic ketones to lactones is frequently encountered with furanoid compounds (Scheme 37) (J.M.J. Tronchet and J.Tronchet, Helv., 1970, 53, 1174). It is less prevalent with 6-membered

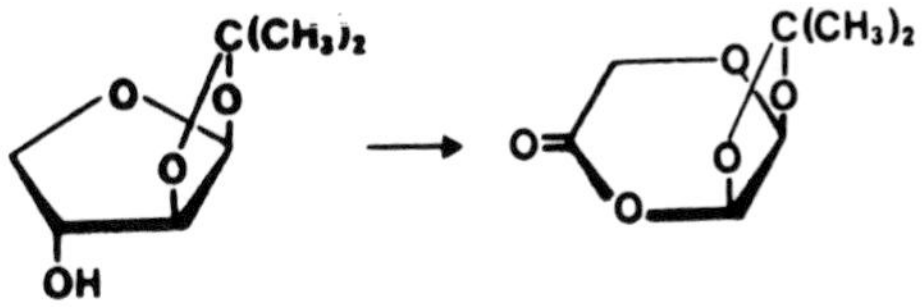

SCHEME 37

ketones although, with *m*-chloroperbenzoic acid, such compounds can be converted into ring-enlarged lactones (Scheme 38) (K. Heyns *et al.*, Tetrahedron Letters, 1972, 5081).

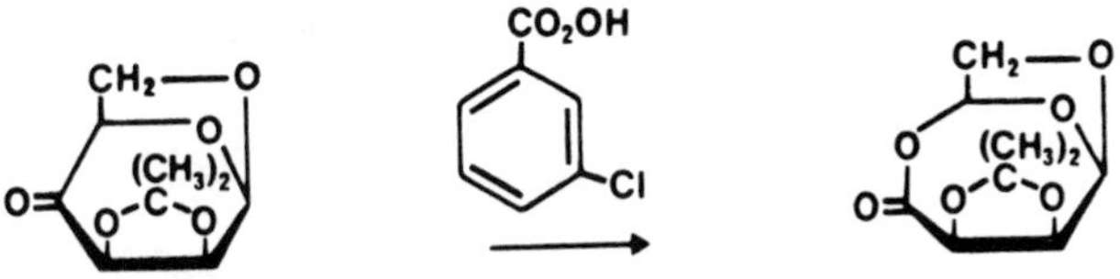

SCHEME 38

(3) *Oxidations with other inorganic reagents.* Chromium trioxide in pyridine can oxidise primary alcohols to carboxylic acids, and has often been used to generate carbohydrate ketones from acyclic, furanoid or pyranoid precursors although its activity is very sensitive to steric factors (R.F.Butterworth and S. Hanessian, Synthesis, 1971, 70). Chromium trioxide-dipyridine complex used in dichloromethane offers convenient means of oxidising 1,2:3,4-di-*O*-isopropylidene-α-D-galactopyranose to the 6-aldehydo derivative (R.E. Arrick, D.C. Baker and D. Horton, Carbohydrate Res., 1973, 26, 441), and freshly prepared lead acetate in pyridine has also been advocated for the same conversion (D.J. Ward, W.A. Szarek and J.K.N. Jones, *ibid.*, 1972, 21, 305). Newer reagents which have been used successfully to prepare carbohydrate ketones are pyridinium chlorochromate and pyridinium dichromate in reactions which are catalysed by molecular sieves (J. Herscovici and K. Antonakis, Chem.Comm., 1980, 561).

Chromium trioxide in acetic acid, although used on occasion to oxidise secondary alcohols to ketones, also causes oxidation of acetals to ketoesters; its use is referred to on pp. 75 and 103.

Selective oxidation can be effected at allylic sites with activated manganese dioxide in chloroform (B. Fraser-Reid, *et al.*, Canad.J.Chem., 1970, 48, 2877) or with silver carbonate on Celite (Scheme 39) (J.M.J. Tronchet *et al.*, Helv., 1970, 53, 1489). D-Glucal is thus oxidised much more effi-

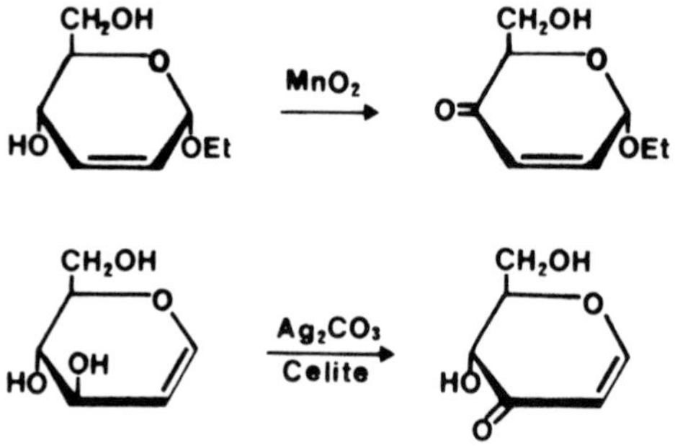

CH_2OH ... HO ... OEt —MnO_2→ CH_2OH ... O= ... OEt

CH_2OH ... HO OH —Ag_2CO_3 / Celite→ CH_2OH ... HO ... O

SCHEME 39

ciently than catalytically with platinum oxide (K. Heyns and H. Gottschalck, Ber., 1966, 99, 3718). This last reagent normally oxidises primary hydroxyl groups preferentially, and axial secondary groups are oxidised faster than are equatorial groups (K. Heyns and H. Paulsen, Adv.carbohydrate Chem., 1962, 17, 169), but exceptions are know to these generalisations (K. Heyns *et al.*, Ber., 1975, 108, 3611).

(4) *Other methods.* Carbonyl-containing derivatives may be produced from primary amines by treatment with ninhydrin (K.N. Slessor *et al.*, Canad.J.Chem., 1974, 52, 3905) or by oxidative deamination of secondary amines by way of imines (Scheme 40) (B. Lengstad and J. Lönngren, Carbohydrate Res., 1979, 72, 312).

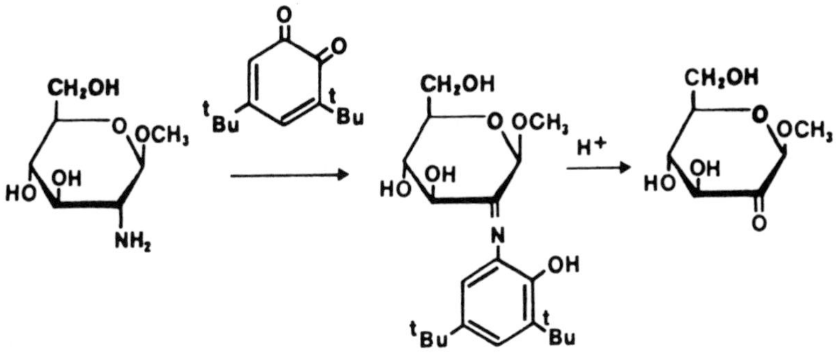

SCHEME 40

A related process involves the preparation of ketones from enol esters which afford deoxyoximes on treatment with hyroxylamine hydrochloride (Scheme 41) (F.W. Lichtenthaler *et al.*, Tetrahedron Letters, 1980, 21, 1425, 1429).

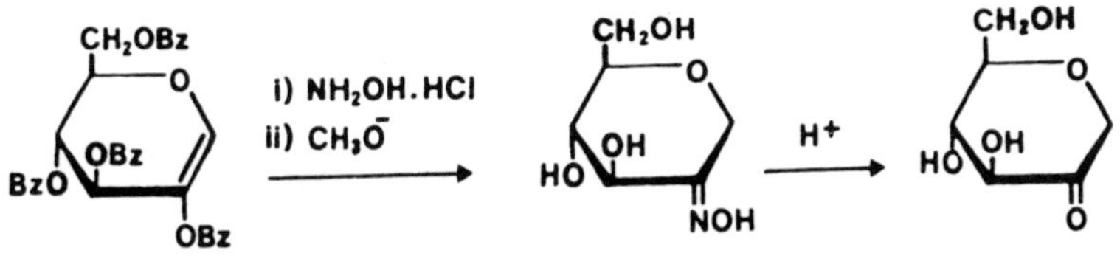

SCHEME 41

Photochemical methods have been found useful, and of particular value for obtaining chemically sensitive products such as those with ester groups β- to the carbonyl functions. Primary azidodeoxy compounds give aldehydes, and secondary analogues give ketones somewhat less efficiently on photolysis followed by mild treatment with acid (D.C. Baker and D.Horton, Carbohydrate Res., 1972, 21, 393). Radiation of pyruvate esters proceeds efficiently to give aldehydes or ketones in yields which compare well with those obtained by purely chemical methods (Scheme 42) (R.W. Binkley, J.org.Chem., 1977, 42, 1216).

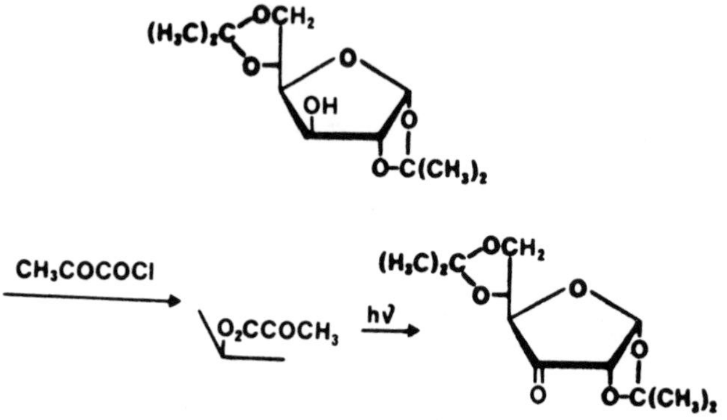

SCHEME 42

Biochemical methods can be used to a limited extent, methyl α- and β-D-xylopyranoside giving the corresponding glycosid-4-uloses on oxidation with suspensions of *Acetobacter suboxydans* (W.A. Szarek and G.W. Schnarr, Carbohydrate Res., 1977, 55, C5), and methyl β-D-galactopyranoside undergoing specific oxidation at the primary alcohol site on treatment in buffered solution (pH 7) with D-galactose oxidase to give the aldehydic product (A. Maradufu and A.S. Perlin, *ibid.*, 1974, 32, 127).

Main interest in the aldehydes and ketones obtainable by the above methods centres around their value as starting materials for the preparation of a wide range of carbohydrate derivatives - notably extended-chain and branched-chain compounds. Although many uneventful carbonyl reactions have been reported, the propensity of the compounds to undergo such reactions as rearrangement, epimerisation and β-elimination has to be guarded against. Thus, the two ketones derived by oxidation at $C_{(2)}$ or $C_{(3)}$ of methyl α-D-glucopyranoside afford the same cationic magnesium complexes (J. Defaye, H. Driguez and A. Gadelle, *ibid.*, 1974, 38, C4), pyranosid-4-ulose derivatives can invert at C-5 on treatment with base allowing access to rare compounds (C.L. Stevens *et al.*, J.org.Chem., 1973, 38, 4311), and oxidation of methyl 4,6-*O*-benzylidene-2-deoxy-α-D-*lyxo*-hexopyranoside with chromium trioxide in pyridine gives an enone product (Scheme 43) (W.G. Overend *et al.*, J.chem.Soc.(C), 1966, 1131).

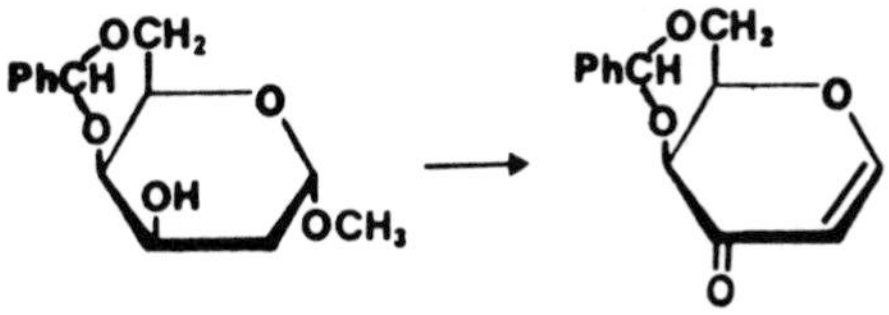

SCHEME 43

Hydroxycarbonyl derivatives may dimerise, and thus the products of acetylation of methyl β-D-*galacto*-hexodialdo-1,5-pyranoside contain both a product of elimination and one of acetylation of a dimer formed by acetal-like condensation between an aldehydic group of one monomer and the 4,6-hydroxyaldehydo system of another (Scheme 44) (A. Maradufu and A.S. Perlin, *loc.cit.*). Hydroxyketones similarly may dimerise and

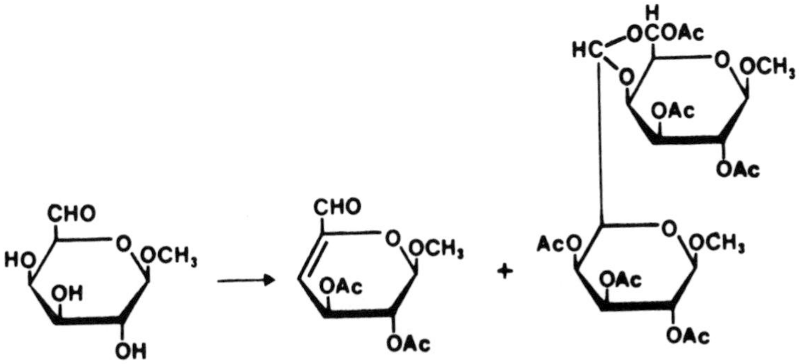

SCHEME 44

give derivatives of the dimers after substitution (P. Köll and J. Kopf, Carbohydrate Res., 1979, 68, 189).

Aldol dimerisation can be induced under basic conditions (Scheme 45) (P.M. Collins and R. Iyer, *ibid.*, 1976, 46, 277).

Enol esters can be produced from carbonyl-containing derivatives by treatment with acetic anhydride and potassium carbonate (S.L. Cook and J.A. Secrist, *ibid.*, 1976, 52, C3).

A novel way of descending a sugar series by extrusion of a non-terminal carbon atom is afforded by the photolysis of some pyranoid ketonic compounds. Methyl 6-deoxy-2,3-*O*-isopropyli-

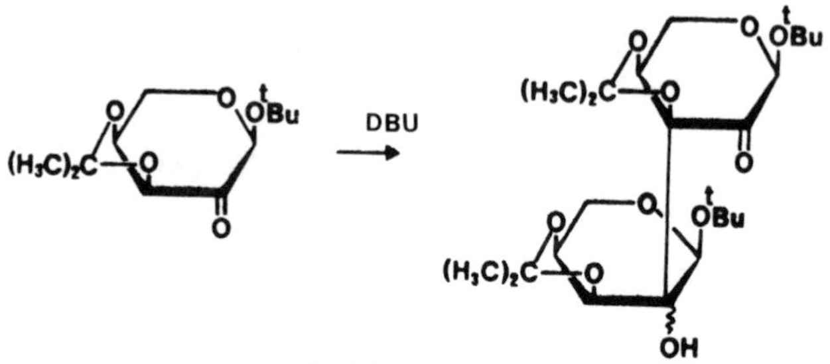

SCHEME 45

dene-α-L-*lyxo*- and β-D-*ribo*-hexopyranosid-4-ulose on ultraviolet irradiation lose carbon monoxide and give mainly, by way of a diradical, the same mixture of furanoside derivatives, with the β-D-riboside predominating (Scheme 46) (P.M.

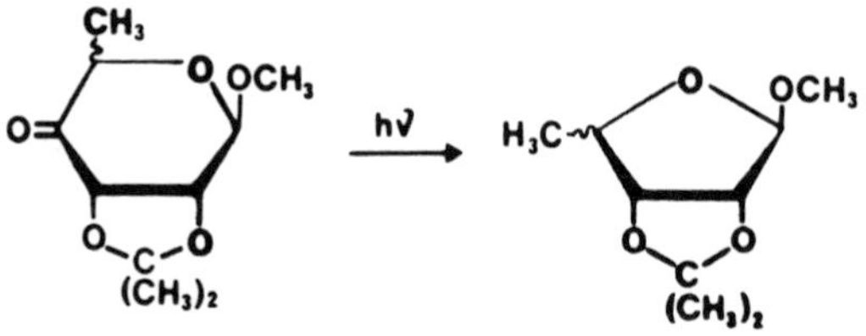

SCHEME 46

Collins and P. Gupta, J.chem.Soc.(C), 1971, 1960, 1965). Although other ketones also react mainly in this fashion, the reactions are very sensitive to several variables, and a wide range of products can be formed (P.M. Collins *et al.*, J.chem. Soc.Perkin I, 1972, 1670; J.chem.Res.(S), 1978, 446). Nevertheless, this photochemical approach affords a means of preparing D-*erythro*-pentulose from D-fructose by extrusion of $C_{(3)}$ (*idem.*, J.chem.Soc.Perkin I, 1980, 277). Use of methanol as solvent can lead primarily to the production of branched-chain compounds (Scheme 47) (*idem.*, *ibid.*, 1977, 2423).

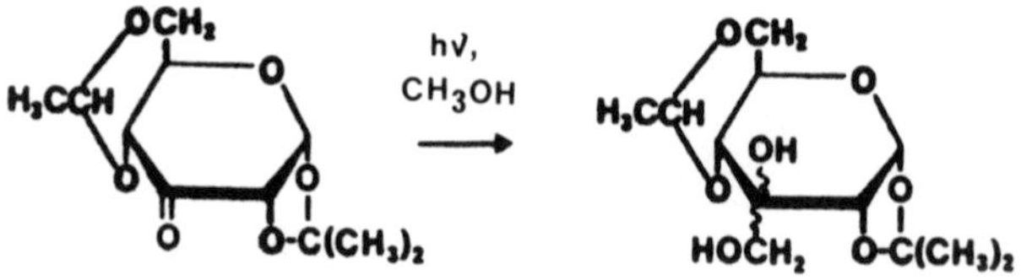

SCHEME 47

(v) *Aldulosonic acids*

Derivatives of hex-4- and 5-ulosonic acids are obtainable from acetylated alkyl hexo-furanosides and -pyranosides, respectively, in highly efficient and selective manner by oxidation with chromium trioxide in acetic acid. While with the furanosides the anomeric configuration is immaterial, in the pyranoside series the aglycone must be equatorial (Scheme 48) (S.J. Angyal and K. James, Austral.J.Chem., 1970, 23, 1209).

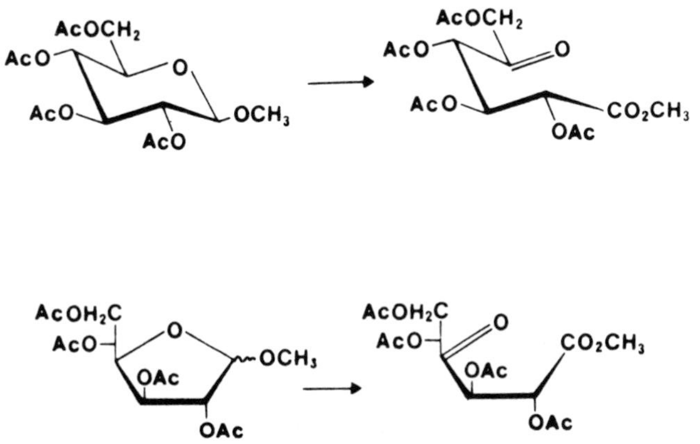

AcOCH$_2$ AcO AcO AcO O OCH$_3$ → AcOCH$_2$ AcO O AcO OAc CO_2CH_3

AcOH$_2$C AcO O OAc OCH$_3$ OAc → AcOH$_2$C AcO O OAc CO_2CH_3 OAc

SCHEME 48

L-*xylo*-Hex-2-ulosonic acid, an intermediate in the synthesis of L-ascorbic acid, is obtainable by oxidation of the primary hydroxyl group of 2,3:4,6-di-*O*-isopropylidene-L-sorbofuranose, and chromium trioxide in ether - dichloromethane in the presence of Celite is effective for the reaction (S.J. Flatt, G.W.J. Fleet and B.J. Taylor, Synthesis, 1979, 815).

3-Deoxy-D-*arabino*-hept-2-ulosonic acid 7-phosphate, formed from phosphoenolpyruvate and D-erythrose 4-phosphate, is the first metabolic intermediate in the common biosynthetic pathway to aromatic compounds in plants and bacteria, and the biosynthetic route to the acid can be applied chemically (Scheme 49) (K.M. Herrmann and M.D. Poling, J.biol.Chem., 1975, 250, 6817).

$$\begin{array}{l} CO_2H \\ | \\ CO \\ | \\ CH_2CO_2H \\ CHO \\ \vdash OH \\ \vdash OH \\ CH_2OPO_3H_2 \end{array} \xrightarrow{Co^{2+}} \begin{array}{l} CO_2H \\ | \\ CO \\ | \\ CH_2 \\ \sim OH \\ \vdash OH \\ \vdash OH \\ CH_2OPO_3H_2 \end{array}$$

SCHEME 49

3-Deoxy-D-*manno*-oct-2-ulosonic acid occurs in the endotoxin lipopolysaccharides of Gram negative bacteria and in bacterial capsular antigens (H.J. Jennings *et al.*, Biochem.biophys.Res. Comm., 1974, 61, 489). During acidic hydrolysis of such materials the acid is converted into anhydro-derivatives involving the hydroxyl group at $C_{(4)}$ (which can be lost by β-elimination) and $C_{(7)}$ and $C_{(8)}$ (W.A. Volk *et al.*, J.biol.Chem., 1972, 247, 3881). Salts of the acid exist at equilibrium in aqueous solution as the α-pyranose (60%), β-pyranose (11%) and furanoses (20 and 9%), and its 5-deoxy analogue has been synthesised by condensation between 2-deoxy-D-*erythro*-pentose and oxalacetic acid (R. Cherniak, R.G. Jones and D.S. Gupta, Carbohydrate Res., 1979, 75, 39). L. Szabó and coworkers have effected a range of chemical syntheses of this type (see e.g. R.S. Sarfati, M. Mondange and L. Szabó, J.chem.Soc.Perkin I, 1977, 2074 for the synthesis of 3-deoxy-D-*manno*-oct-2-ulosonic acid).

3-Deoxy-L-*glycero*-pent-2-ulosonic acid occurs as the acidic component of the capsular polysaccharide of *Klebsiella* type 38 (B. Lindberg, K. Samuelsson and W. Nimmich, Carbohydrate Res., 1973, 30, 63).

(vi) *Ascorbic acid*

^{13}C N.M.R. studies of L-ascorbic acid (S. Berger, Tetrahedron, 1977, 33, 1587) indicate that the dominant species present in solutions of increasing pH values are (28)-(30), the ions being stabilised by delocalisation. When the monoanion is involved in reactions (as in methylation with diazomethane and β-D-glucosylation with tetra-*O*-acetyl-α-D-glucopyranosyl bromide, W.A. Szarek and K.S. Kim, Carbohydrate Res.,

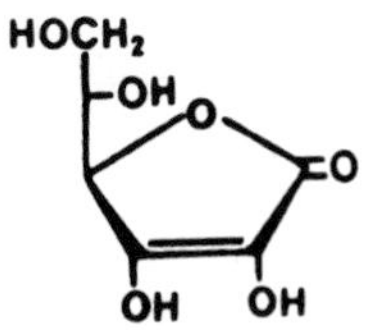

(28) 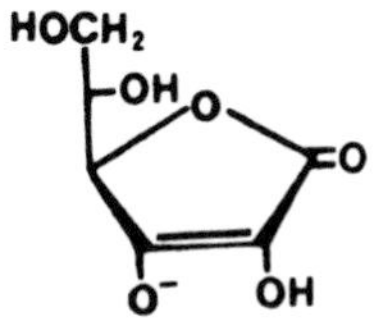 (29) 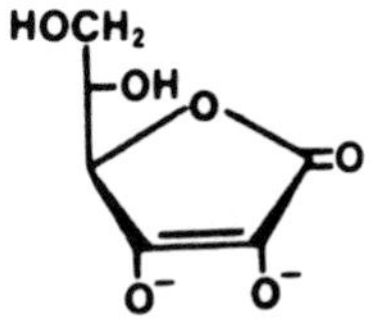 (30)

1978, 67, C13), substitution occurs at $O_{(3)}$, whereas under more basic conditions (as used during the preparation of L-ascorbic acid monophosphate and monosulphate, T. Radford, J.G. Sweeny and G.A. Iacobucci, J.org.Chem., 1979, 44, 658), reaction takes place at the more basic $O_{(2)}$ position.

The synthesis of L-ascorbic acid requires the establishment from a hexonic acid derivative of a compound with a carbonyl group at $C_{(2)}$ [or $C_{(3)}$] and an L-*threo*-diol at $C_{(4)}$, $C_{(5)}$; (usually L-*xylo*-hex-2-ulosonic acid (31) is produced as an intermediate) and several recent new approaches have been reported and some rationales are outlined on p.106. (See T.C. Crawford and S.A. Crawford, Adv.carbohydrate Chem.Biochem., 1980, 37, 79 for a review).

(vii) *Oxidation of monosaccharides with glycol-cleaving agents*

This reaction continues to be used widely for both analytical and synthetic purposes; for further reviews see B. Sklarz, Quart.Rev., 1967, 21, 3, A.J. Fatiadi, Synthesis, 1974, 229, and A.S. Perlin in "The Carbohydrates", Vol. 1B, eds. W. Pigman and D. Horton, Academic Press, New York, 1980, p.1167.

2. Functional derivatives of monosaccharides

(a) Glycosides

Recent years have revealed the occurrence in Nature of a numberless variety of compounds containing sugars bonded glycosidically to other carbohydrates and to non-carbohydrates. The plant kingdom provides glycosides with a wide array of aromatic, terpenoid and steroidal substances with simple sugars attached singly, individually at more than one point, and as oligosaccharides. Micro-organisms, on the other hand,

Synthesis of L-Ascorbic acid

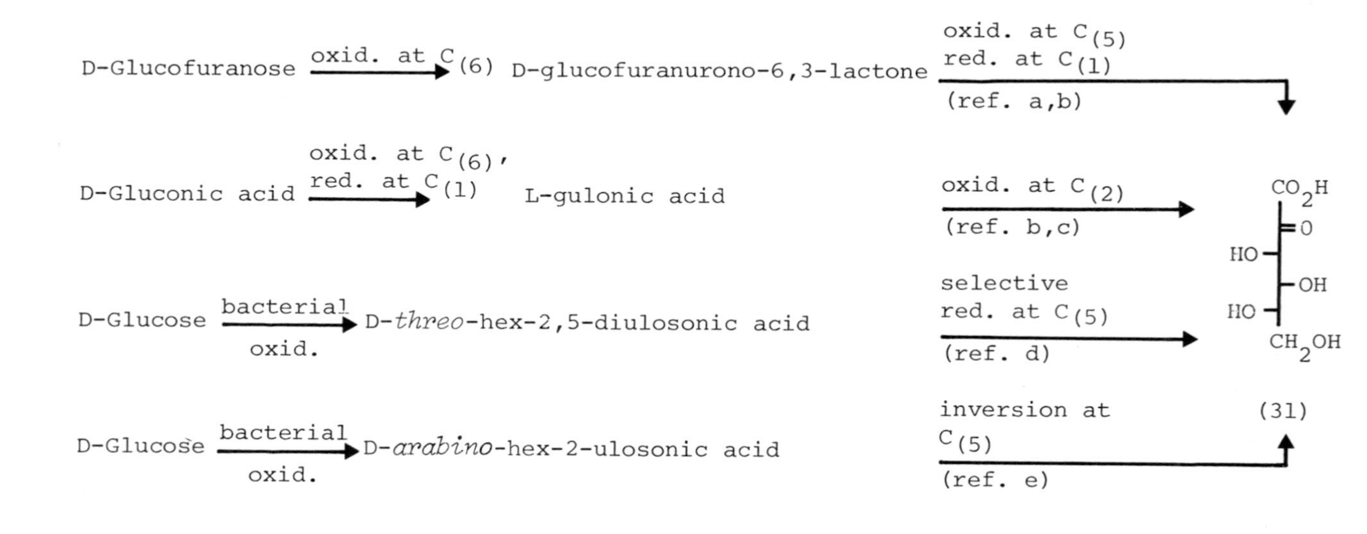

[References: a, J. Bakke and O. Theander, Chem.Comm., 1971, 175; b, T.C. Crawford and R. Breitenbach, *ibid.*, 1979, 388; c, R.J. Ferrier and R.H. Furneaux, *ibid.*, 1977, 332; d, G. Andrews *et al.*, *ibid.*, 1979, 740; e, T. Ogawa *et al.*, Carbohydrate Res., 1976, 51, C1].

produce a plethora of compounds containing highly modified sugars. Many such substances have been investigated in the course of searches for antibiotic materials, and while the structural modifications to the sugar components often involve deoxy-, amino- and branched-chain features (J.S. Brimacombe, Angew. Chem.internat.Edn., 1971, 10, 236), compounds containing unsaturated monosaccharides and even one containing a fluorine atom have been encountered. The finding of such a range of naturally occurring glycosides, and the recognition of the biological significance of many of these has been paralleled by significant developments in chemical methods for the specific glycosidic bonding of monosaccharides. The main challenges provided by natural products of aminal origin are complex oligosaccharides such as those which determine blood group specificities, and new glycosidation methods have permitted major developments in this field also.

Traditional methods of synthesis of *O*-glycosides are well suited to the preparation of compounds with the *trans*-relationship between the aglycone and ring substituent at $C_{(2)}$, but the preparation of *cis*-related anomers has continued as a special challenge (S. Umezawa *et al.*, Bull.chem.Soc.Japan, 1969, 42, 529; J.S. Brimacombe and L.C.N. Tucker in "Annual Reports B", Vol.70, The Chemical Society, London, 1973, p.431). Several important new procedures have been developed (see below). See also "Chemistry of the *O*-Glycosidic Bond: Formation and Cleavage", A.F. Bochkov and G.E. Zaikov, Pergamon Press, Oxford, 1979.

Reference is made to *S*-glycosides on p.194; *C*-glycosides, which occur in plants and micro-organisms, are noted on p.120.

(i) Synthesis of O-*glycosides*

Reviews on the synthesis of glycosides have appeared as follows:

R.J. Ferrier, Fortschr.Chem.Forsch., 1970, 14, 289; W.G. Overend in "The Carbohydrates" Vol. IA, eds. W. Pigman and D. Horton, Academic Press, New York, 1972, p.279; G. Wulff and G. Röhle, Angew.Chem.internat.Ed., 1974, 13, 157; S. Hanessian and J. Banoub, Adv.Chem.Ser., 1976, 39, 36; A.F. Bochkov and G.E. Zaikov (*loc.cit.*); P.M. Collins in "Annual Reports B", Vol.74, The Chemical Society, London, 1977, p.343.

(1) *From reducing sugars*. Glycosides can be prepared from

sugars with unprotected anomeric centres in two ways (Scheme 50); in the presence of acids, alcoholysis occurs by attack of the alcohol at these centres (Fischer glycosidation); secondly, they are derivable by direct substitution of the anomeric hydroxyl groups:

H^+ ; ROH ; RX ; $\overset{+}{O}H_2$; OH ; OR

SCHEME 50

Although detailed studies have been carried out on the methanolysis of sugars (Ferrier, *loc.cit.*) a complete understanding of the reactions has not yet been gained. The general pathway sugar → furanosides → pyranosides has, however, been fully substantiated for the aldoses and also for 2-amino-2-deoxy-D-glucose (S. Umezawa *et al.*, Carbohydrate Res., 1979, 77, 267), D-fructose and L-sorbose (G.S. Bethell and R.J. Ferrier, *ibid.*, 1973, 31, 69), and D-galacturonic acid (K. Larsson and G. Petersson, *ibid.*, 1974, 34, 323). In the last case, and in the cases of D-arabinose and D-galactose (D.D. Heard and R. Barker, J.org.Chem., 1968, 33, 740) and D-xylose and D-glucose (R.J. Ferrier and L.R. Hatton, Carbohydrate Res., 1968, 6, 75), small proportions of dimethyl acetals (detected by radiochemical and chromatographic methods), are present amongst the initial reaction products. They appear to be formed concurrently with the furanosides rather than as their precursors as Emil Fischer had proposed. Analogous ketals were not located during the methanolysis of hex-2-uloses (Bethell and Ferrier, *loc.cit.*) or pent-2-uloses (K. Linek *et al.*, *ibid.*, 1973, 35, 242). In the latter case furanosides alone are therefore produced.

Various metal ions bind selectively to particular configurations of polyhydroxy systems (p. 37), and they can be used to disturb the equilibria formed during Fischer glycosidation reactions and to provide access to products usually formed in low proportions. Judicious selection of reaction conditions thus, for example, permits the isolation in good yields of all four methyl D-allosides formed by methanolysis of the free sugar. Direct, kinetically controlled reaction gives 61% (isolated) of the β-furanoside; kinetically controlled reaction carried out in the presence of strontium chloride gives the α-furanoside (67%, which is selectively complexed to the metal ion); similar reaction under thermodynamic control affords the α-pyranoside (47%) which is also bound to the metal ion, and the β-pyranoside is obtainable in good yield (*ca.* 60%) from equilibrated mixtures of the glycosides produced without added ions (M.E. Evans and S.J. Angyal, *ibid.*, 1972, 25, 43). Metal complexing has been used in this way to facilitate the preparation of many methyl glycosides (S.J. Angyal *et al.*, Austral.J.Chem., 1975, 28, 1541; 1977, 30, 1259).

Usually compounds which are protected at all positions other than the anomeric centre are used for glycoside preparation by the direct substitution procedure (Ferrier, *loc.cit.*), and although these are liable to undergo mutarotation and produce mixtures of products, good selectivity can be achieved. Preferential methylation of 2-acetamido-4,6-*O*-benzylidene-2-deoxy-β-D-glucose thus occurs to give the methyl β-glycoside in 70% yield when dimethyl sulphate is used in aqueous alkali, while the α-anomer is produced in 86% yield when dimethyl sulphoxide is the solvent (W. Roth and W. Pigman, J.Amer.chem. Soc., 1960, 82, 4608). Metal ion complexing can again be used profitably to control specificity since anomerically discrete complexes can be formed, and these act as specific nucleophiles in substitution reactions. Use of complexes (32) and (33) gives β- and α-linked D-ribofuranosyl compounds (including disaccharides) in good yield (R.R. Schmidt and M. Reichrath, Angew.Chem.internat.Edn., 1979, 18, 466). Markedly

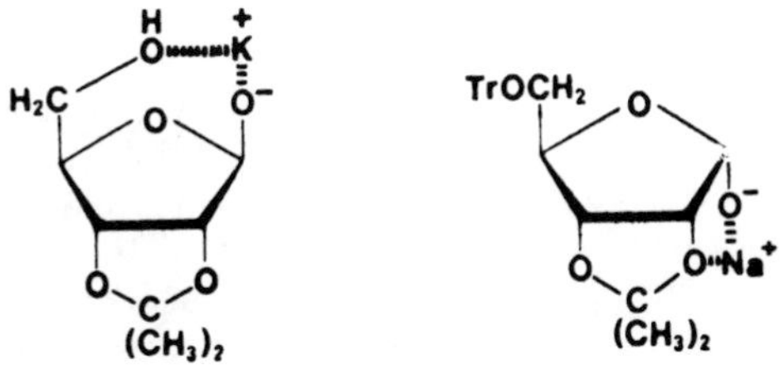

(32) (33)

different ratios obtained on methylating some 1-hydroxy compounds with the aid of silver oxide or sodium hydride can no doubt be attributed to the participation of such complexes

(A.H. Haines and K.C. Symes, J.chem.Soc.(C), 1971, 2331).

Reaction of D-glucose in mixed methanol-acetone (1:4) in the presence of concentrated sulphuric acid (4%) gives several products from which, by selective extraction and column chromatography, methyl α- and β-D-glucoseptanoside can both be obtained in practicable yields as their 2,3:4,5-di-isopropylidene acetals (J.D. Stevens, Carbohydrate Res., 1972, 21, 490).

Alcohols and vinyl ethers, from which very stable carbonium ions are produced, are suitable for the direct substitution of sugar derivatives under acid conditions, and benzhydrol and dihydropyran give diphenylmethyl and tetrahydropyranyl glycosides, respectively, in this way. Vinyl glycosides cannot be made by the Fischer strategy and are produced using acetylene under basic conditions or by transvinylation procedures (Ferrier, *loc.cit.*). Aryl glycosides can be prepared by direct condensation of phenols with the anomeric hydroxyl groups of sugars using diethyl azodicarboxylate and triphenylphosphine in tetrahydrofuran (G. Grynkiewicz, Carbohydrate Res., 1977, 53, C 11).

(2) *From glycosyl halide derivatives*. Extensive data on the reaction of glycosyl halide derivatives with alcohols have been reviewed (Wulff and Röhle, *loc.cit.*; K. Igarashi, Adv. carbohydrate Chem.Biochem., 1977, 34, 243), and earlier generalisations have been confirmed that halides with participating ester groups at $C_{(2)}$ readily form 1,2-acyloxonium ions which then may react with alcohols (a) at the anomeric centres to give 1,2-*trans*-related glycoside derivatives, or (b) at the acyloxonium ion carbon atom to afford cyclic 1,2-orthoesters which themselves can be used in the preparation of this type of glycoside (pp.112,142). Innumerable variants have been advocated for the direct preparation of 1,2-*trans*-related glycosides. These include (a) the use of silver trifluoromethanesulphonate as reaction catalyst and 1,1,3,3-tetramethylurea as proton acceptor in dichloromethane (S. Hanessian and J. Banoub, Carbohydrate Res., 1977, 53, C13), (b) the use of crown ethers to solubilise silver nitrate as catalyst (A. Knöchel, G. Rudolph and J. Thiem, Tetrahedron Letters, 1974, 551, 3739), (c) the use of mercury(II) cyanide in benzene-nitromethane (H.M. Flowers, Methods Carbohydrate Chem., 1972, 6, 474), and (d) the "activation" of the alcohols used as nucleophiles by their conversion into various ethers which readily cleave to give stable carbonium ions.

These include t-butyl, trityl, trimethylsilyl, 2,3-diphenyl-2-cyclopropenyl and tributylstannyl ethers (T. Ogawa and M. Matsui, Carbohydrate Res., 1976, 51, C13).

Although 1,2-*cis*-related glycosides can be obtained from peracylated glycosyl halides - especially from 1,2-*trans*-halides when conditions favour transition states with strong S_N2 character (non-polar solvents, insoluble silver salts, glycosyl chlorides) - this approach is unsatisfactory as a general means of obtaining such glycosides because of competing anomerisation of the halide and the intrusion of S_N1 character to the displacement reactions. Likewise, the use of mixed mercury(II) cyanide and mercury(II) bromide in polar solvents, although it can be reasonably efficient, does not offer a satisfactory and general 1,2-*cis*-glycoside synthesis.

Several useful syntheses of 1,2-*cis*-glycosides have relied upon glycosyl halide derivatives with groups at $C_{(2)}$ which cannot participate in the displacement of the halide from the anomeric centre (e.g. nitrate, chlorosulphate, trichloroacetyl esters and benzyl ethers), but the major development in this approach followed the recognition that, if conditions are controlled to permit rapid anomerisation of the glycosyl halides, the equatorial anomers will react preferentially with alcohols to give products with inverted configuration. (R.U. Lemieux *et al.*, J.Amer.chem.Soc., 1975, 97, 4056). Tetra-*O*-benzyl-D-glucopyranosyl bromide, for example, exists as a syrup and mainly as the α-anomer, but in the presence of tetraethylammonium bromide this isomer rapidly equilibrates with the β-compound which, although never present in large proportions, is relatively reactive towards alcohols in the presence of bases. This approach (Scheme 51) represents a

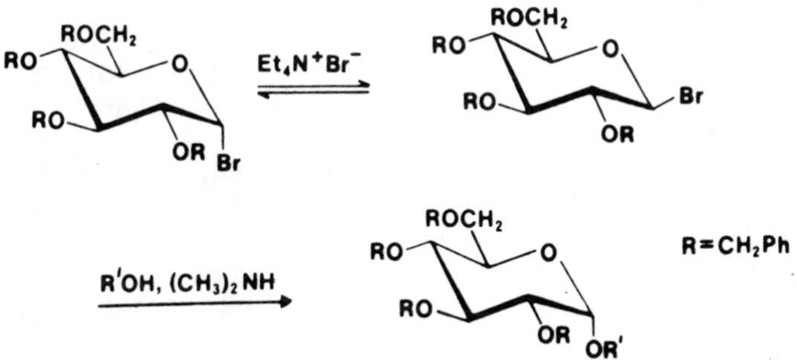

SCHEME 51

major development in the synthesis of α-D-glucopyranosides and other 1,2-*cis*-glycosides.

For the synthesis of specific glycosides of 2-amino-2-deoxy sugars, 2-azido-2-deoxyglycosyl halide derivatives of defined anomeric configuration are the reagents of choice (H.Paulsen and W. Stenzel, Angew.Chem.internat.Edn., 1975, 14, 558).

Frequently, glycosyl halide derivatives are produced from substituted glycosyl esters by use of hydrogen halide with which thermodynamically stable anomers are normally obtained. Dihalogenomethyl methyl ethers can be used as alternative sources of halogen, and with zinc chloride as catalyst, they too gave access to the more stable halides. However, when these ethers are used with 1,2-*trans*-related sugar peresters and with boron trifluoride etherate as catalyst, 1,2-*trans*-glycosyl halides are produced, and these are frequently the thermodynamically less stable anomers. (R. Bognár *et al.*, Carbohydrate Res., 1976, 48, 136). Otherwise, halogens can be introduced at the anomeric centre by displacement of hydroxyl groups using such reagents are triphenylphosphine in carbon tetrachloride (J.B. Lee and T.J. Nolan, Tetrahedron, 1967, 23, 2789) or dichlorocarbene in a phase transfer system (P. Di Cesare and B. Gross, Carbohydrate Res., 1977, 58, C1). This latter method has particular appeal for furanosyl compounds since substituted aldono-γ-lactones offer good access to the initial 1-hydroxy compounds. The use of sulphonic acid anhydrides in the presence of halide ion and a base also affords glycosyl halides from, for example, 2,3,4,6-tetra-*O*-benzyl-D-glucose, and these can be used in "one-pot" syntheses of glycosides (J. Leroux and A.S. Perlin, *ibid.*, 1978, 67, 163).

A further approach to glycosyl halide derivatives uses additions to glycals (p.208), and, in particular, nitrosyl chloride adducts have afforded an alternative synthesis of 1,2-*cis*-related glycosides. Added to tri-*O*-acetyl-D-glucal the reagent gives a dimeric product with the α-D-*gluco*-configuration which reacts in DMF with alcohols and phenols to give 2-oximino-α-glycosides with high specificity (R.U.Lemieux *et al.*, Canad.J.Chem., 1973, 51, 7). These, on deoximation by treatment with aldehydes or ketones under acidic conditions, afford ulose derivatives which give α-D-glucopyranosides on reduction with sodium borohydride (*idem.*, *ibid.*, 1973, 51, 27) (Scheme 52). Since hydrogenation of the oximino-inter-

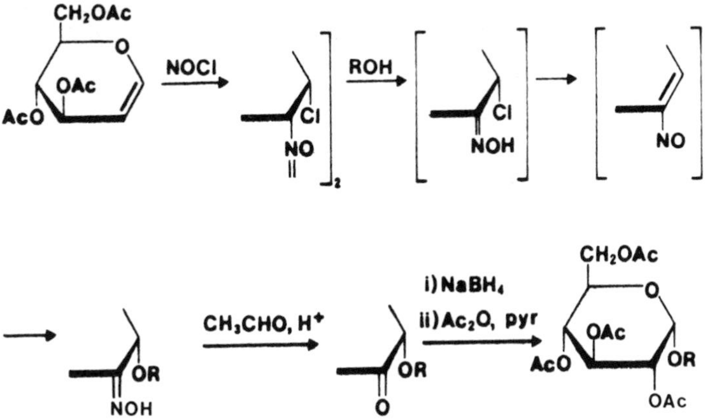

SCHEME 52

mediate is also highly stereoselective, this approach provides a route to α-pyranosides of 2-amino-2-deoxy-D-glucose as well as of D-glucose itself (*idem., ibid.*, 1973, 51, 33).

(3) *From other glycosyl derivatives*. Several glycosyl compounds with other leaving groups at the anomeric centre have found use in glycoside synthesis. Applications of some of the main examples are illustrated below together with notes of the reactions involved. Many applications of these processes have been to the synthesis of di- and higher saccharides.

Glycosyl imidates, formed from corresponding glycosyl chloride derivatives and secondary amides, react with alcohols in the presence of acids to give high yields of glycosides formed with good stereoselectivity. This represents a good alternative α-glucopyranoside synthesis. Non-participating groups at $C_{(2)}$ are required (Scheme 53) (P. Sinaÿ, Pure appl.Chem., 1978, 50, 1437).

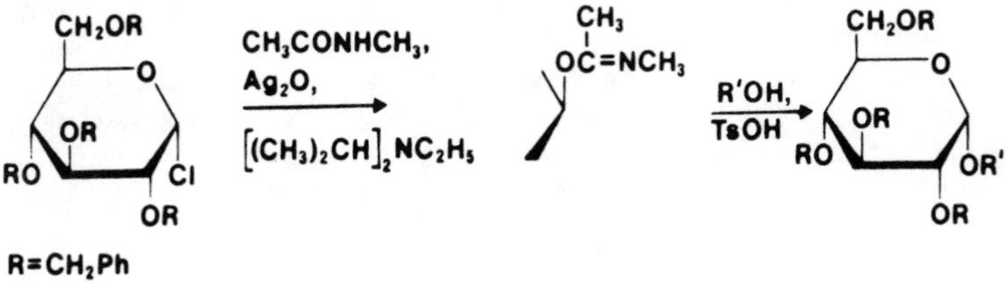

SCHEME 53

Glycosyl tosylates react with alcohols to give glycosides, and the presence of the *N*-phenylcarbamoyl group at $C_{(6)}$ affects the reactions in favour of the production of α-glucosides (Scheme 54) (R. Eby and C. Schuerch, Carbohydrate Res., 1974, 34, 79).

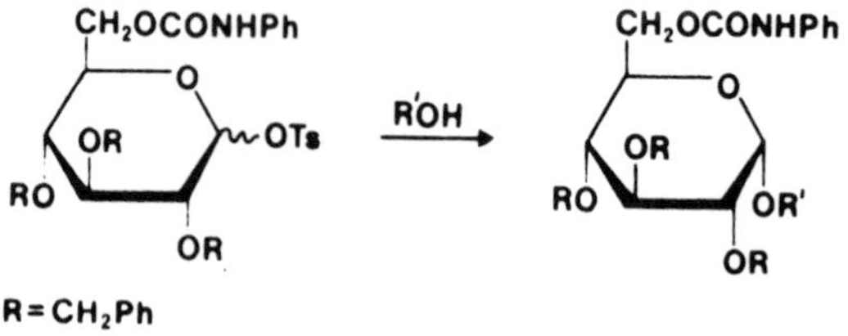

SCHEME 54

Quatenary ammonium salts produced from benzylated α-halogeno-compounds have the β-configuration which is favoured by the reverse anomeric effect (p. 23). They undergo dispacements with alcohols to give 1,2-*cis*-related glycosides (Scheme 55) (A.C. West and C. Schuerch, J.Amer.chem.Soc., 1973, 55, 1333).

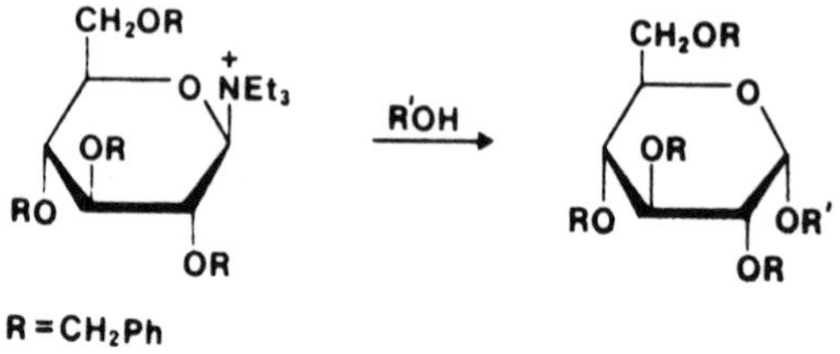

SCHEME 55

Glycosyl acetates have frequently been used together with Lewis or protonic acids in glycoside syntheses, but often mixtures of products are obtained following acid catalysed anomerisations. The use of tin(IV) chloride and 1,2-*trans*-related peresters and N,N-dimethylformamide dialkyl acetals in equimolar proportions gives high yields of *trans*-glycosides specifically. This is particularly useful in the furanoid series since unstable glycosyl halide derivatives are thus avoided (Scheme 56) (S. Hanessian and J. Banoub, Tetrahedron Letters, 1976, 657, 661).

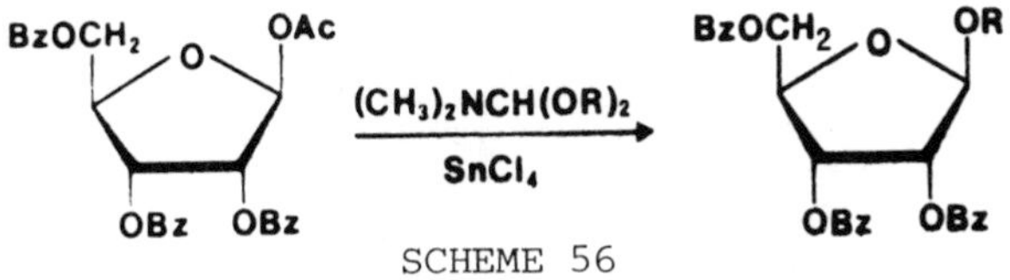

SCHEME 56

Acylated glycosyl halides, on treatment with simple alcohols (ROH) in boiling ethyl acetate in the presence of lead carbanate and anhydrous calcium sulphate, give 1,2-orthoesters (N.K. Kochetkov, A.J. Khorlin and A.F. Bochkov, Tetrahedron, 1967, 23, 693) which can also be prepared conveniently from the halides by use of *N,N*-dimethylformamide dialkylacetals (S. Hanessian and J. Banoub, Carbohydrate Res., 1975, 44, C14). These derivatives afford 1,2-*trans*-related glycosides with alcohols (R^1OH) in boiling nitromethane in the presence of mercury(II) bromide, or in boiling chlorobenzene in the presence of catalytic 2,6-dimethylpyridinium perchlorate. However, when orthoesters of lower alcohols are treated in 1,2-dichloroethane with higher alcohols and catalytic toluene-*p*-sulphonic acid and with removal of the simple alcohol, transesterification reactions occur, to give new orthoesters. With further acid and pyridinium perchlorate these isomerise usually to 1,2-*trans*-glycosides. This latter step therefore proceeds without liberation of an alcohol as by-product. (Scheme 57) (N.K. Kochetkov, O.S. Chizhov and A.F. Bochkov, in M.T.P. International Review of Science, Vol.7, ed. G.O. Aspinall, Butterworths, London, 1973, p.147). Factors such as

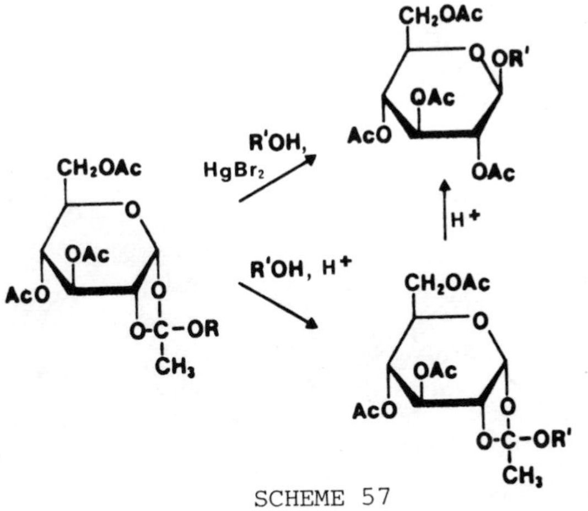

SCHEME 57

the basicity of the alcohol (R^1OH), however, influence the reaction, and under conditions in which ethyl 1,2-orthoesters rearrange to give only β-glycosides, 2-chloro-, 2,2-dichloro- and 2,2,2-trichloro-ethyl analogues afford 16, 50 and 67% respectively, of the α-anomers (P.J. Garegg and I. Kvarnström, Acta chem.Scand., 1976, B30, 655).

A reaction analogous to the orthoester glycoside synthesis affords 1,2-*trans*-related glycosides of *N*-acyl-2-aminosugars from oxazoline derivatives (Scheme 58) (K.L. Matta, E.A. Johnson and J.J. Barlow, Carbohydrate Res., 1973, 26, 215).

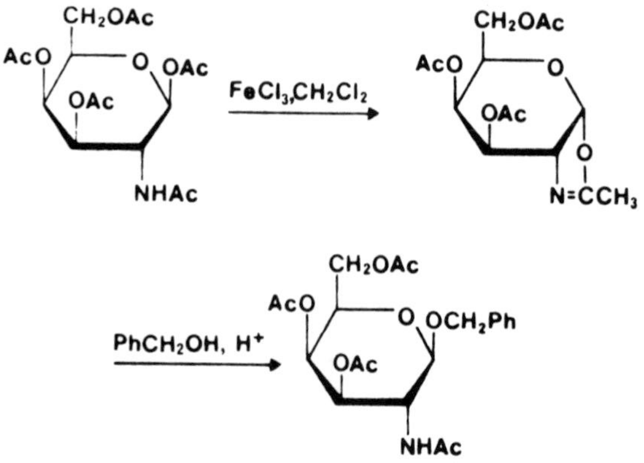

CH_2OAc, AcO, O, OAc, OAc, NHAc — $FeCl_3,CH_2Cl_2$ → CH_2OAc, AcO, O, OAc, O, $N{=}CCH_3$ — $PhCH_2OH, H^+$ → CH_2OAc, AcO, O, OCH_2Ph, OAc, NHAc

SCHEME 58

Co-ordination between mercury(II) ions and the sulphur and nitrogen atoms of pyridin-2-yl 1-thio-β-D-glucopyranoside generates a reactive leaving group which is readily displaced by alcohols to give glucosides. Initial results indicate that the stereoselectivity of the reaction is not high (Scheme 59) (S. Hanessian, C. Bacquet and N. Lehong, *ibid.*, 1980, 80, C17).

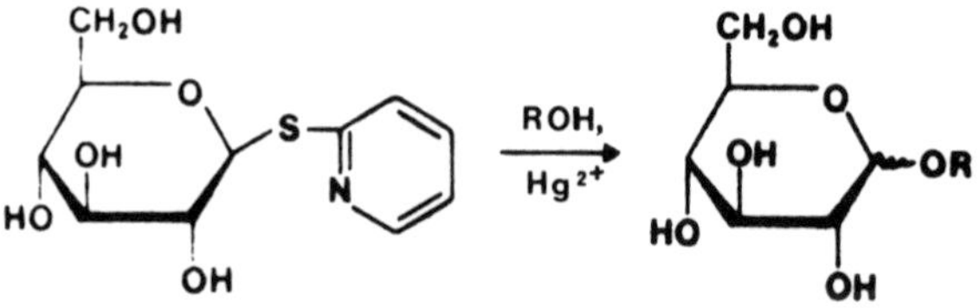

CH_2OH, O, S, N, OH, HO, OH — ROH, Hg^{2+} → CH_2OH, O, OR, OH, HO, OH

SCHEME 59

This procedure extends the use of phenyl 1-thioglycosides and mercury(II) ions in glycoside synthesis (R.J. Ferrier, R.W. Hay and N. Vethaviyasar, *ibid.*, 1973, 27, 55) which itself was a development of the reaction by which, in the presence of such ions, aldose dithioacetals are converted into alkyl glycofuranosides (J.W. Green, Adv.carbohydrate Chem., 1966, 21, 95).

(4) *From unsaturated sugar derivatives.* Some addition reactions of glycals afford glycosidic products directly (p.288), and some, for example the addition of nitrosyl chloride (p.112), give glycosyl derivatives of value in glycoside synthesis. Furthermore, glycal esters and 2-hydroxyglycal esters react with alcohols in the presence of Lewis acids to afford 2,3-unsaturated compounds which also can be used to prepare specific glycosides (see p.214).

(5) *From more accessible glycosides.* A further approach to the synthesis of specific glycosides depends upon the modification of available glycosidic compounds, and acid-catalysed anomerisations of thermodynamically unfavoured isomers represents the most commonly used procedure of this type (W.G. Overend in "The Carbohydrates", Vol.1A, eds. W. Pigman and D. Horton, Academic Press, New York, 1972, p.279). Alternatively, transglycosylation of a sugar moiety from one aglycone to another under acid conditions can be used, and enzymic methods can also bring this about.

Otherwise, starting with readily available compounds, configurations at specific centres of the sugars can be altered to give less accessible compounds. β-D-Mannopyranosides, which are difficult to prepare by ordinary methods, can be made from β-D-glucopyranoside analogues by suitable protection of hydroxyl groups at $C_{(3)}$, $C_{(4)}$ and $C_{(6)}$, oxidation to the 2-ulosides, and reduction with sodium borohydride (E.E. Lee, G. Keaveney and P.S. O'Colla, Carbohydrate Res., 1977, 59, 268). Inversion at $C_{(4)}$ by nucleophilic displacement of sulphonyloxy groups by acetate anions allows the conversion of glucopyranosides to galactopyranosides (M. Petitou and P. Sinaÿ, *ibid.*, 1975, 42, 180).

(ii) Reactions of O-*glycosides*

For a review of the mechanisms of enzymic and chemical hydrolyses of glycosides see B. Capon, Chem.Rev., 1969, 69, 407.

Kinetic evidence confirms that methyl α- and β-D-glucopyranoside hydrolyse in alkaline conditions by mechanisms which involve participation by anions at $O_{(6)}$ and $O_{(2)}$, respectively (Y.Z. Lai, Carbohydrate Res., 1972, 24, 57). In concurrence with this, methylation of the hydroxyl group at $C_{(2)}$ of *p*-nitrophenyl β-D-xylopyranoside reduces the rate of alkaline hydrolysis by a factor of 10^3, and reaction occurs by nucleophilic attack on the aromatic ring (C.K. De Bruyne, F. Van Wijnendaele and H. Carchon, *ibid.*, 1974, 33, 75). In the case of *p*-nitrophenyl α-D-glucopyranoside such attack occurs intramolecularly to give a Meisenheimer complex (C.S. Tsai and C. Reyes-Zamora, J.org.Chem., 1972, 37, 2725) and, by way of this, migration of the aglycone occurs to give the 2-ether (and its 2-epimer). Subsequent migration of the aryl substituent to $O_{(3)}$ gives a product which readily undergoes β-elimination and formation of saccharinic acids (Scheme 60) (D. Horton and A.E. Luetzow, Chem.Comm., 1971, 79).

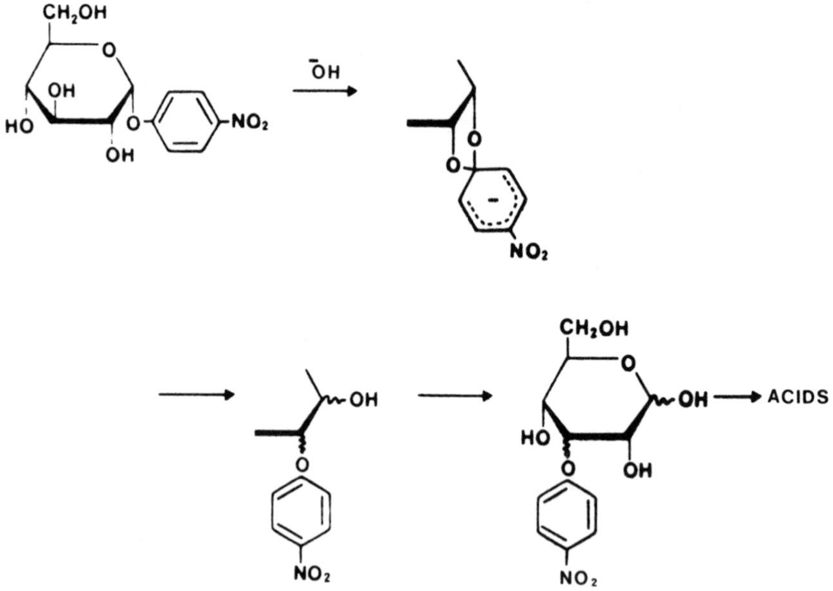

SCHEME 60

As a reaction of importance in biochemistry, the acid catalysed hydrolysis of glycosides has been extensively studied (for a review see J.N. BeMiller, Adv.carbohydrate Chem., 1967,

22, 25), and, in particular, the significance of intramolecular catalytic factors has become apparent. o-Carboxyphenyl β-D-glucopyranoside hydrolyses at pH 4.55 with a rate constant 1.3 X 10^4 times greater than that for the *p*-carboxyisomer in consequence of the general acid catalysis provided by the o-carboxy group (B. Capon *et al.*, J.chem.Soc.(B), 1969, 1038). In the case of the 2-acetamido-2-deoxy analogue of the o-substituted glycoside, hydrolysis occurs by a mechanism involving concerted general acid catalysis of the above kind and intramolecular acetamido group nucleophilic catalysis. The enzyme lysozyme which catalyses the hydrolysis of analogous glycosidic compounds may well also utilise reactions which proceed in analogous fashion (D. Piszkiewicz and T.C. Bruice, J.Amer.chem.Soc., 1968, 90, 2156).

Addition of copper(II) ions to solutions of 8-quinolyl β-D-glucopyranoside gives a complex which is hydrolysed 10^5 times faster than the uncomplexed glycoside in the pH range 5.5-6.2 (Scheme 61). This striking factor signifies that traces of the metal are as effective as strong acid in cleaving this glycoside, and correlates with the discovery that some glycosidases require metal ions for the full expression of their activity (R.R. Clark and R.W. Hay, J.chem.Soc.Perkin II, 1973, 1943).

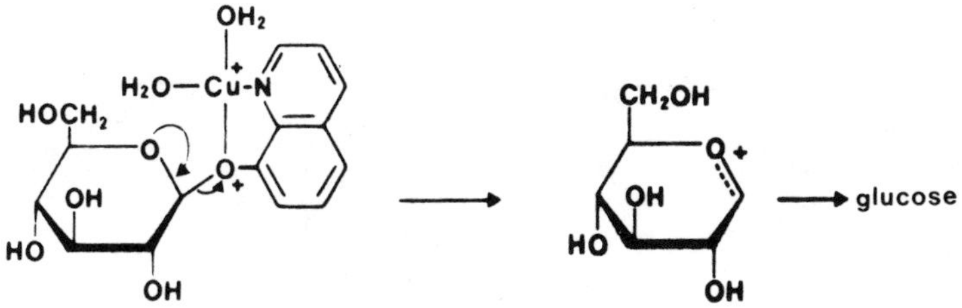

SCHEME 61

The cleavage of aryl glycosides (and other aspects of their chemistry and biochemistry) have been reviewed specifically (S.M. Hopkinson, Quart.Rev., 1969, 23, 98). They are markedly more stable than alkyl glycosides to radiolysis since energy can be transferred from the carbohydrate to the aglycone. Glycosidic bond cleavage does occur, however, both in the solid phase and in aqueous solution (G.O. Phillips, *et al.*, Carbohydrate Res., 1971, 16, 79, 89). o-Nitrobenzyl glycosides are readily cleaved photochemically (U. Zehavi and A.

Patchornik, J.org.Chem., 1972, 37, 2285).

Glucuronide linkages can also be cleaved photochemically (p. 90), and the complex situation relating to the acid-catalysed hydrolysis of such compounds has been resolved by the recognition that the rates comprise terms for the hydrolysis of both the glucuronides and the ionised species (B. Capon and B.C. Ghosh, J.chem.Soc.(B), 1971, 739). This accounts for the finding, for example, that 2-naphthyl β-D-glucopyranoside hydrolyses 45 times faster than the corresponding glucuronide in 1M-hydrochloric acid but 35 times slower at pH 4.79.

(iii) C-Glycosides

Considerable attention has been given to this class of compounds because of their wide occurrence in Nature and the biological activities exhibited by members of the group. Initially, they were found in plant sources and bonded to phenolic compounds - often anthracenes or flavones - and barbaloin (34) and vitexin (35) are amongst the best known of this group. L.J. Haynes has reviewed earlier work (Adv.carbohydrate

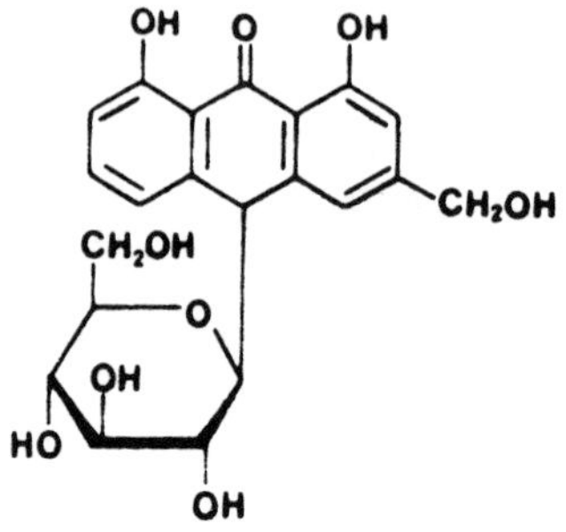

(34) 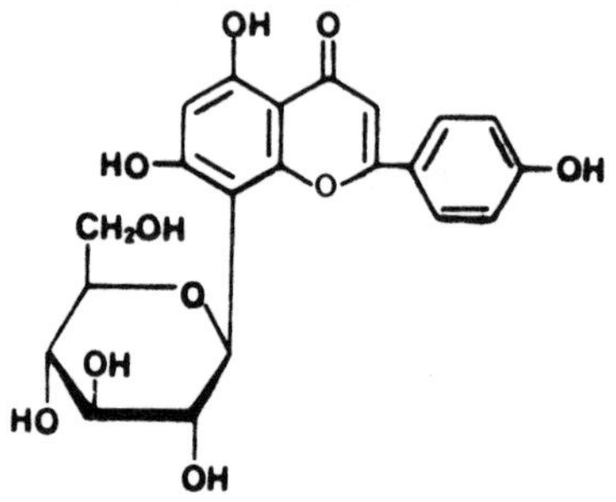 (35)

Chem., 1963, 18, 227; 1965, 20, 357), and a later review on *C*-glycosylflavones has appeared (J. Chopin and M.L. Bouillant, Flavonoids, 1975, 2, 631).

Work in this area has been greatly stimulated by the finding of pseudouridine (36) as a minor component of transfer ribonucleic acids, and of such other *C*-nucleosides as showdomycin (37)(which exhibits antibacterial and antitumour activity) and formycin (38) (which shows antiviral action) amongst microbiological fermentation products. S. Hanessian and A.G. Pernet have reviewed the syntheses of such compounds and of other *C*-glycosides (Adv.carbohydrate Chem.Biochem., 1976, 33, 111).

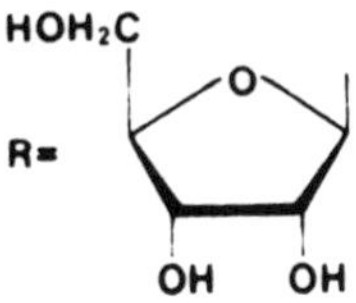

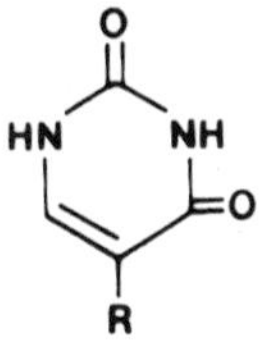

HOH₂C R= OH OH

(36) 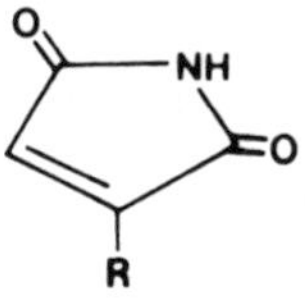 (37) 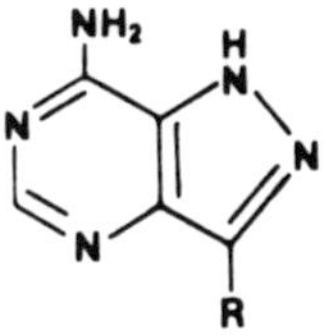 (38)

There are two approaches to the synthesis of glycosyl compounds of this type. Firstly, anhydride ring closures of acyclic derivatives can be employed (e.g. Scheme 62) (J.G. Buchanan *et al.*, J.chem.Soc.Perkin I, 1975, 1191; 1976, 68).

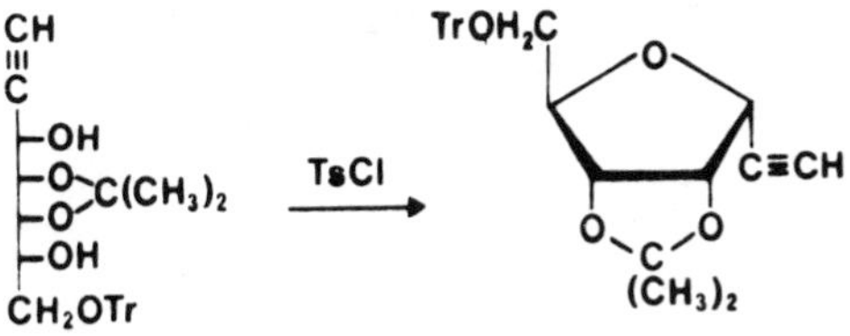

SCHEME 62

Secondly, displacement reactions with carbon nucleophiles at anomeric centres of glycosylating agents can be utilised. Frequently, the aglycones are then elaborated from simple $C_{(1)}$ substituents (Scheme 63) (T. Ogawa, A.G. Pernet and S. Hanessian, Tetrahedron Letters, 1973, 3543).

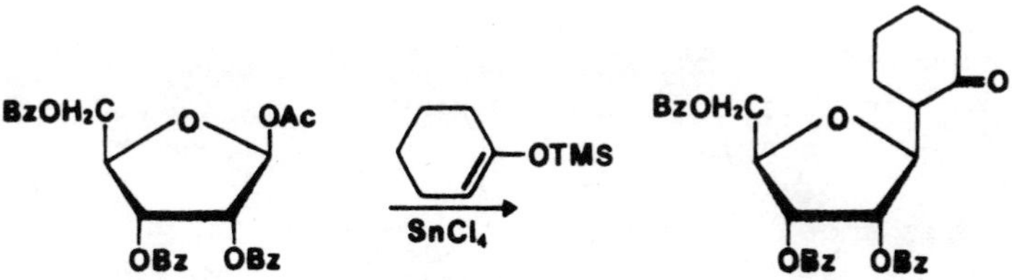

SCHEME 63

(b) Monosaccharide ethers

Methyl ethers retain their significance as labelling groups for structural analytical work particularly as they are amenable to treatment in the gas chromatograph and mass spectrometer; several new methylation methods have been developed. Other ethers serve mainly as protecting groups during synthesis.

(i) Methyl ethers

A considerable number of ethers of simple sugars and of modified sugars have been isolated from polysaccharides and from microbiological sources, respectively; many specific syntheses have been reported.

While the traditional methylation methods are still frequently used, the Kuhn modification (DMF as solvent) of the Purdie method has found particular favour, and the diazomethane-boron trifluoride procedure (J.O. Deferrari, E.G.Gros and I.M.E. Thiel, Methods carbohydrate Chem., 1972, 6, 365) is often adopted for the methylation of compounds having base-labile groups - particularly ester functions which may migrate under the conditions of other methylation reactions. An interesting variant of this method involves the use of diazomethane in methanol-dichloromethane with catalytic tin(II) chloride which causes, in some cases, highly selective monomethylation of diols. Applied to methyl 4,6-*O*-benzylidene-α-D-glucopyranoside (M. Aritomi and T. Kawasaki, Chem.pharm. Bull.Tokyo, 1970, 18, 677) and benzyl 4,6-*O*-benzylidene-β-D-talopyranoside (G.J.F. Chittenden, Carbohydrate Res., 1976, 52, 23) the method gives 93 and 89% of the 3-ethers, but 91% of the 2-substituted product is obtained from benzyl 4,6-*O*-benzylidene-β-D-galactopyranoside (*idem.ibid.*, 1975, 43, 366).

A methylation method which has the virtue of giving rapid and complete substitution of polyhydroxy compounds involves the use of the dimethylsulphinyl carbanion as base. This is prepared by dissolving sodium hydride in DMSO, and the carbohydrate is added in this solvent prior to the further addition of methyl iodide [S. Hakomori, J.Biochem.(Tokyo), 1964, 55, 205]. Partial methylation by this method (with limited proportions of methyl iodide) of methyl β-D-xylopyranoside gives the reactivity order of the hydroxyl groups as 4>2>3, whereas the sequence is 2>4>3 for methylation by the Haworth, Purdie or Kuhn methods (Y.S. Ovodov and E.V. Evtushenko, Carbohydrate Res., 1973, 27, 169).

An entirely different approach to the synthesis of methyl ethers also uses DMSO which, together with acetic acid and acetic anhydride converts primary, secondary or tertiary alcohols into methylthiomethyl ethers. These are then reduced with "nickel boride" (P.M. Pojer and S.J. Angyal, Austral.J. Chem., 1978, 31, 1031):

$$R\text{-}OH \xrightarrow{DMSO, Ac_2O, AcOH} R\text{-}OCH_2SCH_3 \xrightarrow{NiCl_2,\ NaBH_4} ROCH_3$$

Other effective newer methods involve the use of methyl halide in phase transfer conditions with aqueous sodium hydroxide and tetrabutylammonium bromide as catalyst (P.Di Cesare and B. Gross, Carbohydrate Res., 1976, 48, 271), and methyl trifluoromethanesulphonate and hindered bases e.g. 2,6-di-tert-butylpyridine (B. Lindberg *et al.*, *ibid.*, 1975, 44, C5; J.M. Berry and L.D. Hall, *ibid.*, 1976, 47, 307).

Demethylation can be effected by use of chromium trioxide in acetic acid, the ethers being oxidised to formate esters (S.J. Angyal and K. James, *ibid.*, 1970, 12, 147). Acyl esters survive the oxidation step but are then cleaved concurrently with the formates. Lithium in diethylamine also causes demethylation without, surprisingly, cleaving glycosidic bonds or ketal rings (C. Monneret *et al.*, Tetrahedron Letters, 1971, 1935).

While chemical inertness is the primary characteristic of methyl ethers, occasionally they participate in nucleophilic displacement reactions. One of the products obtained on base treatment of methyl 6-*O*-methanesulphonyl-2,3-di-*O*-methyl-β-D-galactopyranoside contains a 3,6-anhydro ring implying that the sulphonyloxy group has been displaced by $O_{(3)}$ (J.S. Brimacombe and O.A. Ching, Carbohydrate Res., 1969, 9, 287). Much more frequently, methoxyl groups migrate from the anomeric centre during such reactions, and although this does not represent a further example of methyl ether cleavage, it does reveal how methyl ether groups can arise in the products.

(ii) Trityl ethers

A new procedure for making trityl ethers can be employed successfully with secondary alcohols and uses triphenylmethylium perchlorate (made from triphenylcarbinol and acetic anhydride followed by perchloric acid) and a sterically hindered base e.g. 2,4,6-tri-*tert*-butylpyridine (Y.V. Wozney and N.K. Kochetkov, *ibid.*, 1977, 54, 300).

A modification of the normal trityl ether as a protecting group is derived by the use of insoluble polymers which contain trityl chloride as part of their structures. Friedel Crafts benzoylation of a polystyrene resin, followed by Grignard reaction and chlorination, gives such materials.

$$\text{(P)} \xrightarrow{PhCOCl, AlCl_3} \text{(P)}-COPh \xrightarrow{PhMgBr} \text{(P)}-C(OH)Ph_2 \xrightarrow{AcCl} \text{(P)}-C(Cl)Ph_2$$

Reaction of the polymer in pyridine with methyl α-D-glucopyranoside followed by benzoyl chloride gives an insoluble product from which all excess of reagents can be washed. Passage of hydrogen bromide through a suspension of the resin then releases methyl glucoside 2,3,4-tribenzoate in high yield. This approach represents a very convenient way of carrying out specific substitution reactions (J.M.J. Fréchet and K.E. Haque, Tetrahedron Letters, 1975, 3055). Whereas the above treatment for cleavage of the trityl ether bond is normal, trityl ethers can also be removed conveniently using silica gel columns (J. Lehrfeld, J.org.Chem., 1967, 32, 2544).

The conversion of an alcohol to a trityl ether activates it towards glycosylation reagents (p.111).

(iii) Benzyl ethers

Several of the new methylating reactions can be adapted for the preparation of benzyl ethers, for example, the use of benzyl trifluoromethanesulphonate (R.U. Lemieux and T. Kondo, Carbohydrate Res., 1974, 35, C4) and phase transfer catalysts (S. Czernecki *et al.*, Tetrahedron Letters, 1976, 3535). Benzyl ethers also result from the reductive ring opening with lithium aluminium hydride-aluminium chloride of benzylidene acetals (Scheme 64) (S.S. Bhattacharjee and P.A.J. Gorin, Canad.J.Chem., 1969, 47, 1195).

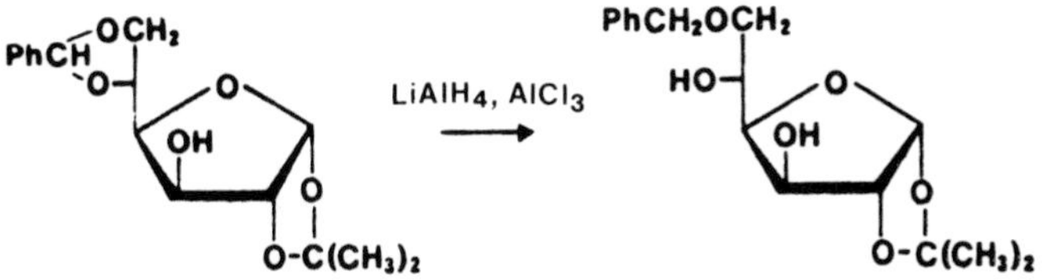

$\xrightarrow{LiAlH_4,\ AlCl_3}$

SCHEME 64

A good range of debenzylation methods have become available. These include free radical bromination (photochemical with bromine) at the benzyl centres followed by mild hydrolysis (J.N. BeMiller, R.E. Wing and C.Y. Meyers, J.org.Chem., 1968, 33, 4292), oxidation to benzoates with chromium troxide in acetic acid (S.J. Angyal and K. James, Carobhydrate Res., 1970, 12, 147), treatment with boron trifluoride in a volatile thiol (thiolysis occurs concurrently at anomeric centres, H.G. Fletcher and H.W. Diehl, *ibid.*, 1971, 17, 383), treatment with sodium in liquid ammonia (S.A. Holick and L. Anderson, *ibid.*, 1974, 34, 208).

(iv) Allyl ethers

Allyl ethers have proved of value as protecting groups since they are relatively stable but rearrange readily and completely on treatment with potassium *tert*-butoxide in DMSO into *cis*-prop-1-enyl ethers which, being vinyl ethers, can be removed readily in several ways including by treatment with dilute acid or with mercury(II) salts:

$$R\text{-}OCH_2\text{-}CH{=}CH_2 \longrightarrow R\text{-}OCH{=}CHCH_3 \longrightarrow ROH$$

Substitution of methyl groups into the allyl radicals has significant effect on these reactions. For example, the conditions used to isomerise allyl ethers remove 3-methylallyl groups. The considerable opportunities offered by these characteristics have been exploited mainly by Gigg (P.A. Manthorpe and R. Gigg, Methods carbohydrate Chem., 1980, 8, 305).

(v) Silyl ethers

While the trimethylsilyl ether group has proved to be excellent for analytical work involving gas chromatography and mass spectrometry, it tends to be too susceptible to solvolysis in protic media for use in synthesis. However, the *tert*-butyldimethylsilyl analogues which are 10^2-10^3 times more stable have provided suitable alternatives. These ethers are prepared by use of the silyl chloride in DMF with imidazole as catalyst, and are cleaved specifically with tetrabutylammonium fluoride in tetrahydrofuran (E.J. Corey and A. Venkateswarlu, J.Amer.chem.Soc., 1972, 94, 6190). Under these conditions, however, ester groups can migrate (G.H. Dodd, B.T. Golding and P.V. Ioannou, Chem.Comm., 1975, 249). Methyl α-D-glucopyranoside gives the 6-ether in virtually quantitative yield when treated with a molar equivalent of *tert*-butyldimethylsilyl chloride in pyridine (F. Franke and R.D. Guthrie, Austral.J.Chem., 1977, 30, 639).

The *tert*-butyldiphenylsilyl ether group is similar in its selectivity, is even more stable, and also cleaves on treatment with fluoride ion (S. Hanessian and P. Lavallee, Canad. J.Chem., 1975, 53, 2975).

(c) Cyclic acetals and ketals

Reviews have appeared on the cyclic acetals of ketoses (R.F. Brady, Adv.carbohydrate Chem.Biochem., 1971, 26, 197) and aldoses and aldosides (A.N. de Belder, *ibid.*, 1977, 34, 179).

(i) Formation

Various adaptations of existing methods have extended the accessibility and range of cyclic acetals obtainable from free sugars and their derivatives. Use of iron(III) chloride and acetone offers a simple means of producing such acetals as 1,2:5,6-di-*O*-isopropylidene-α-D-glucofuranose which is the normal main product of acid catalysed condensation of D-glucose with acetone (H.C. Srivastava *et al.*, Tetrahedron Letters, 1977, 439). Alternatively, condensation with acetone in DMF solution in the presence of copper(II) sulphate, gives, at intermediate stages in the reaction, significant proportions of the 4,6- and 5,6-acetals (S. Morgenlie, Acta chem. Scand., 1975, B29, 278). More surprisingly, methods have been developed for the isolation of 1,2:3,4-di-*O*-isopropylidene-α- and 2,3:4,5-di-*O*-isopropylidene-D-glucoseptanose (39) and (40) from the products of condensation of the sugar with acetone in the presence of concentrated sulphuric acid. This has led to detailed studies of septanose derivatives (J.D.Stevens, Austral.J.Chem., 1975, 28, 525).

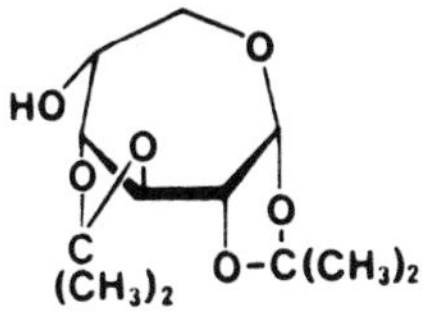

(39) 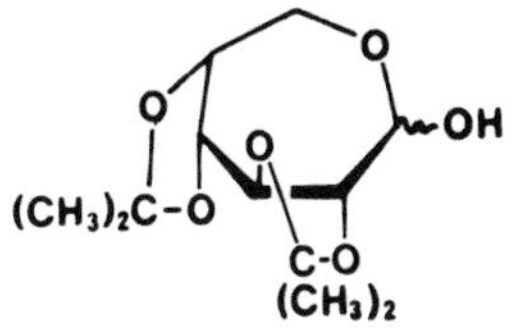 (40)

In analogous fashion, the anomeric methyl glycosides of this second acetal (40) are obtainable from the several products formed on reaction of D-glucose in acetone-methanol in the presence of the same catalyst. Approximately 50 g of each of these crystalline anomers can be isolated from the products obtained from 1 Kg of the sugar, the procedures involving initial distribution between an aqueous and an organic phase followed by column chromatography (*idem.*, Carbohydrate Res., 1972, 21, 490).

The introduction of methods which avoid the generation of water during acetal formation represents a significant development, and acetal exchange reactions can be used to force the ring closure of strained acetals. Isopropylidene derivatives of suitable diols are readily prepared using 2,2-dimethoxypropane in DMF in the presence of catalytic toluene-*p*-sulphonic acid. D-Glucose, under these conditions, affords the 4,6-acetal at room temperature, and at elevated temperatures the 5,6-acetal and 1,2:5,6-di-*O*-isopropylidene-α-D-glucofuranose are obtained as well as some acyclic products. The absence of 1,2-*O*-isopropylidene-α-D-glucofuranose suggests that the diacetal may be derived by way of the 5,6-substituted monoacetal (M. Kiso and A. Hasegawa, *ibid.*, 1976, 52, 87). High yields of the 4,6-acetal are obtainable by use of these procedures with methyl α-D-glucopyranoside (M.E. Evans, F.W. Parrish and L. Long, *ibid.*, 1967, 3, 453), and the application of higher boiling acetals permits full substitution. Methyl 2,3:4,6-di-*O*-cyclohexylidene-α-D-glucopyranoside can be produced in near quantitative yield by use of 1,1-dimethoxycyclohexane at elevated temperatures

(Scheme 65) (F.H. Bissett, M.E. Evans and F.W. Parrish, *ibid.*, 1967, 5, 184).

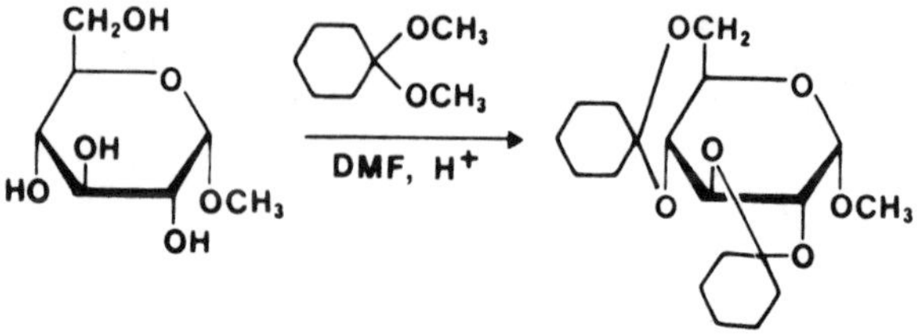

SCHEME 65

An alternative synthesis of cyclic acetals uses vinyl ethers in DMF with catalytic acid. Methyl α-D-glucopyranoside with ethyl vinyl ether gives, initially, a mixed acetal at $C_{(6)}$ by an acid-catalysed addition process, and this undergoes intramolecular trans-acetalation to afford the 4,6-*O*-ethylidene product (Scheme 66) (M.L. Wolfrom *et al.*, J.org.Chem., 1968, 33, 1067).

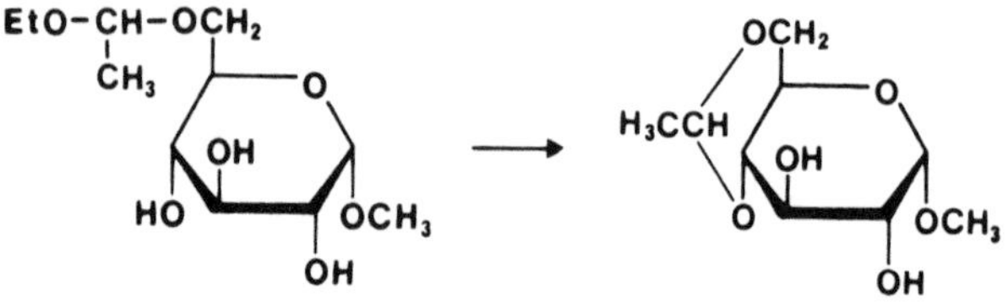

SCHEME 66

Isopropylidene acetals are similarly available through the use of ethyl isopropenyl ether (*idem.*, Carbohydrate Res., 1974, 35, 87).

Geminal dihalides offer means of synthesising cyclic acetals under basic conditions, and strained rings can be formed in this way. Thus, methyl 4,6-*O*-benzylidene-α-D-glucopyranoside treated in DMF with sodium hydride followed by methylene dichloride gives the 2,3-methylidene acetal (A.B. Foster *et al.*, J.chem.Soc.(C), 1967, 2404), and the same condensation can be effected in aqueous solution by use of methylene dibromide in the presence of tetrabutylammonium bromide under phase transfer conditions (K.S. Kim and W.A. Szarek, Synthesis, 1978, 48). An alternative route to cyclic methylene acetals involves treatment of diols with *N*-bromosuccinimide in DMSO, the CH_2 group being derived from the solvent (S. Hanessian, P. Lavallee and A.G. Pernet, Carbohydrate Res., 1973, 26, 258). Benzal chloride or, better, benzal bromide in refluxing pyridine reacts with suitable diols to give cyclic benzylidene acetals and, somewhat surprisingly, the thermodynamically preferred stereoisomers are favoured. This permits the formation of these acetals in the presence of acid sensitive groups such as trityl ethers (P.J. Garegg *et al.*, Acta chem.Scand., 1973, 26, 518, 3895).

A bis-sulphonium derivative of benzaldehyde ethylene dithioacetal offers means of preparing benzylidene acetals under alternative, non-acidic conditions (Scheme 67) (R.M. Munavu and H.H. Szmant, Tetrahedron Letters, 1975, 4543).

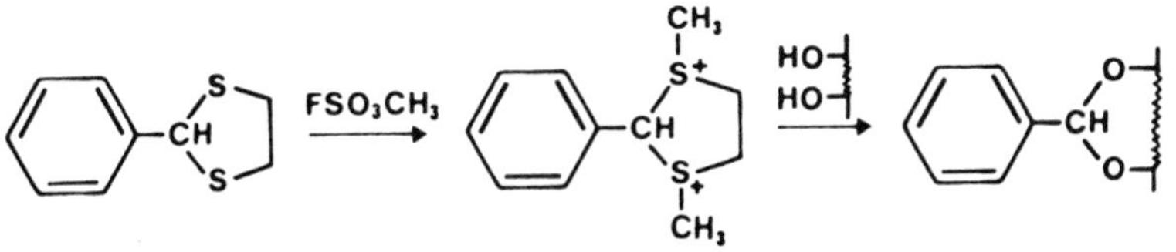

SCHEME 67

Insoluble resins containing benzaldehyde groups form substituted benzylidene acetals and offer means of carrying out solid phase partial substitution reactions. Methyl α-D-glucopyranoside, for example, can be converted into its 2- or 2,3-substituted derivatives in this way (S. Hanessian *et al.*, Carbohydrate Res., 1974, 38, C15).

Other approaches to the synthesis of carbohydrate cyclic acetals involve the reaction between acylated glycosyl halides and dialkyl cadmiums (giving 1,2-acetals, R.G. Rees, A.R. Tatchell and R.D. Wells, J.chem.Soc.(C), 1967, 1768), the acid-catalysed exchange reaction undergone by 1,2-orthoesters with ketones (R.U. Lemieux and D.H. Detert, Canad.J.Chem., 1968, 46, 1039) and the reduction of acyloxonium ions derived from orthoesters (J.G. Buchanan and A.R. Edgar, Carbohydrate Res., 1976, 49, 289).

(ii) Reactions of cyclic acetals and ketals

The general acid-catalysed hydrolysis of acetals has been

reviewed, although no special reference is made to cyclic derivatives (T.H. Fife, Acc.chem.Res., 1972, 5, 264). de Belder (*loc.cit.*) has briefly outlined the reactions of cyclic acetals of carbohydrates.

A reaction which has proved very useful occurs when benzylidene acetals are treated in refluxing carbon tetrachloride with *N*-bromosuccinimide in the presence of barium carbonate. With 4,6-*O*-benzylidene derivatives of hexopyranosyl compounds the preponderant products are the analogous 6-bromo-6-deoxyhexopyranosyl 4-benzoates (Scheme 68), and the reaction is frequently referred to as the Hanessian-Hullar process (S. Hanessian, Methods carbohydrate Chem., 1972, 6, 183). The high

SCHEME 68

selectivity of the reaction and the functionality introduced have made it particularly useful for the synthesis of 6-deoxyhexosyl compounds. It proceeds by hydrogen abstraction from the acetal centre followed either by bromine radical attack at the more accessible primary position, or attack at the acetal carbon atom to give an intermediate which readily loses bromide to give a benzoxonium ion which is then attacked at $C_{(6)}$ by bromide ion. Benzylidene acetals formed from vicinal secondary hydroxyl groups (for example at the 2,3-positions of ribofuranose derivatives) react similarly to give *trans*-related bromodeoxybenzoates; usually both possible products are formed, and stereochemical inversion occurs at the centres which undergo bromination (S. Hanessian and N.R. Plessas, J. org.Chem., 1969, 34, 1053).

A similar acetal-opening reaction, but one giving hydroxy esters, occurs on photolysis of 2-nitrobenzylidene acetals (P.M. Collins *et al.*, J.chem.Soc.Perkin I, 1975, 1695, 1700). 4,6-*O*-Derivatives of methyl α- and β-D-glucopyranoside both give the corresponding 4- and 6-(o-nitroso)benzoates in the ratio 3:7.

A further ring opening reaction occurs when benzylidene

acetals are treated with the hydride abstracting reagent trityl tetrafluoroborate. Hydrolysis of the derived benzoxonium ions again gives monohydroxy monobenzoates, 2,3-*O*-benzylidene-D-ribofuranose acetals giving equal proportions of the *cis*-related 2- and 3-monoesters (S.Hanessian and A.P.A. Staub, Tetrahedron Letters, 1973, 3551). Alternatively, treatment of the ionic intermediates with nucleophiles affords means of preparing *trans*-related products (C. Pedersen *et al.*, Acta chem.Scand., 1977, B31, 359, 365). Benzylidene acetals also give hydroxy monobenzoates by ozonolysis (P. Deslongchamps *et al.*, Canad.J.Chem., 1975, 53, 1204) and deoxybenzoates on treatment with di-t-butyl peroxide (p.173), but the specificities of both these reactions is not good.

Reductive ring opening is effected when acetals are treated with aluminium chloride and lithium aluminium hydride, and hydroxy monoethers are formed. 4,6-Benzylidene derivatives of hexopyranosyl compounds normally open to give the 4-benzyl ethers with high selectivity (Scheme 69), and this provides a useful means of producing compounds with free primary hydroxyl groups (A. Lipták, *et al.*, Carbohydrate Res., 1975, 44, 1). Complexing of the Lewis acid occurs at $O_{(6)}$ to direct

SCHEME 69

ring opening. In other cases the reaction can also be highly selective, the *exo*- and *endo*-diastereoisomers of benzyl 2,3-*O*-benzylidene-α-L-rhamnopyranoside giving the 3-benzyl and the 2-benzyl ether, respectively, with almost complete specificity (*idem.ibid.*, 1976, 51, C19).

Although acetals frequently do not react with Grignard reagents, in benzene as solvent the combination of the nucleophilicity of the carbon and coordination of the magnesium to oxygen is sufficient to cause ring opening. 1,2:5,6-Di-*O*-cyclohexylidene-α-D-glucofuranose reacts preferentially at the 5,6-acetal with methylmagnesium iodide to give a 5-hydroxy 6-ether as main product (Scheme 70) (M. Kawana and S. Emoto,

Bull.chem.Soc.Japan, 1980, 53, 230).

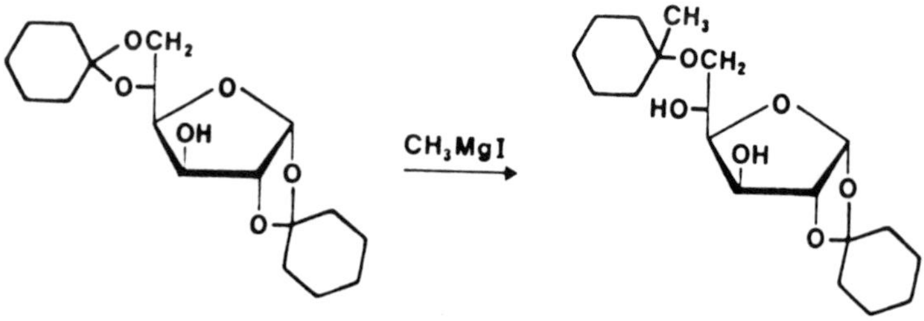

SCHEME 70

Treated with strong proton abstracting reagents carbohydrate acetals give unsaturated products (including enols), but frequently the isolated yields are not good (G. Rodemeyer and A. Klemer, Ber., 1976, 109, 1708). However, on occasions the reaction can be of considerable synthetic value. The 5-membered ring of methyl 2,3:4,6-di-*O*-benzylidene-α-D-mannopyranoside (mixed isomers) opens specifically when the diacetal is treated with butyllithium in tetrahydrofuran at -40°C, and a 3-uloside is produced in very high yield (Scheme 71) (D. Horton and W. Weckerle, Carbohydrate Res., 1975, 44, 227).

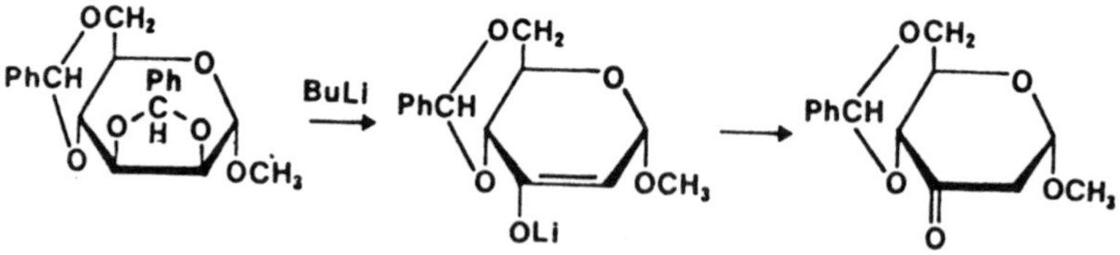

SCHEME 71

(d) *Anhydro derivatives of monosaccharides*

For a review of all anhydro-sugars see R.D. Guthrie in "The Carbohydrates", Vol.1A., eds. W. Pigman and D. Horton, Academic Press, New York, 1972, p.423.

(i) The glycosans

(1) *1,6-Anhydrides and related compounds.* (For reviews see M. Černý and J. Staněk, Fortschr.Chem.Forsch., 1970, 14, 526; Adv.carbohydrate Chem.Biochem., 1977, 34, 23). 1,6-Anhydro-β-D-glucopyranose is readily isolated from the pyrolysate of cellulose; it is formed together with smaller amounts of the furanose analogue, 1,4:3,6-dianhydro-α-D-glucopyranose, and several unsaturated compounds including 1,5-anhydro-4-deoxy-D-*glycero*-hex-1-en-3-ulose (41) (F. Shafizadeh *et al.*, Carbohydrate Res., 1978, 67, 433). During pyrolyses carried out in the presence of catalytic phosphoric acid, significant proportions of "levoglucosenone" (42) are formed (*idem.ibid.*, 1978, 61, 519).

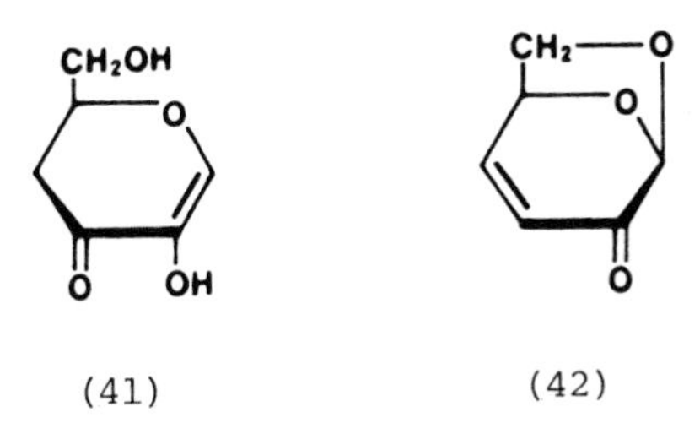

(41) (42)

It is now well established that aldohexoses equilibrate with their 1,6-anhydropyranose derivatives in acid solution, the proportions of anhydrides varying substantially according to their relative thermodynamic stabilities (p.80). In similar fashion the aldoheptoses equilibrate with their 1,6-anhydropyranose derivatives, and the equilibria lie more towards the anhydrides than is the case with the comparable (homomorphous) aldohexoses (S.J. Angyal and R.J. Beveridge, *ibid.*, 1978, 65, 229). Small amounts of 1,6-anhydro-β-D-glucopyranose are, surprisingly, produced by a transferase acting on maltose (J.H. Pazur *et al.*, *ibid.*, 1978, 61, 279). Hexuloses cannot form strictly analogous anhydrides, but heptuloses give 2,7-anhydro-derivatives (S.J. Angyal and K. Dawes, Austral.J.Chem., 1968, 21, 2747), as do octuloses (N.K. Richtmyer, Carbohydrate Res., 1972, 23, 319) and nonuloses (S.K. Gross *et al.*, *ibid.*, 1975, 41, 344).

When treated with acid in DMF under dehydrating conditions aldohexoses give large proportions of 1,6-anhydrofuranoses (p. 80). D-Galactose, for example, affords 87% of this derivative (43), and vacuum pyrolysis of this sugar likewise gives this anhydride, together with some of the 1,6-anhydro-β-D-pyranose isomer (44) and also some 1,5-anhydro-α-D-galactofuranose (45). This last product is the same as 1,4-

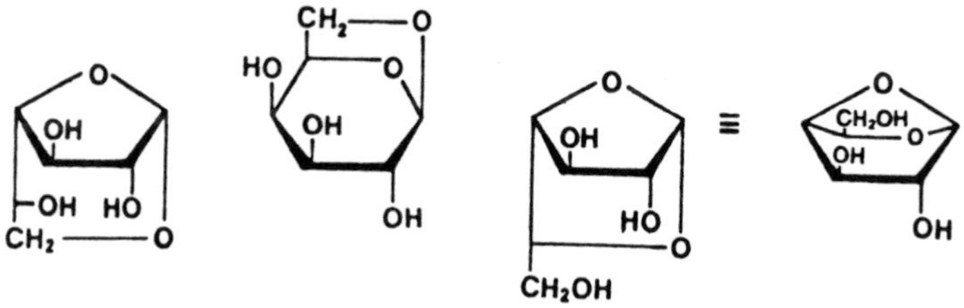

(43) (44) (45)

anhydro-β-D-galactopyranose, and its mode of formation is not known; however it is also present in small amounts when the free sugar equilibrates in acidic solution (P.Köll, Ber., 1973, 106, 3559).

While this type of 1,5-anhydrofuranose ring system is uncommon, the parent compounds are produced in small amounts on vacuum pyrolysis of the aldopentoses, and the derivatives of arabinose and xylose (*trans*-hydroxyl groups at $C_{(2)}$, $C_{(3)}$) are also present in tiny proportions (0.03%) in equilibrated acidic solutions of the free sugars (P. Köll, S. Deyhim and K. Heyns, *ibid.*, 1973, 106, 3565). Analogous 2,6-anhydrohexulofuranoses (identical to 2,5-anhydrohexulopyranoses) are known, the D-fructose derivative being produced by high temperature, high pressure hydrogenolysis of sucrose in the presence of copper chromium oxide (H.R. Goldschmid and A.S. Perlin, Canad.J.Chem., 1960, 38, 2178). All of the hexuloses give small proportions on vacuum pyrolysis (P. Köll, S.Deyhim and K. Heyns, Ber., 1978, 111, 2909).

Anhydro rings of the above type can be opened under acidic conditions. Some of the newly discovered features of these compounds which do not involve such opening reactions are as follows: selective tosylation of 1,6-anhydro-β-D-glucofuranose occurs at $O_{(5)}$, and the ester, with base, gives 1,6:3,5-dianhydro-α-L-idose (P.Köll and J. Schulz, Tetrahedron Letters, 1978, 49); radical bromination of tri-*O*-acetyl-1,6-anhydro-β-D-glucopyranose occurs specifically at $C_{(6)}$, and the product (46) is subject to displacement reactions (Scheme 72) (R.J. Ferrier and R.H. Furneaux, Austral.J.Chem., 1980,

33, 1025);

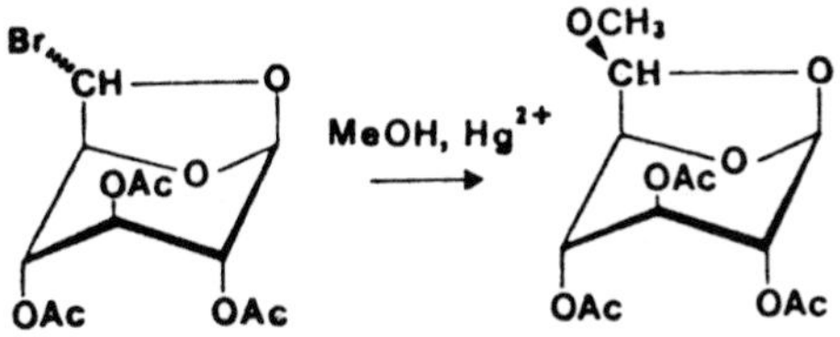

(46)

SCHEME 72

diazotisation of a 2-amino-2-deoxy-1,6-anhydro-D-altropyranose ester gives a 2,5-anhydroalloseptanose derivative by a carbonium ion rearrangement process (Scheme 73) (C.-D. Chang and T.L. Huller, Carbohydrate Res., 1977, 54, 217).

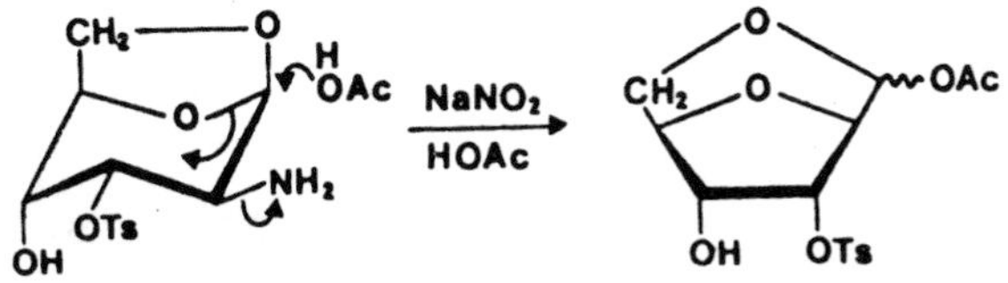

SCHEME 73

(2) *1,2-, 1,3- and 1,4-Anhydrides.* Although tri-*O*-acetyl-1,2-anhydro-α-D-glucopyranose can be used for the synthesis of both α- and β-D-glucopyranosides (W.G. Overend in "The Carbohydrates", Vol.1A, eds. W. Pigman and D. Horton, Academic Press, New York, 1972, p.279), other approaches are now favoured (see p.107). 1,2-Anhydrides are nevertheless important reaction intermediates in, for example, the displacement, under alkaline conditions, of leaving groups from the anomeric centres of β-glucopyranosyl compounds. Conversely, nitrous acid deamination of 2-amino-2-deoxy-D-mannose involves the intermediacy of 1,2-anhydro-D-glucose (J.W. Llewellyn and J.M. Williams, Carbohydrate Res., 1975, 42, 168).

The only known 1,3-anhydropyranose derivative is obtained from 2,4,6-tri-*O*-benzyl-α-D-glucopyranosyl chloride by treatment with sodium hydride in tetrahydrofuran (Scheme 74) (C. Scheurch *et al.*, *ibid.*, 1980, 86, 193).

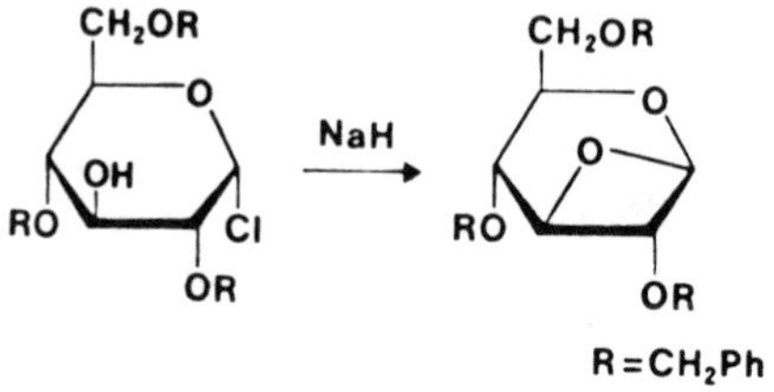

SCHEME 74

1,4-Anhydro-compounds can be produced under basic conditions from 2,3,6-trisubstituted pyranose 4-sulphonates by nucleophilic attack of $O_{(1)}$ at $C_{(4)}$. If these esters also carry an acyl group at the anomeric centre they can undergo related nucleophilic displacement reactions under neutral conditions (for example, Scheme 75) (J.S. Brimacombe, J. Minshall and L.C.N. Tucker, J.chem.Soc.Perkin I, 1973, 2691).

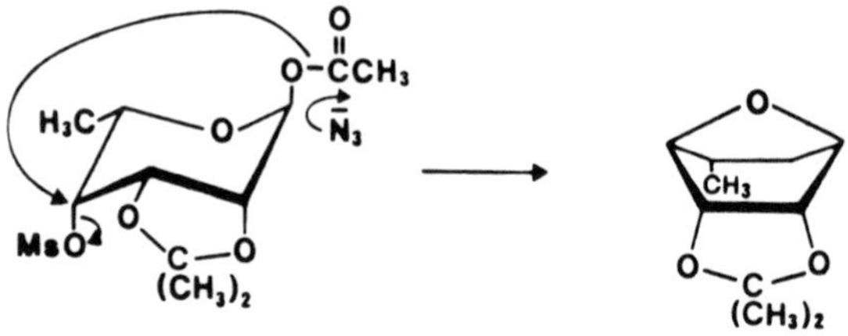

SCHEME 75

2,3,6-Trisubstituted-D-glucoses may give similar derivatives under acid catalysed, dehydrating conditions (F. Micheel *et al.*, Ann., 1974, 124). Some compounds afford 1,4-anhydrides very readily, 3,6-anhydro-D-glucal existing preferentially as the anhydro-compounds formed by addition of the alcohol group to the double bond (Scheme 76) (J.S. Brimacombe *et al.*, Carbohydrate Res., 1973, 27, 254).

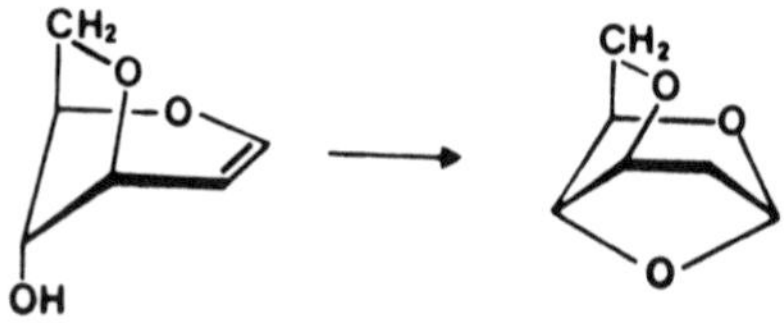

SCHEME 76

(ii) Epoxides

The general methods of synthesis of sugar epoxides, their reactions and spectroscopic properties have been reviewed by N.R. Williams (Adv.carbohydrate Chem.Biochem., 1970, 25, 109).

Usually these oxiranes are made under basic conditions from hydroxy-compounds with vicinally and *trans*-related leaving groups such as sulphonyloxy- or halogeno-groups; otherwise, in some cases, the diazotisation of α-hydroxyamines affords epoxides. Methyl 2-amino-4,6-*O*-benzylidene-2-deoxy-α-D-glucopyranoside, however, on deamination gives 2,5-anhydro-4,6-*O*-benzylidene-D-mannose by a rearrangement reaction involving attack at $C_{(2)}$ by the ring oxygen atom (Williams *loc.cit.*). A convenient method of preparing methyl 2,3-anhydro-4,6-*O*-benzylidene-α-D-mannopyranoside directly from methyl 4,6-*O*-benzylidene-α-D-glucopyranoside uses sodium hydride in dry DMF followed by *N*-tosylimidazole. (D.R. Hicks and B. Fraser-Reid, Synthesis, 1974, 203).

Alternatively, epoxides can be made by epoxidation of unsaturated sugars, the oxygen atom being directed to enter in *cis*-relationship to allylic hydroxyl groups but *trans* to substituted allylic hydroxyl groups (p.225).

Although many epoxide ring opening reactions have been reported which conform with the Fürst-Plattner generalisation, several do not, and give mainly either *trans*-diequatorial products (N.R. Williams *loc.cit.*) or even *cis*-related compounds. Stereochemical, electronic, mechanistic or isomerisation factors can account for this; for example, the finding of the *cis*-compound methyl 3-cyano-3-deoxy-β-D-ribopyranoside in the products of treatment of methyl 2,3-anhydro-β-D-ribopyranoside with cyanide ion at pH 8 is attributable to isomerism of the expected *trans-xylo*-product by way of the $C_{(3)}$ carbanion (N.R. Williams, Chem.Comm., 1967, 1012). R.D. Guthrie and J.A. Liebmann (Carbohydrate Res., 1974, 33, 355) have systematically studied the ring opening of methyl 2,3-anhydro-4,6-*O*-benzylidene-D-hexopyranosides with azide ion and the factors controlling the direction of ring opening. Neighbouring hydroxyl and acyloxy groups can participate in ring openings of sugar epoxides and control the nature of the products formed (J.G. Buchanan *et al.*, *ibid.*, 1974, 35, 151; J.chem.Soc.(C), 1971, 1515).

Other functional groups may control the mode of ring opening reactions. For example, double bonds ensure that allylic C-O bonds are broken specifically on attack by nucleophiles (Scheme 77) (H.W. Pauls and B. Fraser-Reid, J.Amer.chem.Soc., 1980, 102, 3956).

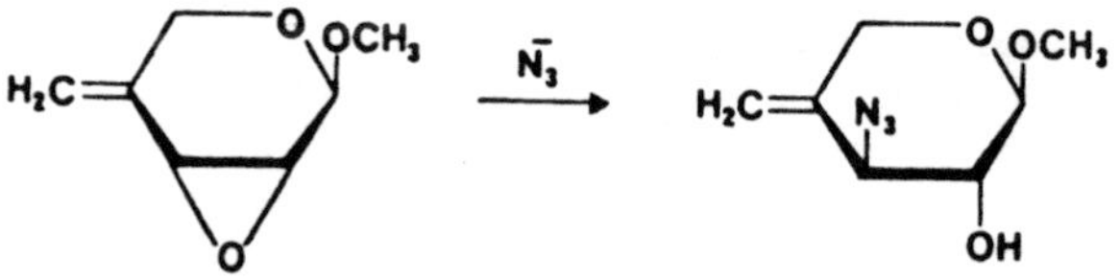

SCHEME 77

Leaving groups attached to epoxide rings similarly control the site of attack (Scheme 78) (S. Kumazawa *et al.*, Angew. Chem.internat.Edn., 1973, 12, 921).

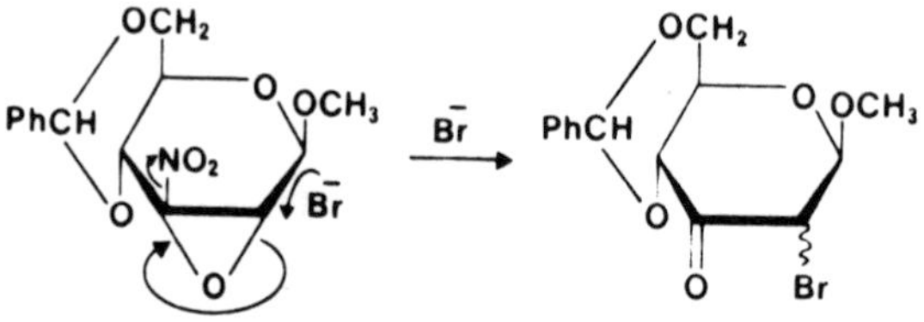

SCHEME 78

(iii) Other anhydro-sugars

The chemistry of 2,5-anhydro-sugars is now well developed and has been reviewed (J. Defaye, Adv.carbohydrate Chem. Biochem., 1970, 25, 181). They can be formed by deamination of 2-amino-2-deoxyaldoses (p.135) and related compounds, or by intramolecular nucleophilic displacements of leaving groups at $C_{(2)}$ or $C_{(5)}$. Some ring closures occur surprisingly readily, and selective tosylations of dialkyldithioacetals of ribose, xylose and lyxose (but not arabinose) result in the isolation of the 2,5-anhydrides of the acetals (J. Defaye and D. Horton, Carbohydrate Res., 1970, 14, 128). Hydrolysis at the reducing centres gives the furanoid aldehydes which are of value for the synthesis of tetrofuranosyl *C*-glycosides (p.121).

Treatment of methyl 2,3-di-*O*-mesyl-α-D-glucopyranoside with base gives methyl 2,6:3,4-dianhydro-α-D-altropyranoside, and stronger conditions then cause selective opening of the oxirane ring to afford methyl 2,6-anhydro-α-D-mannopyranoside (P. Köll, H. Komander and J. Kopf, Ber., 1980, 113, 3919).

Ring closure of suitable compounds with leaving groups at either $C_{(3)}$ or $C_{(6)}$ to give 3,6-anhydro-derivatives is a common reaction; ring opening of the products is less so. Methyl 3,6-anhydro-α-D-glucopyranoside, treated in dichloromethane with boron trichloride at low temperatures, gives the 6-chloro-6-deoxyglucoside as main product, but other compounds are also formed following additional reaction at $C_{(1)}$ (A.B. Foster *et al.*, Carbohydrate Res., 1967, 4, 195).

Four-membered anhydro rings fused to furanoid rings are well known (e.g. L. Hough and B.A. Otter, *ibid.*, 1967, 4, 126), but are less commonly found bonded to pyranoid systems. The latter are, however, obtainable by standard procedures (e.g. Scheme 79), and are chemically relatively stable (R.J. Ferrier and N. Vethaviyasar, J.chem.Soc.Perkin I, 1973, 1791; C.R. Hall and T.D. Inch, Carbohydrate Res., 1977, 53, 254).

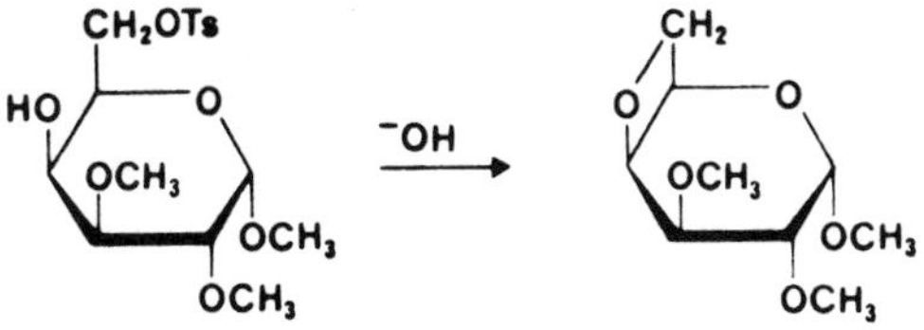

SCHEME 79

(e) Esters

For a review see M.L. Wolfrom and W.A. Szarek in "The Carbohydrates", Vol.1A, eds. W. Pigman and D. Horton, Academic Press, New York, 1972, p.217.

(i) Carboxylic esters

Anhydrous phosphoric acid added to acetic anhydride is an effective and mild reagent for acetylating all types of carbohydrate hydroxyl groups including those which are sterically hindered; other acylations can be effected similarly (A.J. Fatiadi, Carbohydrate Res., 1968, 6, 237). Tetrabutylammonium fluoride also catalyses the reaction of acyl anhydrides with alcohols, and, in tetrahydrofuran, permits ready substitution of compounds which are unreactive towards the anhydrides in pyridine (S.L. Beaucage and K.K. Ogilvie, Tetrahedron Letters, 1977, 1691). Likewise, 4-(dimethylamino)-pyridine catalyses acylation reactions, tertiary alcohols being reactive towards benzoyl chloride in pyridine in its presence (H. Redlich, H.-J. Neumann and H. Paulsen, Ber., 1977, 110, 2911). Benzoyl cyanide in acetonitrile with triethylamine as catalyst is also a convenient system for effecting benzoylations (A. Holý and M. Souček, Tetrahedron Letters, 1971, 185).

The complex question of relative reactivities of hydroxyl groups towards acylating agents has been discussed in detail by A.H. Haines (Adv.carbohydrate Chem.Biochem., 1976, 33, 11). While initial investigations on this matter examined the products of monoacylation of polyhydroxy compounds, later work gained information by analysis of products containing several ester groups. Thus, tribenzoylation of methyl α-D-glucopyranoside with benzoyl chloride in pyridine gives the 2,3,6-triester in yields greater than 50%, and the other partially substituted products indicate that the reactivity order of the hydroxyl groups is 6>2>3>4 (J.M. Williams and A.C. Richardson, Tetrahedron, 1967, 23, 1369). Many factors complicate a full analysis of such reactions - not least, details of the mechanisms involved (are hydroxyl groups or oxyanions the nucleophilic species?), and the influence of introduced esters on the reactivities of the unsubstituted hydroxyl groups. In the case of methyl α-D-mannopyranoside the hydroxyl group at $C_{(3)}$ is more reactive than that at $C_{(2)}$ (which is axial and expected to be less reactive on this account) confirming the finding that the latter hydroxyl group is not always the most reactive of the secondaries. From work with alkyl hexopyranoside 4,6-acetals it appears that the reactivity of equatorial hydroxyl groups is enhanced by neighbouring axial alkoxy groups perhaps in consequence of intramolecular hydrogen bonding or related solvation effects. Thus $HO_{(2)}$ in α-D-glucopyranosides is more reactive than in the β-anomers, and selective reaction at this site is therefore more pronounced in the former series. With the galactosides $HO_{(3)}$, with the adjacent axial $O_{(4)}$, is more reactive than $HO_{(2)}$ despite the fact that α-compounds also have $O_{(1)}$ axial.

Sometimes these selectivities can be enhanced by use of appropriate reagents, *N*-benzoylimidazole giving exclusively, or almost so, the 2-benzoate from methyl 4,6-*O*-benzylidene-α-D-glucopyranoside (F.A. Carey and K.O. Hodgson, Carbohydrate Res., 1970, 12, 463), the 3-ester from benzyl 4,6-*O*-benzylidene-β-D-galactopyranoside (G.J.F. Chittenden, *ibid.*, 1971, 16, 495) and, less predictably, the 2-substituted product from methyl 4,6-*O*-benzylidene-α-D-altropyranoside (N.L.Holder and B. Fraser-Reid, Synthesis, 1972, 83).

Similar high selectivity is induced in some instances by use of stannyl derivatives which serve to enhance the reactivity of hydroxyl groups. Treatment of methyl 4,6-*O*-benzylidene-α-D-glucopyranoside with equimolar dibutyl-tin oxide gives a crystalline cyclic 2,3-*O*-dibutylstannylene derivative which, with benzoyl chloride in dioxane in the presence of triethylamine, gives the 2-benzoate of the initial acetal in 86% yield. The β-anomer reacts less selectively as does the α-*manno*-acetal, but the α-*allo*-analogue gives 91% of the 2-ester. More surprisingly, methyl α-D-glucopyranoside also seemingly gives a 2,3-cyclic stannylene derivative, because selective benzoylation after treatment with dibutyltin oxide gives methyl 2-*O*-benzoyl-α-D-glucopyranoside in >80% yield, and no 6-ester is formed (Scheme 80). Again, results are

CH₂OH HO O HO OH OCH₃ → CH₂OH HO O O Sn (Bu)₂ O OCH₃ → CH₂OH HO O HO BzO OCH₃

SCHEME 80

different with methyl β-D-glucopyranoside which gives the 6-benzoate in similar yield (R.M. Munavu and H.H. Szmant, J. org.Chem., 1976, 41, 1832). When derived from *cis*-diols on 6-membered rings the stannylene compounds lead to highly selective monoesterification at the equatorial oxygen atoms (M.A. Nashed and L. Anderson, Tetrahedron Letters, 1976, 3503). An adaptation of this procedure uses tributylstannyl derivatives, and offers means of obtaining methyl 2,6-di-*O*-benzoyl-α-D-glucopyranoside and methyl 3,6-di-*O*-benzoyl-β-D-galactoside and -α-D-mannopyranoside in excellent yields (T. Ogawa and M. Matsui, Carbohydrate Res., 1977, 56, C1).

J. Staněk *et al.* (Carbohydrate Res., 1974, 36, 273) have examined kinetic aspects of the acylation (and alkylation) of diols taking into account effects on the reactivity of one hydroxyl group of substituting the other.

Anomeric selectivity can be controlled during acylations at the anomeric centre by use of metal complexing effects (*cf.* p.109). 2,3,4,6-Tetra-*O*-benzyl-D-glucose treated in tetrahydrofuran at -30°C with butyllithium gives an anion which reacts with acyl halides to give 90% of α-products, and when these procedures are carried out in benzene at 60°C, β-products are obtained in similar yields (Scheme 81) (P.E. Pfeffer, E.S. Rothman and G.G. Moore, J.org.Chem., 1976, 41, 2925). Conceivably, under the milder conditions, a specific

BuLi, C_6H_6, 60°, R'COCl

BuLi, THF, -30°, R'COCl

R = CH_2Ph

SCHEME 81

metal complex of the α-modification of the sugar is involved, whereas at elevated temperatures, rapid equilibration of the anions can occur and the equatorial product is obtained under kinetic control.

Glycosyl acetates are frequently prepared by the acetolysis with acetic acid, acetic anhydride and acids (often sulphuric acid) of glycosides and other glycosyl compounds (R.D.Guthrie and J.F. McCarthy, Adv.carbohydrate Chem., 1967, 22, 11), but reaction may not just involve acid-catalysed nucleophilic displacement of the aglycones (or other groups) at the anomeric centres. Furanose compounds with *cis*-related groups at $C_{(2)}$, $C_{(3)}$ undergo stereochemical inversion at $C_{(2)}$ (p.74), and products formed by displacement of the ring-oxygen atoms can predominate. Thus, pentopyranose compounds frequently give acyclic products, methyl α-D-xylopyranoside with acetic anhydride and boron trifluoride giving a 1:1 mixture of the acyclic 1-methoxyalditol penta-acetates followed by the acyclic hexa-acetate (Scheme 82). (F.W. Lichtenthaler, J. Breunig and W. Fischer, Tetrahedron Letters, 1971, 2825).

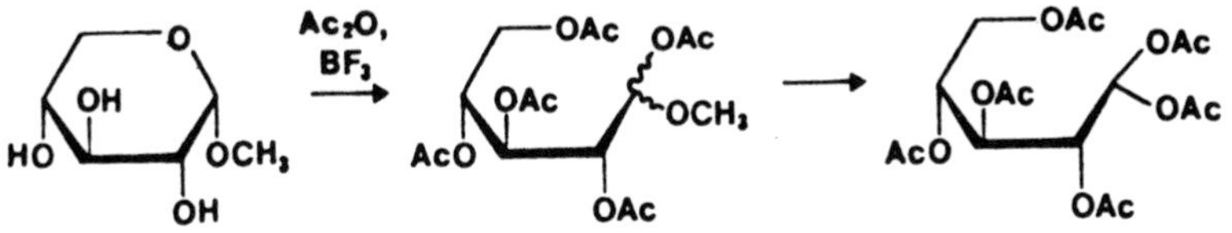

SCHEME 82

In related fashion, acetolysis of 3,5-di-*O*-acetyl-1,2-*O*-isopropylidene-α-D-xylofuranose initially gives an acyclic 1-acetoxy-1,2-*O*-isopropylidene perester which then affords the acyclic hexa-acetate (A. Magnani and Y. Mikuriya, Carbohydrate Res., 1973, 28, 158).

The chemistry of carbohydrate acyl esters is closely linked to that of acyloxonium ions which has been reviewed by H. Paulsen (Adv.carbohydrate Chem.Biochem., 1971, 26, 127; Pure appl.Chem., 1975, 41, 69). These ions take part as intermediates in many carbohydrate reactions and frequently lead to products derived by a variety of rearrangement processes.

The conformational analysis of esterified sugar derivatives has been studied in considerable detail (see p.26; P.L. Durette and D. Horton, Adv.carbohydrate Chem.Biochem., 1971, 26, 49).

Many substituted acyl esters are known. Trichloroacetyl groups are very readily removed under basic conditions such as with ammonia in dioxane solution or even with pyridine; selective removal can be effected in the presence of acetate groups (G. Excoffier, D.Y. Gagnaire and M.R. Vignon, Carbohydrate Res., 1976, 46, 201). Likewise, chloroacetates can be cleaved specifically from polyesters containing acetyl and benzoyl esters by treatment with thiourea (M. Bertolini and C.P.J. Glaudemans, *ibid.*, 1970, 15, 263). Trifluoroacetyl groups, being strongly electron withdrawing, stabilise neighbouring glycosidic bonds. In consequence, compounds with free hydroxyl groups at $C_{(2)}$, on treatment with trifluoroacetic acid and its anhydride, are esterified and hence stabilised towards trifluoroacetolysis, whereas analogues which are initially substituted at this position are subject to glycosidic cleavage under these conditions (L.-E. Franzén and S. Svensson, *ibid.*, 1979, 73, 309).

Several substituted-acetyl derivatives of sugars occur in Nature. Notably, 1-*O*-(indole-3-acetyl)-β-D-glucopyranose has been found in plants and has a stronger growth promoting effect than has the free acid, and 2-, 3-, 4- and 6-bonded isomers have also been detected (A. Ehmann, *ibid.*, 1974, 34, 99).

(ii) Orthoesters

The use of sugar orthoesters in synthesis has been reviewed (N.K. Kochetkov and A.F. Bochkov, Recent Develop.Chem.Nat. Carbon Compounds, 1971, 4, 75).

Newer syntheses of cyclic 1,2-orthoesters are referred to on p.115, and they may otherwise be prepared from acylated glycosyl halides by condensations in tetrahydrofuran with alcohols in the presence of silver salicylate (G. Wulff and W. Schmidt, Carbohydrate Res., 1977, 53, 33) or in acetonitrile in the presence of silver nitrate and 2,4,6-trimethylpyridine - a process which involves intermediate 1,2-*trans*-related glycosyl nitrates (S.E. Zurabyan *et al.*, *ibid.*, 1973, 26, 117). Cyclic orthoesters of simple alcohols may be converted to analogues involving complex alcohols, e.g. sterols, without the use of a catalyst (N.I. Uvarova *et al.*, *ibid.*, 1975, 42, 165). A further route to cyclic orthoesters depends upon the generation of cyclic acyloxium ions derived from alternative sources. Epoxy-compounds with adjacent ester groups treated with acids afford such a source and, for example, methyl 2-*O*-acetyl-3,4-anhydro-α-D-arabinoside, treated with antimony pentachloride and then sodium methoxide, gives the *lyxo*-2,3-orthoesters with the *exo*-methoxy isomer predominating (Scheme 83) (Buchanan and Edgar, *loc.cit.*).

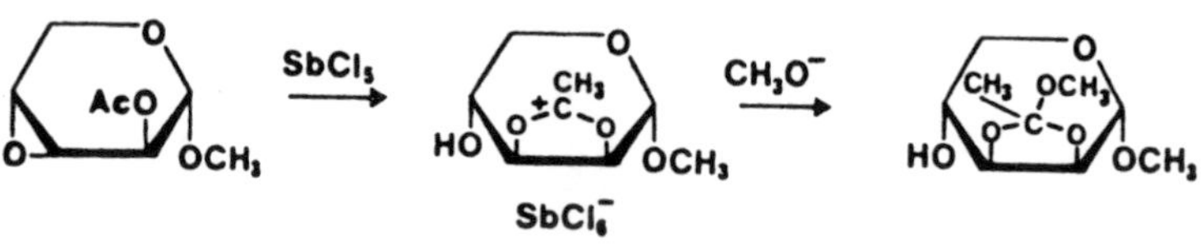

SCHEME 83

The question of the stereochemistry at the orthoester centres is discussed in this paper.

The main interest in cyclic orthoesters stems from their value as glycosylating agents, because they can be induced to react with alcohols to give 1,2-*trans*-glycosides directly, or they can be made to isomerise to these compounds (p.115). Nucleophiles other than alcohols can be used, however, and glycosyl phosphate derivatives are obtainable with dibenzyl hydrogen phosphate (L.V. Volkova, L.L. Danilov and R.P. Evstigneeva, *ibid.*, 1974, 32, 165).

Reduction of orthoesters in the presence of Lewis acids allows the conversion to cyclic acetals (J.G. Buchanan and A.R. Edgar, *ibid.*, 1976, 49, 289), and acid-catalysed hydrolysis, which also would be expected to proceed through cyclic acyloxonium ions, results in the formation of monohydroxy monoesters. In this latter reaction, when the orthoester groups are fused to conformationally fixed six-membered rings, the axially oriented ester products predominate (J.F. King and A.D. Allbutt, Canad.J.Chem., 1970, 48, 1754), so that 1,2-substituted α-D-glucopyranose and β-D-mannopyranose compounds often give mainly 1- and 2-*O*-acyl derivatives, respectively. This has been rationalised on stereoelectronic grounds (P. Deslongchamps *et al.*, Canad.J.Chem., 1975, 53, 1601; Tetrahedron, 1975, 31, 2463). In the case of 1,2-orthoesters, however, the ratios of the two possible esters can depend upon the conditions used for the hydrolysis (L.R. Schroeder, D.P. Hultman and D.C. Johnson, J.chem.Soc.Perkin II, 1972, 1063) which emphasises the kinetic nature of the above generalisation.

(iii) Sulphonate esters

(1) *General.* The displacement of sulphonyloxy groups, both intermolecularly and intramolecularly, has become one of the most widely used reactions in synthetic carbohydrate chemistry offering means of introducing an extensive range of functional groups, effecting many rearrangement reactions and giving access to both saturated and unsaturated products. All aspects of the synthesis of sulphonates and their reduction and displacement reactions have been described in two reviews by D.H. Ball and F.W. Parrish (Adv.carbohydrate Chem., 1968, 23, 233; Adv. carbohydrate Chem.Biochem., 1969, 24, 139). Displacement reactions, and particularly aspects of them which

involve neighbouring group participation, have been further reviewed by L. Goodman (Adv.carbohydrate Chem., 1967, 22, 109), J.S. Brimacombe (Fortsch.Chem.Forsch., 1970, 14, 367), J.G. Buchanan (M.T.P.International Rev. of Science, Organic Chem.Ser.1, 1973, 7, 31), and R.J. Ferrier (*ibid.*, Ser.2, 1976, 7, 35).

Most work continues to be carried out on mesylates and tosylates, but the desirability of using esters with even better leaving properties has been recognised and, in particular, trifluoromethanesulphonates (triflates) are important in this respect. In one example the rate enhancement factor relative to the tosylate is 4000 (L.D. Hall and D.C. Miller, Carbohydrate Res., 1976, 47, 299). The triflate (47) prepared using trifluoromethanesulphonic anhydride in dichloroethane containing pyridine at 0°C, reacts with sodium benzenethiolate in DMF at 5°C (Scheme 84) (T.H. Haskell, P.W.K. Woo and D.R. Watson, J.org.Chem., 1977, 42, 1302).

(47) SCHEME 84

The selectivity exhibited during monosulphonylation of polyhydroxy systems has been discussed by Ball and Parrish (*loc.cit.*) and by A.H. Haines (Adv.carbohydrate Chem.Biochem., 1976, 33, 11), and correlates largely with results of selective acylations (p.140). It can be enhanced by use of more sterically demanding sulphonylating reagents, for example, 2,4,6-trimethylbenzenesulphonyl chloride (S.E. Creasey and R.D. Guthrie, J.chem.Soc.Perkin I, 1974, 1373) or *N*-tosylimidazole (D.R. Hicks and B. Fraser-Reid, Synthesis, 1974, 203). When used in phase transfer conditions this latter reagent gives high yields of the 2-tosylate of methyl 4,6-*O*-benzylidene-α-D-glucopyranoside as expected (p.140) but, contrary to expectations, also the 2-ester from the α-D-*manno* analogue (P.J. Garreg, T. Iversen and S. Oscarson, Carbohydrate Res., 1977, 53, C5). Tosyl chloride in pyridine reacts preferentially at $O_{(3)}$ of this latter compound, and this reagent generally shows selectivity for equatorial hydroxyl groups in keeping with the principles of conformational analysis (S.E. Creasey and R.D. Guthrie, *ibid.*, 1972, 22, 487).

While sodium amalgam, Raney nickel and lithium aluminium hydride are still commonly used for the reductive cleavage of sulphonates, the following offer suitable alternatives (at least for tosylates) and are less prone to lead to "anomalous" products such as anhydrides or deoxy-compounds: sodium in liquid ammonia (M.A. Miljković *et al.*, *ibid.*, 1970, 15, 162), sodium naphthalene in tetrahydrofuran (J.K.N. Jones *et al.*, Canad.J.Chem., 1973, 51, 1767), and photolysis in methanol in the presence of a base (sodium methoxide) (A.D. Barford, A.B. Foster and J.H. Westwood, Carbohydrate Res., 1970, 13, 189). This last procedure, however, applied to α-ketotosylates results in the removal of the tosyloxy group (W.A. Szarek and A. Dmytraczenko, Synthesis, 1974, 579).

(2) *Nucleophilic displacement reactions*. A range of nucleophiles are commonly used in dipolar, aprotic solvents especially hexamethylphosphoric triamide (HMPT), dimethyl sulphoxide (DMSO) and *N*,*N*-dimethylformamide (DMF), and their efficiencies as media for displacement reactions decrease in that order (at least for the displacement of primary sulphonyloxy groups by the azide anion). Tosyloxy groups are better leaving groups than mesyloxy groups, and variation of the substituents on the aryl rings have the influences expected on the basis of their electronic effects. Thus, benzenesulphonates are more reactive than tosylates, and *p*-bromobenzenesulphonates are still more so. These last esters, however, undergo competing reactions within the aromatic rings (L. Anderson *et al.*, J.org.Chem., 1974, 39, 3014).

A.C. Richardson has published in full the rationale he outlined in The Second Edition to account for the markedly different rates at which sulphonates are displaced from different sites on monosaccharides (Carbohydrate Res., 1969, 10, 395). Using the concept of destabilising interactions between permanent dipoles within molecules and those generated in S_N2 transition states, he is able to account for the stability towards charged nucleophiles of ester groups at $C_{(2)}$ of aldopyranosyl compounds [although β-D-*gluco*- and β-D-*manno*-compounds can be reactive (M. Miljković, M. Gligorijević and D. Glišin, J.org.Chem., 1974, 39, 3223)], $C_{(1)}$ of hexulopyranoses, and $C_{(6)}$ of galactopyranoses. Axial, electronegative groups adjacent to esters and in 1,3-relation to them also impede displacements with charged nucleophiles.

Neighbouring groups, as well as participating in displacement reactions by providing nucleophilic assistance at the carbon atom involved (see below), can assist the departure of the leaving group. 5-Deoxy-1,2-*O*-isopropylidene-3-*O*-tosyl-α-D-xylofuranose does not react with potassium thiocyanate in DMF (the nucleophile having to attack from the *endo*-direction), whereas the 5-hydroxy analogue does undergo reaction, conceivably because hydrogen bonding facilitates departure of the tosyloxy group (Scheme 85) (J. Defaye and J. Hildesheim, Carbohydrate Res., 1967, 4, 145).

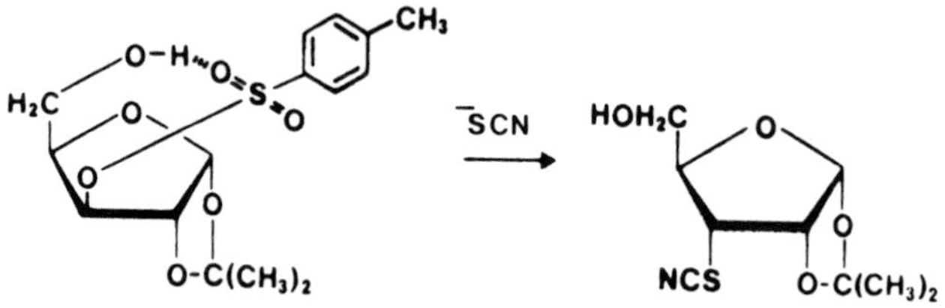

SCHEME 85

Displacement reactions may occur competitively with elimination processes - particularly if there are electrostatic factors impeding the former and if eliminations are favourable. Methyl 6-deoxy-4-*O*-mesyl-2,3-di-*O*-methyl-α-L-talopyranoside (48) has an axial methoxyl group impeding S_N2 sub-

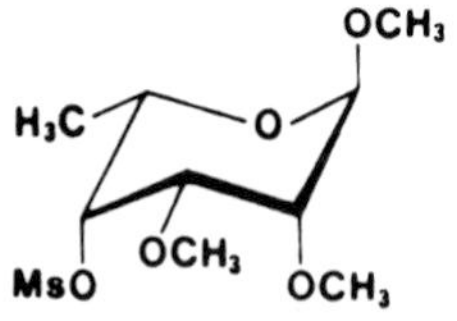

(48)

stitution in the favoured chair conformation and is suitably oriented for E_2 eliminations; in the alternative chair form there are two *vicinal*-axial substituents. With sodium azide in DMF it affords both possible alkenes (3-enose 50%, 4-enose 10%), together with the product of substitution (40%) (A.K.

Al-Radhi, J.S. Brimacombe and L.C.N. Tucker, *ibid.*, 1972, 22, 103).

The participation of intramolecular groups during nucleophilic displacements of sulphonyloxy groups is commonplace, and has been dealt with at length by Goodman, Brimacombe, Buchanan and Ferrier (*loc.cit.*). Schemes (86)-(89) illustrate the phenomenon with oxygen, sulphur and hydrogen as participating nucleophiles; an example of nitrogen participation is given on p. 162.

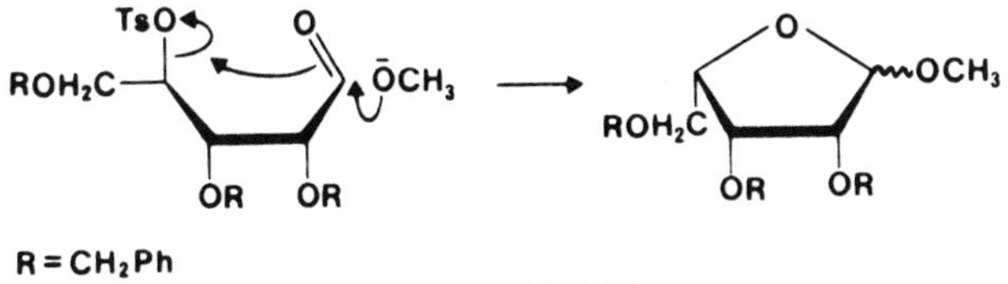

$R = CH_2Ph$

SCHEME 86

(N.A.Hughes and P.R.H.Speakman, J.chem.Soc.(C), 1967, 1186).

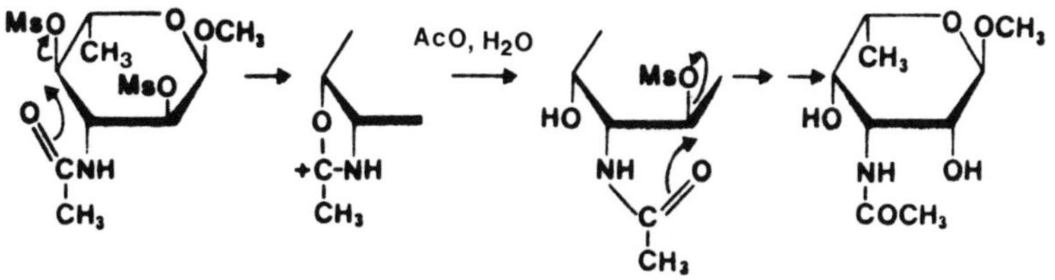

SCHEME 87

(A.C.Richardson and K.A.McLauchlan, J.chem.Soc., 1962, 2499).

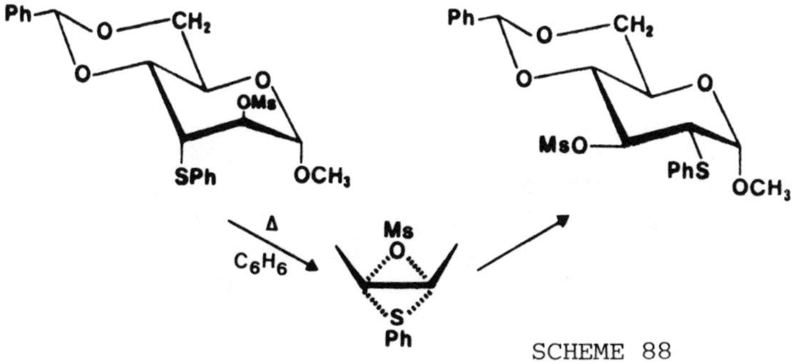

SCHEME 88

(S.Hanessian and A.P.A.Staub, Carbohydrate Res., 1970, 14, 424)

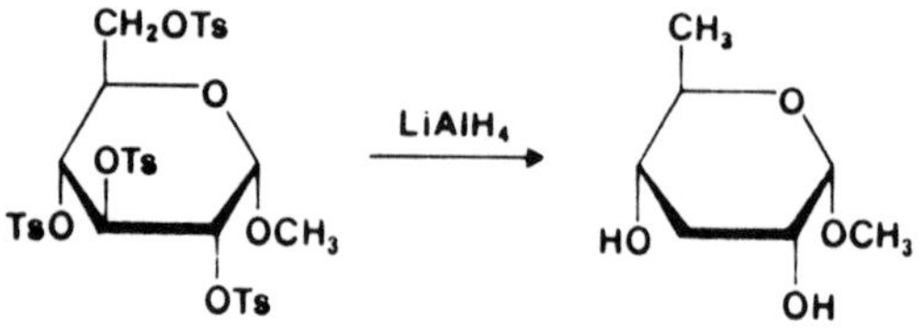

SCHEME 89

(G.Ekborg and S.Svensson, Acta chem.Scand., 1973, 27, 1437).

This last reaction is assumed to proceed in normal fashion at $C_{(6)}$ to give the deoxy group, and at $C_{(2)}$ and $C_{(4)}$ to give secondary alcohols. These react with aluminium hydride ions to afford aluminium alkoxides and thus provide nucleophilic hydrogen suitably oriented for attack at $C_{(3)}$; this leads to the product of anomalous C-O bond fission at this centre.

Elimination reactions involving sulphonates and displacements at allylic centres are referred to in Section 7.

(iv) Other esters

Sugar 1-phosphates are obtained in aqueous solution by phosphorylation with orthophosphate in the presence of cyanogen showing that these biochemically important compounds could conceivably have been formed under prebiotic conditions (C. Degani and M. Halmann, J.chem.Soc.(C), 1971, 1459). Aldohexose 6-phosphates isomerise to give the epimers and corresponding ketose esters in the presence of divalent metal ions (B.E. Tilley, D.W. Porter and R.W. Gracy, Carbohydrate Res., 1973, 27, 289), and the equilibria adopted by various sugar phosphates in aqueous solution have been determined by spectroscopic methods (G.R. Gray, Acc.chem.Res., 1976, 9, 418).

Monosaccharide boronates [$RB(OR')_2$], and to a lesser extent borinates (R_2BOR'), have found use in many analytical and synthetic aspects of carbohydrate chemistry, and their chemistry has been comprehensively reviewed (R.J. Ferrier, Adv. carbohydrate Chem.Biochem., 1978, 35, 31).

Thiocarbonate derivatives have found use in other specific deoxygenation of monosaccharides (p.170) and in the synthesis of unsaturated sugars (p.220).

Applications of other esters can be located via The Chemical Society's Specialist Periodical Reports.

(f) Functional nitrogenous derivatives of monosaccharides

Basic aspects of the chemistry of these compounds are fully covered by Hough and Richardson in the Second Edition. Developments have been many and varied and can be followed by reference to the Chemical Society's Specialist Periodical Reports. A wide range of heterocyclic compounds derived by reaction of sugars both at the anomeric centres and other centres have been produced; these have been reviewed by H. El Khadem (Adv. carbohydrate Chem.Biochem., 1970, 25, 351). Other relevant topics to have been reviewed are the reactions undergone by free sugars with ammonia (M.J. Kort, *ibid.*, 1970, 25, 311), nitrosugars (H.H. Baer, *ibid.*, 1969, 24, 67; A.C. Richardson, M.T.P.International Review of Science, Organic Chemistry, Ser.1, Vol.7, 1973, p.105; Ser.2, Vol.7, 1976, p.131), glycosylamines, phenylhydrazones and osazones (H. Simon and A. Kraus, Fortsch.Chem.Forsch., 1970, 14, 430; Amer.chem.Soc. Symposium Ser., 1976, No.39 188). The vast topic of the nucleosides is not treated in this survey.

3. Amino-sugars

A monograph in four volumes has been published on this topic ("The Amino Sugars", eds. R.W. Jeanloz and E.A. Balazs, Academic Press, New York, 1969). All aspects of the chemistry and biochemistry of these compounds are covered; the following Chapters are of particular relevance: Vol. 1A, Monosaccharide Amino Sugars (D. Horton), Sialic Acids and Muramic Acids (G. Blix and R.W. Jeanloz); Vol.1B, Naturally Occurring Glycosides of Amino Sugars with Antibiotic Activity (R.U. Lemieux); Vol. 2A, Distribution and Biological Role. See also D. Horton and J.D. Wander in "The Carbohydrates", Vol.1B, eds. W. Pigman and D. Horton, Academic Press, New York, 1980, p.644.

Rapid expansion in the study of natural products - particularly microbiological metabolites with antibiotic activity - has led to the finding of a wide range of new amino-sugars. Many contain common structural modifications such as deoxy-, or *C*-, *N*- or *O*-methyl groups, and others are further modified; a component of sisomycin contains a 2,6-diamino-2,3,4,6-tetradeoxyhex-4-enopyranose moiety (p.203), lincomycin has a 6-amino-6,8-dideoxyoctose component (B. Bannister, J.chem.Soc.

Perkin I, 1974, 360), and a 2-amino-2,6-dideoxyhexose 6-sulphonic acid has been isolated from a bacterial cell wall (R. Reistad, Carbohydrate Res., 1977, 54, 308). The most dramatic finding, however, is of nojirimycin, a *Streptomyces* antibiotic, which is 5-amino-5-deoxy-D-glucose, the only known natural product with a sugar having the ring oxygen atom replaced by another hetero atom (S. Inouye *et al.*, Tetrahedron, 1968, 23, 2125).

Table 7 indicates the structures of many of the naturally occurring amino-sugars which have been given trivial names; their names often denote the antibiotics in which they are found. The syntheses of desosamine, mycaminose, perosamine, daunosamine, garosamine, lincosamine, nojirimycin and neosamine B have been reviewed by J.S. Brimacombe (Angew.Chem. internat.Edn., 1971, 10, 236).

^{15}N N.M.R. studies of 2-amino-2-deoxyhexose hydrochlorides and their *N*-acyl derivatives (R.E. Botto and J.D. Roberts, J. org.Chem., 1977, 42, 2247) have shown that the resonances for the α- and β-pyranoses differ by only about 1 p.p.m. for compounds having equatorial amino groups, but with 2-amino-2-deoxy-D-mannopyranose derivatives, the α-anomers give resonances about 7 p.p.m. downfield from those of the β-compounds. For equilibrated solutions of the D-*gluco*-, D-*galacto*- and D-*manno*-compounds in water, the resonances confirmed the α:β-pyranose ratios determined by ^{1}H N.M.R. methods (D. Horton, J.S. Jewell and K.D. Philips, J.org.Chem., 1966, 31, 4022) which are appreciably different from those established by the analogous hexoses (p. 32). While glucose gives an equilibrium containing 62% β-pyranose, 2-amino-2-deoxy-D-glucose gives *ca.* 35% β-pyranose for both the hydrochloride and the *N*-acetyl derivative. Conversely, mannose has 35% β-pyranose, while the hydrochloride and *N*-acetyl derivatives of the 2-amino-2-deoxy sugar have 57% and 43%, respectively. Therefore, the introduction of the amino functions favour the anomers with the 1,2-*cis*-orientation. This presumably is because of the electrostatic attraction between the substituted nitrogen atoms and the *cis*-hydroxyl groups at the anomeric centres. The 2-amino-sugars themselves do not show this effect, and are calculated to exhibit α,β-ratios which are similar to those of the corresponding aldoses [(T. Taga and K. Osaki, Bull.chem. Soc.Japan., 1975, 48, 3250; this paper also gives calculated free energies for 2-aminohexopyranose derivatives in both chair conformations (*cf.* p. 32)].Replacement of the hydroxyl

Table 7

Some naturally occurring amino-sugars (absolute configurations are not given)

Trivial name	Parent sugar	Amino site(s)	Deoxy site(s)	Other Modifi-cations	Reference: Journal	Reference: Reference
Acosamine	*arabino*-hexose	3	2,3,6		A	1977, <u>58</u>, 125
Actinosamine	*arabino*-hexose	3	2,3,6	4-*O*-Me	B	1973, <u>9</u>, 101
Amosamine	glucose	4	4,6	*N*,*N*-Me_2		2nd Edn. p.493
Angolosamine	*arabino*-hexose	3	2,3,6	*N*,*N*-Me_2	C	1975, <u>31</u>, 2989
Bacillosamine	glucose	2,4	2,4,6		D	1977, <u>161</u>, 103
Bamosamine	glucose	4	4,6	NMe		2nd Edn. p.493
Celestosamine	*erythro-galacto*-octose	6	6,8	7-*O*-Me	A	1974, <u>38</u>, 147
Daunosamine	*lyxo*-hexose	3	2,3,6			2nd Edn. p.487
Desosamine	*xylo*-hexose	3	3,4,6	*N*,*N*-Me_2	E	1971, <u>10</u>, 236
Forosamine	*erythro*-hexose	4	2,3,4,6	*N*,*N*-Me_2	E	1976, <u>10</u>, 236
Garosamine	arabinose	3	3	4-*C*-Me-*N*-Me	E	1976, <u>10</u>, 236
Gentosamine	xylose	3	3	*N*-Me	A	1975, <u>44</u>, 121
Holacosamine	*xylo*-hexose	4	2,4,6	3-*O*-Me	A	1972, <u>24</u>, 297
Holosamine	*ribo*-Hexose	4	2,4,6	3-*O*-Me	C	1970, <u>26</u>, 1695
Kanosamine	glucose	3	3			2nd Edn. p.488
Kasugamine	*arabino*-hexose	2,4	2,3,4,6		C	1973, <u>29</u>, 3141
Lincosamine	*erythro-galacto*-octose	6	6,8		E	1971, <u>10</u>, 236
Lividosamine	*ribo*-hexose	2	2,3		A	1975, <u>45</u>, 323
Mycaminose	glucose	3	3,6	*N*,*N*-Me_2		2nd Edn. p.489
Mycosamine	mannose	3	3,6			2nd Edn. p.491

(continued on p. 154)

Trivial Name	Parent Sugar	Amino site(s)	Deoxy site(s)	Other modifi-cations	Reference Journal	Reference
Nebrosamine	*ribo*-hexose	2,6	2,3,6		F	1973, 710
Neosamine B (paranose)	idose	2,6	2,6			2nd Edn. p.498
Neosamine C	glucose	2,6	2,6			2nd Edn. p.497
Nojirimycin	glucose	5	5		C	1968, 24, 2125
Perosamine	mannose	4	4,6		G	1970, 92, 3160
Polyoxamic acid	xylonic acid	2	2		H	1973, 5051
Prumycin	arabinose	2,4	2,4	4-*N*-(D-alanyl)	H	1975, 1853
Purpurosamine	*erythro*-hexose	2,6	2,3,4,6		I	1975, 979
Rhodosamine	*lyxo*-hexose	3	2,3,6	*N*,*N*-Me_2	I	1973, 1369
Ristosamine	*ribo*-hexose	3	2,3,6		A	1977, 55, 253
Sibirosamine	altrose	4	4,6	3-*C*-Me-*N*-Me	J	1974, 27, 866
Sisosamine	*glycero*-hexose	2,6	2,3,4,6	4-ene	F	1975, 11
Streptolidine	arabinonic acid	2,3,5	2,3,5	2,3-*N*,*N*-Carbimino	K	1976, 49, 3611
Streptozotocin	glucose	2	2	*N*-(*N*-methyl-*N*-nitrosocarbamoyl)	L	1974, 57, 2622
Thomosamine	galactose	4	4,6			2nd Edn. p.492
Tolyposamine	*erythro*-hexose	4	2,3,4,6		C	1973, 29, 3141
Vancosamine	*lyxo*-hexose	3	2,3,6	3-*C*-Me	C	1975, 31, 2989
Viosamine	glucose	4	4,6			2nd Edn. p.493

Journals are as follows: A, Carbohydrate Res.; B, Khim prirod.Soedinenii; C, Tetrahedron; D, Biochem.J.; E, Angew.Chem.internat.Edn.; F, Chem.Comm.; G, J.Amer.chem.Soc.; H, Tetrahedron Letters; I, J.chem.Soc.Perkin I; J, J.Antibiotics; K, Bull.chem.Soc.Japan; L, Helv.

group at $C_{(1)}$ of D-fructose by substituted amino-groups (Amadori reaction products) does not influence the equilibria established in pyridine solution (W. Funcke and A. Klemer, Carbohydrate Res., 1976, 50, 9).

The mutarotation rate of 2-amino-2-deoxy-D-glucose hydrochloride is similar to that of D-glucose (A.P. Fletcher and A. Neuberger, J.chem.Soc.Perkin II, 1972, 12), but the rate of equilibration of the pyranose forms of 2-amino-2-deoxy-D-mannose hydrochloride when the sugar is dissolved in water is so high that mutarotation is not observed under normal circumstances (D. Horton, J.S. Jewell and K.D. Philips, J. org. Chem., 1966, 31, 3843).

(a) Synthesis and methods of preparation

For a review see A.C. Richardson (M.T.P.International Review of Science, Series One, Vol.7, 1973, p.105) and D.Horton (*loc. cit.*).

Many of the naturally occurring compounds noted in Table 7, and many analogues have been synthesised since 1967 by the basic methods outlined by Hough and Richardson, and extensions have led to sugars containing several amino-groups. 2,3,4,6-Tetra-amino-2,3,4,6-tetradeoxy-D-glucose, for example, is obtained from a 2,3-diamino-compound by effecting azide displacements of sulphonyloxy groups at $C_{(4)}$ and $C_{(6)}$. Hydrogenolysis of the tetra-hydrochloride of the benzyl glycoside (49), however, gives, as well as the free sugar, large proportions of a pyrrolidine formed by hydrogenation of an intramolecular Schiff's base (Scheme 90) (W. Meyer zu Reckendorf and N. Wassiliadou-Micheli, Ber., 1974, 107, 1188).

(49)

SCHEME 90

Compounds with three contiguous amino-groups are obtainable by cyclisation of dialdehydes with nitromethane in the presence of primary amines and reduction of the nitrodiamines formed. 1,6-Anhydro-β-D-glucopyranose is convertible in this way into the triamino-D-*ido*-analogue by way of its product of periodate oxidation (Scheme 91) (F.W. Lichtenthaler, T. Nakagawa and A.El-Scherbiney, Angew.Chem.internat.Edn., 1967, 6, 568). Alternatively, by related reactions, secondary nitro-

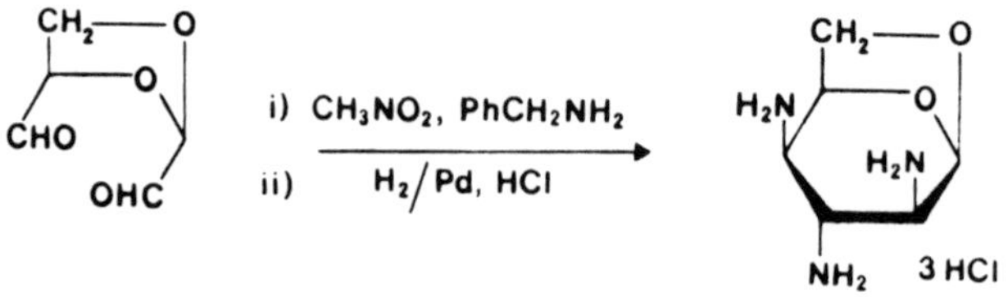

SCHEME 91

compounds with ester groups at both α-positions, on treatment with primary amines, give diaminonitro-products by elimination, addition processes (F.W. Lichtenthaler, P. Voss and N. Majer, *ibid.*, 1969, 8, 211).

Total synthesis (p. 70) represents an alternative route to amino-sugars, and although racemic products are usually obtained, the approach can be modified by use of resolved starting materials to afford compounds in optically pure form (I. Dyong and R. Wiemann, Ber., 1980, 113, 1592). *N*-Acetyl-β-DL-vancosamine has been synthesised (Scheme 92), the starting material being obtained by an allylic amination process (I. Dyong and H. Friege, *ibid.*, 1979, 112, 3273).

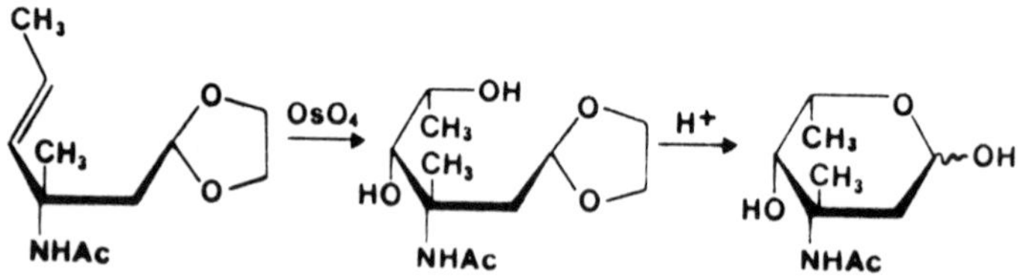

SCHEME 92

The method of introducing an amino-group which depends upon the reduction of the oxime function has gained in importance because of the ease of access to oximes either from ulose derivatives (p. 62) or following the addition of nitrosyl chloride to glycals (p.113). For ketoximes the method of reduction is critically important in determining the ratio of epimeric amino-sugars formed; while hydrogenation of 2-oximino-α-D-*arabino*-hexopyranosides (obtained from the nitrosyl chloride, tri-*O*-acetyl-D-glucal adduct) over palladium in the presence of hydrazine affords a highly specific synthesis of 2-amino-2-deoxy-α-D-glucopyranosides, reductions with lithium aluminium hydride are not highly stereoselective (R.U. Lemieux *et al.*, Canad.J.Chem., 1973, 51, 33).

Other approaches to the synthesis of amino-sugars have also depended upon unsaturated precursors, and the use of the Sharpless reagent (Chloramine T - osmium tetraoxide) has been notable in this respect. It effects *cis*-addition of hydroxyl and *N*-tosylamino groups from the sterically more accessible side of a double bond, but regioselectivity is not always high (see p.223). A simple synthesis of *N*-acetylmycosamine has been effected by its use (Scheme 93) (I. Dyong *et al.*, Carbohydrate Res., 1979, 68, 257).

SCHEME 93

Otherwise, amino-groups may be introduced by the hydroboration-amination method (effectively causing the addition of ammonia to double bonds), and in this way 2-ethoxy-3,4-dihydro-6-methyl-2H-pyran has been converted into the racemic ethyl glycosides of tolyposamine (S. Yasuda *et al.*, Tetrahedron, 1973, 29, 3141).

Ammonia itself will add to activated double bonds, and conversion of nitroalkenes into amino-sugar derivatives has often been effected - particularly for the synthesis of 2-amino-2-deoxyaldoses. Analogous additions to unsaturated aldonic acid derivatives which are obtainable by Wittig processes allow access to 3-amino-2,3-dideoxy compounds (Scheme 94) (I. Dyong and W. Hohenbrink, Ber., 1977, 110, 3655).

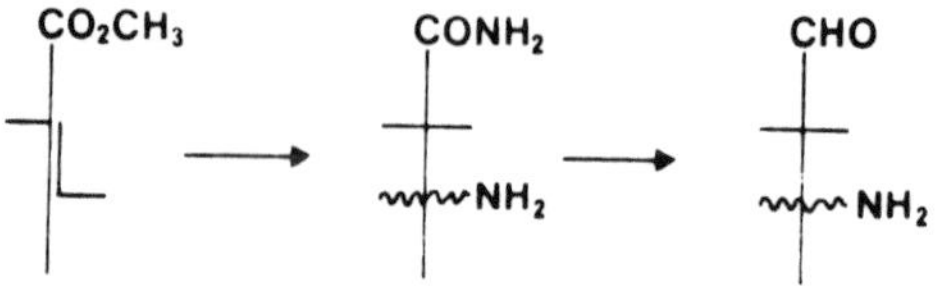

SCHEME 94

Otherwise, hydrazoic acid adds to conjugated enones, and 2,3-unsaturated 4-ulosides in this way give 2-azido-compounds, a synthesis of nebrosamine having been effected in this way (S.D. Gero *et al.*, Tetrahedron, 1977, 33, 965).

Double bonds can be used in quite a different way to introduce amino-groups into sugars since they facilitate nucleophilic displacement reactions at allylic centres; the introduction of azido-groups at $C_{(4)}$ of hex-2-enopyranosides by way of 4-mesyl esters is illustrated on p.229. Not only can the orientation of the azido- (and hence amino-) group be controlled, but these allylic azides offer access to 2-azido-isomers by stereospecific allylic rearrangements. Such processes necessarily give mixtures of 2- and 4-azides, and a better method for using the allylic rearrangement approach to effect the introduction of an amino-group involves the initial use of allylic thiocyanates. A synthesis of a D-epipurpurosamine derivative illustrates the approach (Scheme 95) (R.D. Guthrie and G.J. Williams, Chem.Comm., 1971, 923).

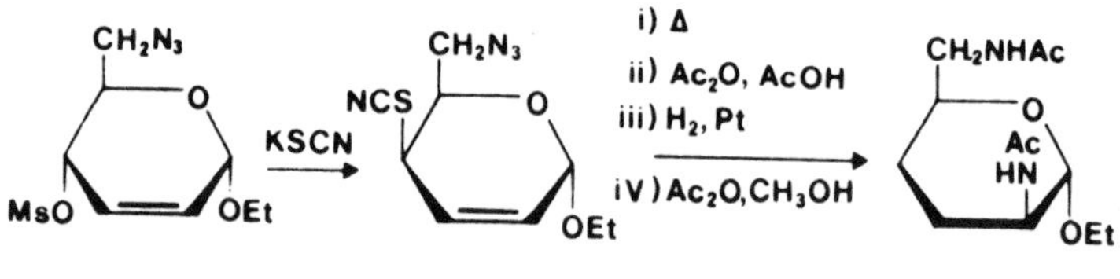

SCHEME 95

A one-step synthesis of deoxyphthalimido derivatives which offers an alternative to the nucleophilic displacement of sulphonyloxy groups as a means of aminating carbohydrates, uses phthalimide, diethyl azodicarboxylate and triphenylphosphine. Yields from primary hydroxylic compounds are reasonable. Some secondary alcohols, e.g. the sterically shielded 1,2:5,6-di-*O*-isopropylidene-α-D-allofuranose, do not react; those which do, give products with inverted configuration (A. Zamojski, W.A. Szarek and J.K.N. Jones, Carbohydrate Res., 1972, 23, 460).

(b) Sugar derivatives with nitrogen as ring hetero-atom

This subject has been reviewed by H. Paulsen and K. Todt (Adv.carbohydrate Chem., 1968, 23, 115).

A free sugar which has an amino- or substituted amino-group at $C_{(4)}$ or $C_{(5)}$ can ring close through nitrogen to give sugar analogues with nitrogen instead of oxygen as the ring hetero atom, and these forms are in equilibrium with the normal furanoses or pyranoses. Because of the high nucleophilicity of nitrogen, ring closure through it is relatively favoured, and 4-amino-4-deoxy-D-glucose and the 5-substituted isomer exist predominantly in the *N*-heterocyclic forms. When, however, the nucleophilicity is reduced by *N*-acetylation, the *O*-cyclic forms are favoured.

While the properties of these *N*-acetyl-derivatives are rather similar to those of the normal sugars - they give glycosides with alcohols in the presence of acid catalysts - the unacetylated compounds are quite different. In particular, in neutral or especially acid solution, they undergo dehydration and rearrangement reactions. Nojirimycin (5-amino-5-deoxy-D-glucose), for example, with dilute acid gives 5-hydroxy-2-hydroxymethylpyridine, and it likewise gives a substituted pyridine derivative with acetic anhydride in pyridine. The α-anomer mutarotates in aqueous solution and then slowly dehydrates to give the cyclic imine by loss of the anomeric hydroxyl group. A synthesis (S. Inouye *et al.*, Tetrahedron, 1968, 24, 2125) is shown in Scheme 96.

i) DMSO, Ac_2O
ii) NH_2OH
iii) H_2, Ni

i) Li, NH_3
ii) H_2SO_3
iii) ^-OH

SCHEME 96

(c) Branched-chain amino-sugars

Interest in amino-sugars having a branching point within the sugar chain emanates from the finding of such compounds in Nature. The amino- and branched-functions may be at the same ring position (e.g. vancosamine) or at different positions (e.g. garosamine); only the former members require specific methods for their synthesis which has been reviewed by F.W. Lichtenthaler (Fortsch.Chem.Forsch., 1970, 14, 556) and A.C. Richardson (M.T.P.International Review of Science, Organic Chemistry, Ser.1, Vol. 7, p.105).

A convenient route to compounds with a branching methyl group at $C_{(3)}$ uses the dialdehyde cyclisation method with nitroethane, and variants use nitroethanol, ethyl nitro-acetate, 3-nitropropanoic acid etc.

Alternatively, ulose derivatives offer suitable starting materials. For example, they can be converted to spiro-epoxides which can be opened with nitrogen-containing nucleophiles. Otherwise, if they are converted to exocyclic alkenes, these can be aminated.

(d) General properties and reactions of amino-sugars

Hough and Richardson covered the reactions of amino-sugars fully in the Second Edition; only a few additional aspects of their chemistry are noted.

Reaction of 2-acetamido-2-deoxyhexoses with alkali results in the formation of 3,6-anhydrides in the same way as 2-deoxyhexoses give 3,6-anhydro-derivatives under these conditions (R.J. Ferrier, W.G. Overend and A.E. Ryan, J.chem.Soc., 1962, 1488). Anhydride formation presumably occurs by addition of the hydroxyl groups at $C_{(6)}$ across the double bonds of 2,3-unsaturated intermediates formed by β-elimination of water. From either the D-*gluco*- or the D-*manno*-compound, mixtures of both their 3,6-anhydrides are obtained (N.K. Kochetkov *et al.*, Carbohydrate Res., 1971, 20, 285). In the case of 2-acetamido-2-deoxy-D-galactose, however, the epimeric D-*gulo*- and D-*ido*-anhydrides are produced because reaction occurs through the furanose form of the unsaturated intermediate, and addition of the primary alcohol group necessarily occurs with configurational inversion at $C_{(3)}$ (*idem.*, *ibid.*, 1973, 26, 201).

Deamination of amino-sugars has been reviewed comprehensively (J.M. Williams, Adv.carbohydrate Chem.Biochem., 1975, 31, 9). While nitrous acid is the normal reagent adopted for this purpose, ninhydrin can be used to convert primary amino-groups into aldehydes, and a method of converting secondary amines into ketones by way of imines is referred to on p. 99. The importance of the participation of adjacent groups during nitrous acid deaminations is illustrated by the reaction of methyl 2-amino-2-deoxy-α-D-arabinofuranoside (Scheme 97) (J.G. Buchanan and D.R. Clark, Carbohydrate Res., 1977, 57, 85).

SCHEME 97

Nucleophilic displacement reactions carried out in the presence of amino-functions are also frequently subject to neighbouring group participation; not only can appended acyl functions take part in reaction (p.105), but the nitrogen atoms themselves can participate, for example, by giving rise to aziridinium intermediates which can lead to products of re-arrangement processes (Schemes 98) (J.Mieczkowski and A. Zamojski, Roczniki Chem., 1976, 50, 2205). This reaction of

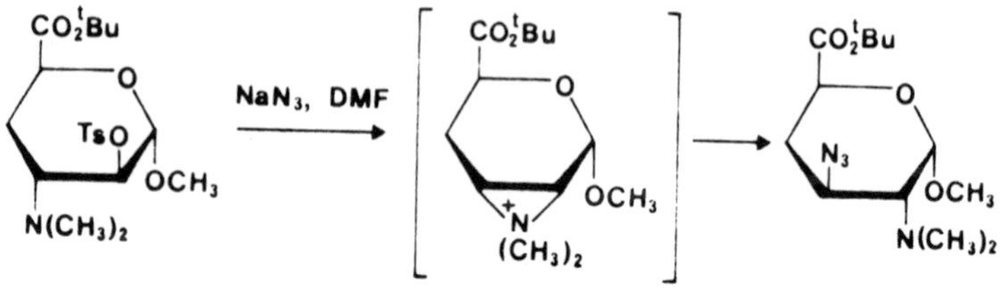

SCHEME 98

the intermediate represents diequatorial ring opening, and although such products can be produced from epimino-sugars also, these, like epoxides, usually react with nucleophiles to give mainly the products of diaxial opening (R.D. Guthrie and G.J. Williams, J.chem.Soc.Perkin I, 1976, 801).

Additional methods of effecting specific *N*-acetylation of amino-sugars include treatment with acetic anhydride in aqueous acetic acid (S. Hirano, Y. Ohe and H. Ono, Carbohydrate Res., 1976, 47, 315) and with *p*-nitrophenyl acetate in DMSO (H. Mukerjee and P.R. Bal, J.org.Chem., 1970, 35, 2042). *N*-Benzoyl derivatives are converted into *N*-acetyl analogues by refluxing in acetic anhydride-acetic acid. Water is then added, and refluxing is continued to partially hydrolyse *N*,*N*-diacetyl derivatives formed initially to some degree (P.A. Gent, R. Gigg and R. Conant, J.chem.Soc.Perkin, I, 1973, 1858). Specific methoxycarbonylation occurs on treatment of 2-amino-2-deoxy-D-glucose with methyl chlorocarbonate in methanol, but the hydrogen chloride liberated then catalyses glycosidation, and good yields of methyl 2-deoxy-2-methoxycarbonylamino-α-D-glucopyranoside are directly obtainable (S. Otani, Bull.chem.Soc.Japan, 1974, 47, 781).

Newer methods of *N*-deacetylation include the use of sodium hydroxide in aqueous DMSO at 100°C (B. Lindberg *et al.*, Carbohydrate Res., 1976, 47, C5), hydrazine in the presence of hydrazine sulphate (B.A. Dmitriev, Yu. A. Knirel and N.K. Kochetkov, *ibid.*, 1973, 29, 451) and phosphorus pentasulphide in pyridine followed by methanolic ammonia (J.J. Fox *et al.*, J.org.Chem., 1965, 30, 2735). Conditions which will cause *O*-alkylation of the amides in the enolic forms to give imidates offer methods of *N*-deacetylating which are effective in the presence of ester groups. Reagents suitable for this purpose include triethyloxonium tetrafluoroborate (S. Hanessian, Tetrahedron Letters, 1967, 1549) and methyl iodide-silver oxide-silver perchlorate (U. Kraska, J.-R. Pougny and P.Sinaÿ, Carbohydrate Res., 1976, 50, 181). The imidates are hydrolysed under mild acidic conditions.

4. Deoxy-sugars

The finding of many examples of this class of sugars - from monodeoxy to trideoxy compounds - in a wide range of natural products has resulted in extensive developments to methods applicable to their synthesis, and this Section will be devoted substantially to these newer procedures. The basic methods were dealt with by Hough and Richardson in the Second Edition. Reviews of various aspects of the chemistry of deoxy-sugars were published as the Proceedings of an American Chemical Society Symposium (Adv.Chem.Ser.No.74, 1968), see also N.R. Williams and J.D. Wander in "The Carbohydrates", Vol.1B, eds. W. Pigman and D. Horton, Academic Press, New York, 1980, p.761.

(a) Synthesis from carbonyl compounds

Free carbonyl groups in sugar derivatives may be used in two general ways to obtain deoxysugars: either they can be reduced directly to methylene groups, or they can be treated with carbanionic reagents which permit the introduction of deoxy-groups in extended-chain or branched-chain products.

Treatment of ketonic compounds with phosphorus pentasulphide in pyridine gives mixtures of cyclic polysulphide products from which the deoxy analogues are obtainable on desulphurisation with Raney nickel (Scheme 99) (P. Köll, R.-W. Rennecke and K. Heyns, Ber., 1976, 109, 2537).

CH_2—O, O, $(CH_3)_2$, O=, O–C–O, P_4S_{10}, S—S, R=, S, =R, Ni, CH_2—O, O, $(CH_3)_2$, O–C–O, R=

SCHEME 99

In related fashion, the Wolff-Kishner reaction can be applied successfully (Scheme 100). However, more complex keto-com-

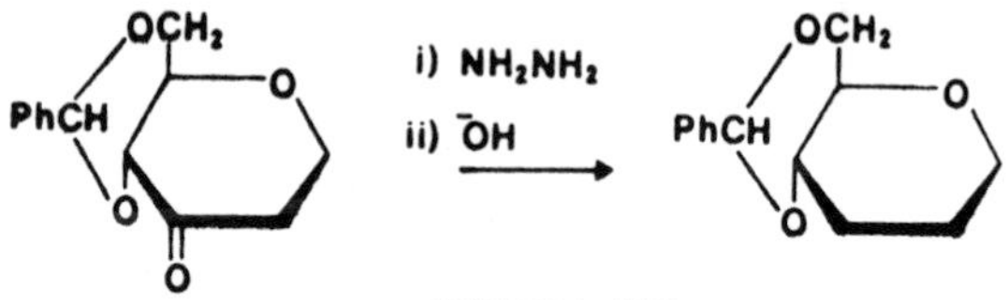

SCHEME 100

pounds, such as related hexopyranosidulose derivatives, give mixtures of products because they are subject to competing, base-catalysed elimination reactions (D. Horton *et al.*, J. chem.Soc.Perkin I, 1977, 1564).

The second approach has given a new synthesis of the biologically important 2-deoxy-D-*erythro*-pentose (Scheme 101) (J.R. Hauske and H. Rapoport, J.org.Chem., 1979, 44, 2472, *c.f.* J.M.J. Tronchet *et al.*, Helv., 1969, 52, 817). In analogous

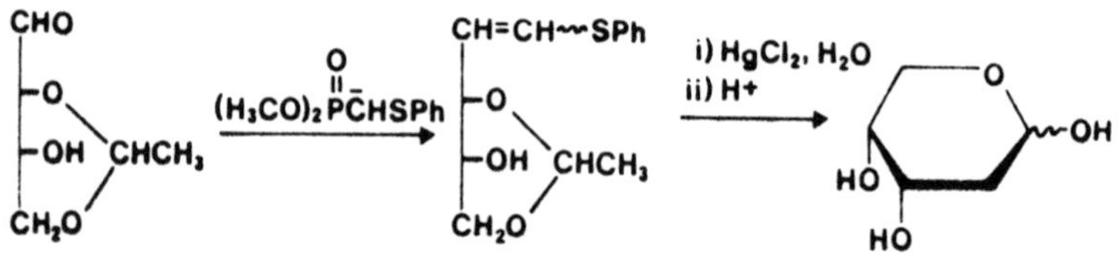

SCHEME 101

fashion methoxymethylene Wittig reagents can be used to obtain 2-deoxyaldoses (Yu.A. Zhdanov and V.G. Alexeeva, Carbohydrate Res., 1969, 10, 184) and, alternatively, dialdose derivatives afford access to sugars with deoxy-groups at the non-reducing ends of their chains (Scheme 102) (e.g. D.E. Kiely *et al.*, Carbohydrate Res., 1973, 31, 387, 397).

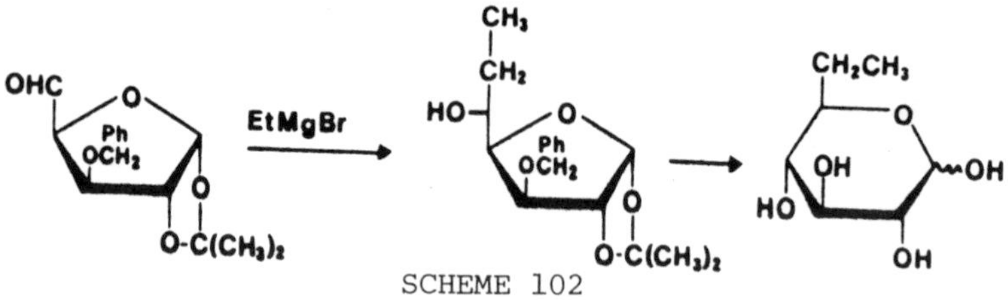

SCHEME 102

Application of the Reformatsky reaction to such aldehydes gives access to 6-deoxyhepturonic acid derivatives (Yu.A. Zhdanov, Yu.E. Alekseev and K.A. Kurdanov, Zhur.obshch.Khim., 1970, 40, 943).

(b) Synthesis from unsaturated compounds

Hydrogenation of unsaturated sugar derivatives affords a useful route to deoxy-compounds, and when unoxygenated double bonds are involved, efficient access to dideoxy analogues is provided. However, many reductions afford mixed products of monodeoxy compounds, the ratios being dependent upon access of the catalyst to the double bonds. While this can depend largely on the stereochemistry at allylic sites, the stereochemistry at more distant sites can, on occasion, be the dominant determining factor. For example, hydrogenation of 6-deoxyhex-5-enopyranosides gives predominantly the 6-deoxyhexosides with the *cis*-relationship between the aglycone and the $C_{(6)}$ methyl group (p.226). In related 5-deoxypent-4-enofuranoses with 1,2-acetal rings, hydrogenation leads predominantly to the isomers with *endo*-methyl groups (Scheme 103) (J. Kiss, R.D'Souza and P. Taschner, Helv., 1975, 58, 311).

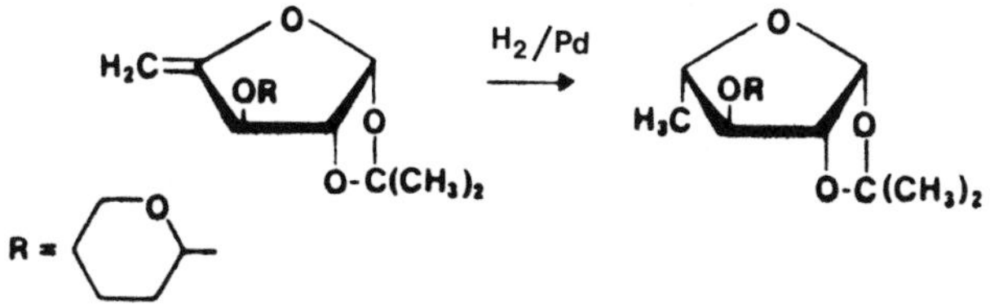

H_2C= OR O O-C(CH₃)₂ H_2/Pd H_3C OR O O-C(CH₃)₂

R =

SCHEME 103

Stereochemical issues are clearly more complex for the reduction of compounds such as enolone esters (p.176), but nevertheless, these can sometimes be used to afford specific deoxy-compounds with good selectivity (Scheme 104) (F.W. Lichtenthaler, U.Kraska and S.Ogawa, Tetrahedron Letters, 1978, 1323).

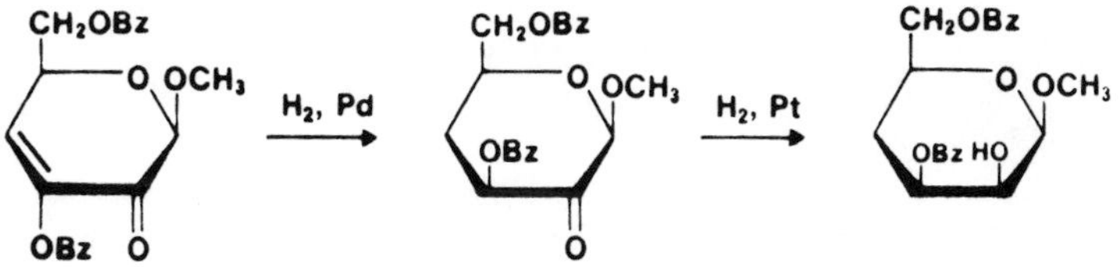

CH_2OBz O OCH_3 OBz O H_2, Pd CH_2OBz O OCH_3 OBz O H_2, Pt CH_2OBz O OCH_3 OBz HO

SCHEME 104

All addition reactions of unsaturated compounds which afford halogeno-compounds give potential access to deoxy-compounds because of the reducibility of carbon-halogen bonds (see next section). 5-Deoxy-D-*xylo*-hexose compounds are, for example, available by this approach (Scheme 105) (W.A. Szarek, R.G.S. Ritchie and D.M. Vyas, Carbohydrate Res., 1978, 62, 89).

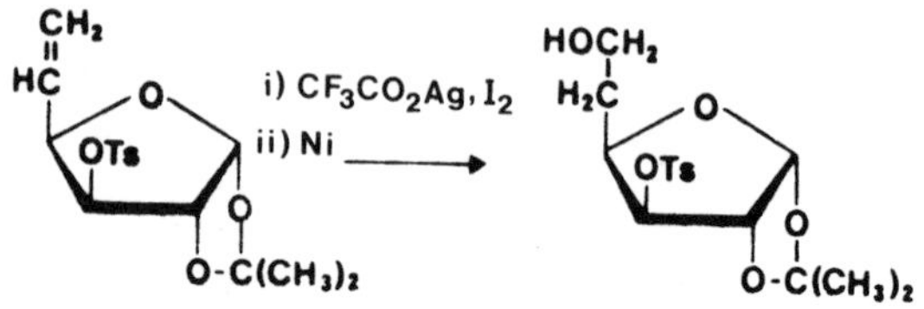

i) CF_3CO_2Ag, I_2
ii) Ni

SCHEME 105

In related fashion, 2-deoxy-2-iodoglycosides, prepared using equimolar proportions of glycal esters, *N*-iodosuccinimide and alcohols (p.210) provide efficient routes to 2-deoxyglycosides, and this approach has been extended to afford elegant syntheses of deoxydisaccharide derivatives related to cardenolide constituents (Scheme 106) (J.Thiem, P. Ossowski and J. Schwentner, Angew.Chem.internat.Edn., 1979, 18, 222).

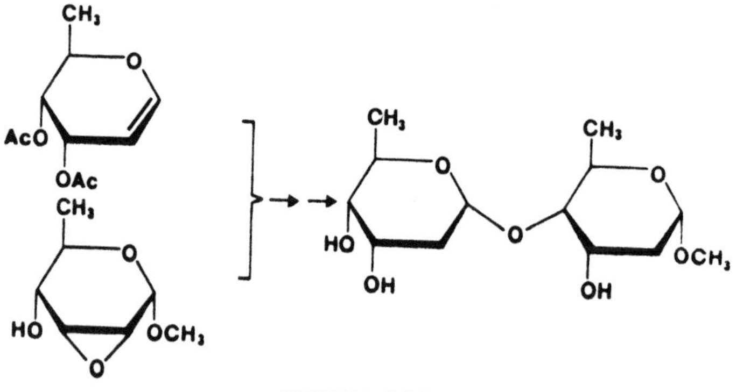

SCHEME 106

A further specific method for preparing 2-deoxyaldoses utilises the base catalysed elimination undergone by aldose dithioacetals which have $O_{(2)}$ involved in, for example, an isopropylidene ring. 2-Deoxy-D-*erythro*-pentose and -D-*threo*-pentose are obtainable in this way from D-arabinose and D-xylose, respectively, (Scheme 107) (M.Y.H. Wong and G.R. Gray, J.Amer.chem.Soc., 1978, 100, 3548).

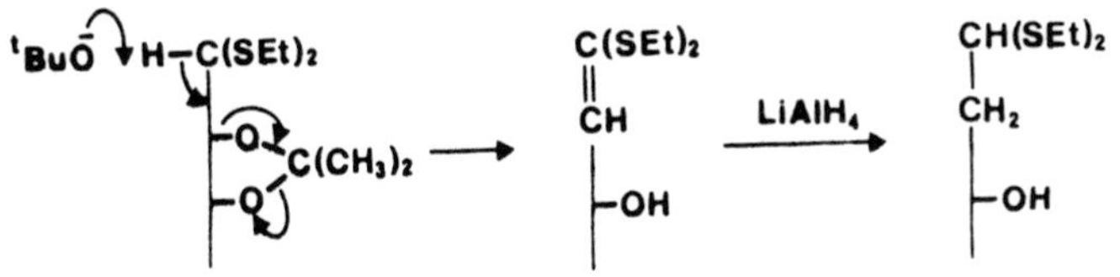

SCHEME 107

(c) Synthesis from halogeno-compounds

A wide range of methods are available for the introduction of halogen atoms into carbohydrate (W.A. Szarek and D.M. Vyas, M.T.P.International Review of Science, Organic Chemistry, Ser. 1, Vol.7, 1973, p.71; Ser.2, Vol.7, 1976, p.89; W.A. Szarek, Adv.carbohydrate Chem.Biochem., 1973, 28, 225; A.B. Foster and J.H. Westwood, Pure appl.Chem., 1973, 35, 147). Because of the facile reduction of carbon-iodine, carbon-bromine and carbon-chlorine bonds (A.R. Pinder, Synthesis, 1980, 425), each represents a means of access to deoxy-compounds. Traditional reagents for effecting the reductive removal of iodine and bromine include Raney nickel, hydrogen with palladium on charcoal, sodium amalgam, lithium aluminium hydride and zinc in acetic acid, however this last reagent can give rise to alkenes when leaving groups are present on the carbon atoms adjacent to the halogenated sites (p.204). More novel reducing agents are tributyltin hydride (E.J. Corey and J.W. Suggs, J.org.Chem., 1975, 40, 2554), nickel boride (nickel chloride and sodium borohydride, J. Thiem and J. Schwentner, Tetrahedron Letters, 1978, 459), triethyl phosphite in the presence of nickel chloride or under light (S. Inokawa *et al.*, Carbohydrate Res., 1973, 26, 230) and chromium(II) acetate in the presence of a thiol (D.H.R. Barton and R.V. Stick, J.chem.Soc.Perkin I, 1975, 1773).

Carbon-iodine bonds can be reductively cleaved by direct photolysis in methanol in the presence of potassium hydroxide (Scheme 108) (W.W. Binkley and R.W. Binkley, Carbohydrate

Res., 1968, 8, 370; 1969, 11, 1).

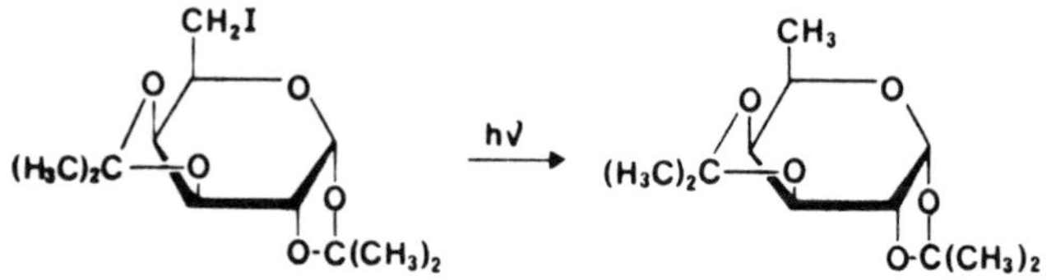

SCHEME 108

While efficient with primary iodo-compounds, this reaction is less so with secondary iodides (D. Horton, *et al.*, *ibid.*, 1977, 58, 109).

Reduction of carbon-chlorine bonds is more difficult, but can be achieved with activated Raney nickel under pressure (B.T. Lawton, W.A. Szarek and J.K.N. Jones, *ibid.*, 1970, 14, 255), and in the presence of potassium hydroxide this reagent allows the complete reduction of methyl 4,6-dichloro-4,6-dideoxy-α-D-galactopyranoside. When, however, this base is replaced by triethylamine, efficient, selective removal of the secondary chlorine atom occurs (*idem.*, *ibid.*, 1970, 15, 397). Alternatively, reaction occurs specifically at $C_{(6)}$ when the dimethyl ether of this dichloride is heated with lithium aluminium hydride in tetrahydrofuran (Scheme 109). While this

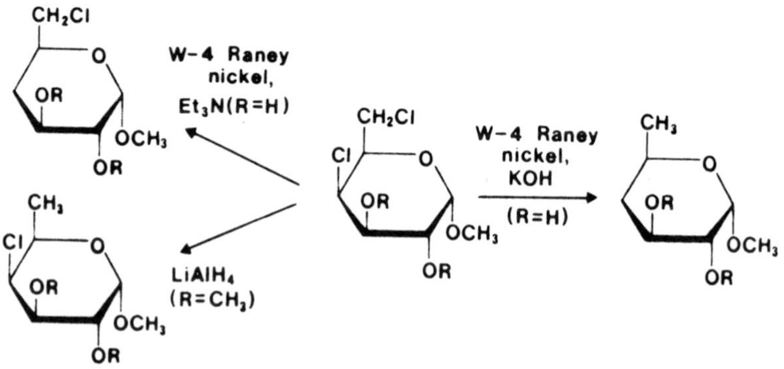

SCHEME 109

latter selectivity is consistent with that expected for a nucleophilic substitution reaction, the former indicates a radical reaction. In agreement with this, tributyltin hydride in the presence of α,α'-azobis-isobutyronitrile (a radical initiator), reductively cleaves secondary carbon-chlorine bonds preferentially and has the advantage of being unreactive towards ester and amide groups (Scheme 110) (H. Arita *et al.*, Bull.chem.Soc.Japan, 1972, 45, 567, 3614).

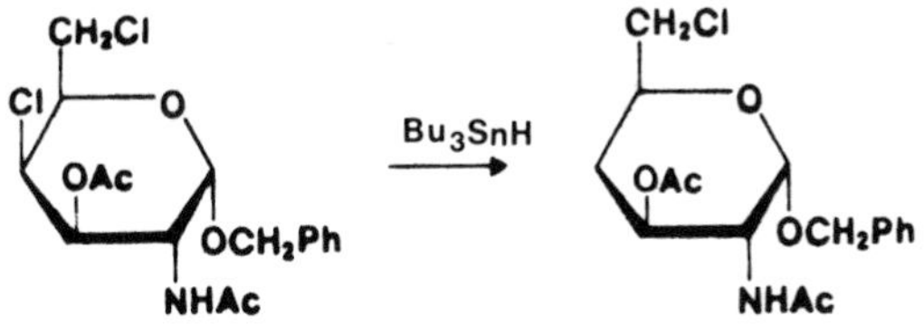

SCHEME 110

Lithium aluminium hydride reduces 3-chloro-3-deoxy-3-deuterio-1,2:5,6-di-*O*-isopropylidine-α-D-glucose with retention of configuration (K.N. Slessor *et al.*, Carbohydrate Res., 1971, 16, 375).

(d) Synthesis from esters

(i) From carboxylate esters

A simple deoxygenation reaction is based on the ultraviolet radiation of acetyl (or pivaloyl) esters in aqueous hexamethylphosphoric triamide. Primary esters and secondary esters on furanoid and pyranoid rings are efficiently cleaved (P.M. Collins and V.R.Z. Munasinghe, Chem.Comm., 1977, 927; T.Kishi, T. Tsuchiya and S. Umezawa, Bull.chem.Soc.Japan, 1979, 52, 3015), but glycosyl acetates may give free sugars following deacylation (J.-P. Pete *et al.*, Synthesis, 1977, 774).

In the special case of tertiary alcohols having a *C*-acyl group at the branching point, deoxygenation can be effected by *O*-acylation followed by reduction with tributyltin hydride in the presence of a radical initiator (Scheme 111) (H. Redlich, H.-J. Neumann and H. Paulsen, Ber., 1977, 110, 2911).

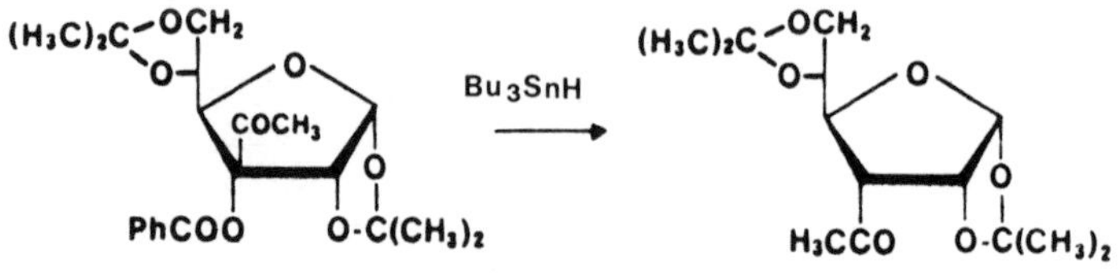

SCHEME 111

(ii) From thiocarboxylate esters

Several reactions have been developed by Barton and his colleagues which allow the conversion of various thio-esters into deoxy-compounds. *O*-Thiobenzoyl esters and *O*-(S-methyl dithiocarbonates) (xanthates) of secondary alcohols, on heating in toluene with tributyltin hydride are converted into the deoxy-analogues in good yield (Scheme 112) (D.H.R. Barton and S.W. McCombie, J.chem.Soc.Perkin I, 1975, 1574). Cyclic thio-

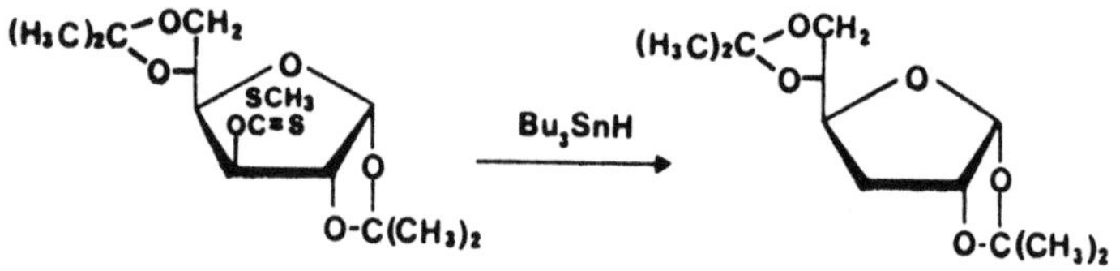

SCHEME 112

carbonates are similarly reduced to give deoxyesters from which deoxysugars are formed on alkaline hydrolysis. From terminal cyclic esters the main products have non-terminal deoxy-groups in keeping with expectations for radical reactions (p.169) (D.H.R. Barton, and R. Subramanian, *ibid.*, 1977, 1718). When, however, these terminal cyclic thiocarbonates are heated with methyl iodide they give (methylthio)carbonyl esters with terminal iodo-groups (D.H.R. Barton and R.V.Stick *loc.cit.*), and so afford access to terminal deoxy-products. In the hexofuranose series, 5- or 6-deoxy derivatives can therefore both be prepared (Scheme 113).

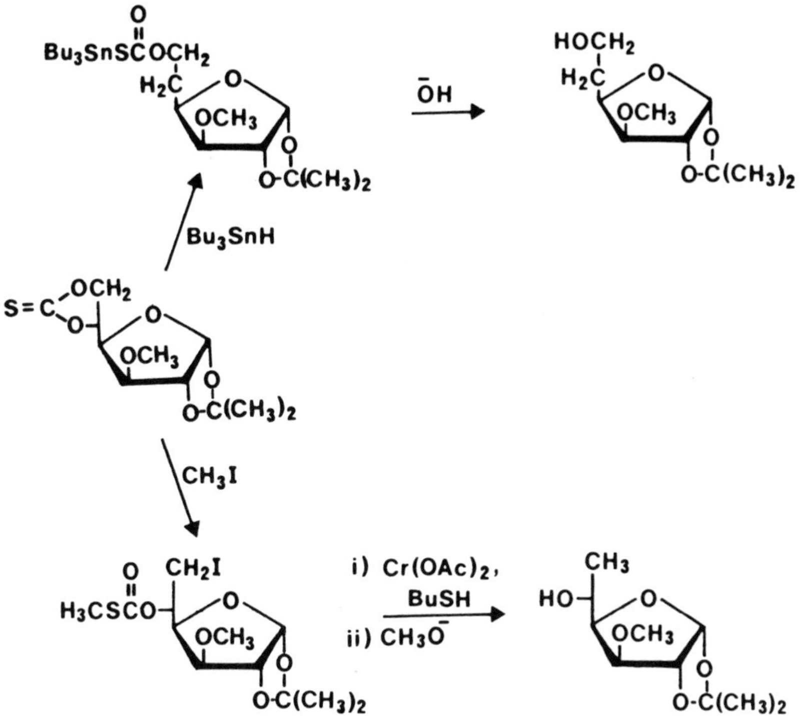

SCHEME 113

Photolysis of primary or secondary dimethylthiocarbamates also leads to deoxy-sugar derivatives, but these products are usually accompanied by the alcohols formed by de-esterification (D. Horton *et al.*, Carbohydrate Res., 1977, 58, 109).

(iii) From sulphonate esters

Generally, primary sulphonates can be reduced with lithium aluminium hydride to give terminal deoxy-compounds while secondary esters cleave with sulphur-oxygen bond fission (D.H. Ball and F.W. Parrish, Adv.carbohydrate Chem., 1968, 23, 233). The former type of reaction can be effected specifically in the presence of secondary sulphonates and acetates by use of sodium borohydride in DMSO (H. Weidmann, N. Wolf and W.Timpe,

Carbohydrate Res., 1972, 24, 184). The presence of neighbouring groups can significantly influence the reduction of secondary esters. For example, methyl 4,6-*O*-benzylidene-2,3-di-*O*-tosyl-α-D-glucopyranoside with lithium aluminium hydride in tetrahydrofuran gives methyl 4,6-*O*-benzylidene-3-deoxy-α-D-*ribo*-hexopyranoside in good yield. (E. Vis and P. Karrer, Helv., 1954, 37, 378). This may occur by way of the product of normal cleavage at $C_{(2)}$ and the derived 2,3-anhydro-allo-pyranoside or, conceivably, by a directed hydride attack at $C_{(3)}$ following de-esterification at $C_{(2)}$. The example of a contiguous trisulphonyloxy compound affording a 1,3-related diol is given on p.150 . Secondary tosyl esters α- to carbonyl functions are cleaved by carbon-oxygen fission on ultraviolet irradiation (W.A. Szarek and A. Dmytraczenko, Synthesis, 1974, 579), and secondary esters adjacent to hydroxy groups can be similarly converted into deoxy-groups on treatment with excess of t-butyl magnesium bromide. This reaction may occur by eliminations to give deoxyuloses, and therefore inversion of configuration may be observed at the alcohol centres (M. Kawana and S. Emoto, Chem.Letters, 1977, 597).

N,*N*-Dimethylsulphamoyl esters of secondary alcohols (prepared using either the sulphamoyl chloride or sulphuryl chloride, pyridine and dimethylamine), on treatment with sodium in liquid ammonia, undergo efficient free radical reactions to give deoxy-compounds (T. Tsuchiya *et al.*, Tetrahedron Letters, 1978, 3365). Many trifluoromethanesulphonates react similarly, either with this reagent or on photolysis (*idem.ibid.*, 1979, 2805), but one such ester has been encountered which gives mainly a ring-contracted deoxy-product (Scheme 114).

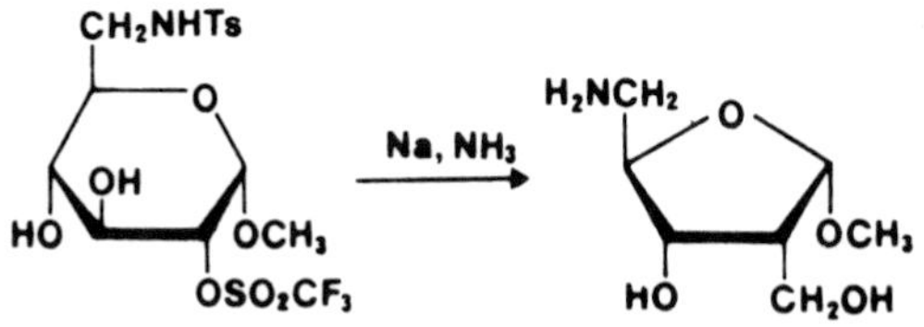

SCHEME 114

Otherwise, deoxygenations may be effected from sulphonates by way of intermediates which are susceptible to reductive cleavage. Trifluoromethanesulphonates, for example, are very susceptible to displacement by sulphur nucleophiles (for example, sodium benzenethiolate), and the products can be desulphurised under standard conditions (T.H. Haskell, P.W.K. Woo and D.R. Watson, J.org.Chem., 1977, 42, 1302). Displacement with cyanide offers means of synthesising, from primary sulphonates, compounds with deoxy-groups adjacent to chain termini. For example, hexopyranose 6-sulphonates can be converted into 6-deoxyheptose derivatives in this way (P.J.Garegg *et al.*, Carbohydrate Res., 1978, 67, 263).

(e) Synthesis from other derivatives

Epoxides (p.137) represent a further general class of compounds from which deoxy-derivatives are readily produced; α-nitroepoxides with sodium borohydride give deoxyketones and thence deoxyalcohols with the hydroxyl groups at the sites of the initial nitro-groups (H.H. Baer and C.B. Madumelu, Canad. J.Chem., 1978, 56, 1177). *N*-Aminoepimines rearrange on acetylation to give α-deoxy-β-ulose *N*-acetylhydrazones (H. Paulsen and M. Budzis, Ber., 1970, 103, 3794).

Benzylidene acetals react with di-t-butyl peroxide to give ring-opened deoxy-benzoates. Hexopyranoside 4,6-acetals lead mainly to 6-deoxy-4-benzoates, but 4-deoxy-isomers can also be produced in significant proportions (C. Pedersen *et al.*, Acta chem.Scand., 1973, 27, 3579).

A further radical reaction resulting in deoxy-derivatives occurs when carbon-mercury bonded compounds are photolysed (D. Horton, J.M. Tarelli and J.D. Wander, Carbohydrate Res., 1972, 23, 440).

Conversion of monohydroxy compounds into tris(dimethylamino)phosphonium salts by treatment with tris(dimethylamino)-phosphine and carbon tetrachloride, and reduction of these with the strongly nucleophilic lithium triethylborohydride, gives a further efficient means for deoxygenation (P. Simon, J.-C. Ziegler and B. Gross, Synthesis, 1979, 951).

5. Branched-chain sugars

(a) General and natural occurrence

Considerable developments have occurred in this area largely in response (a) to the finding of new branched-chain sugars as constituents of antibiotics (H. Grisebach and R. Schmid,

Angew.Chem.internat.Edn., 1972, 11, 159), and of nucleotides (S. Okuda, N. Suzuki and S. Suzuki, J.biol.Chem., 1967, 242, 958; 1968, 243, 6353), and (b) to the fact the nucleosides comprising branched-chain sugars can have cytostatic or antiviral activity. The synthesis of branched-chain sugars has been reviewed by H. Paulsen (Die Stärke, 1973, 25, 389) and the biosynthesis of antibiotic compounds of this class by H. Grisebach (Adv.carbohydrate Chem.Biochem., 1978, 35, 81), see also N.R. Williams and J.D. Wander in "The Carbohydrates", Vol.1B, eds. W. Pigman and D. Horton, Academic Press, New York, 1980, p.761. A further important stimulus has been the recognition that a wide range of non-carbohydrate natural products can be produced synthetically in optically pure form from branched-chain sugars (S. Hanessian, Acc.chem.Res., 1979, 12, 159; B. Fraser-Reid and R.C. Anderson, Prog.Chem.org.nat. Prods., 1980, 39, 1). The complexity and importance of this can be illustrated by the key step in the synthesis of the framework of the macrolide aglycone of the antibiotic erythronolide A from branched-chain monosaccharide derivatives (Scheme 115) (S. Hanessian, G. Rancourt and Y. Guindon, Canad. J.Chem., 1978, 56, 1843).

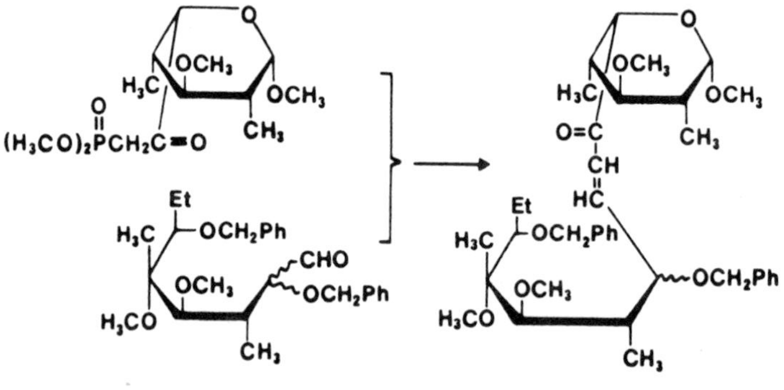

SCHEME 115

Table 8 contains data on the best known naturally occurring sugars which have a methyl branching group. All are antibiotic constituents, and while $C_{(3)}$ is the most common branching point, garosamine and noviose represent sugars which have *C*-methyl groups at $C_{(4)}$ and $C_{(5)}$, respectively. They all are also deoxy-compounds, several have methyl ether groups, and evernitrose and vancosamine are noteworthy having a nitro-

Table 8

Some naturally occurring sugars with methyl branch-groups

Trivial name	Parent sugar	Branch site(s)	Deoxy site(s)	Other modifications	Reference
Arcanose	L-*xylo*-hexose	3	2,6	3-*O*-Me	A
Axenose	L-*xylo*-hexose	3	2,6		B
Chromose B	L-*arabino*-hexose	3	2,6	3-*O*-Ac	A
Cladinose	L-*ribo*-hexose	3	2,6	3-*O*-Me	A
Evermicose	D-*arabino*-hexose	3	2,6		C
Evernitrose	L-*arabino*-hexose	3	2,3,6	4-*O*-Me, 3-nitro	D
Garosamine	L-*arabino*-pentose	4	3	3-NHMe	A
Mycarose	L-*ribo*-hexose	3	2,6		A
Nogalose	L-mannose	3	6	2,3,6-tri-*O*-Me	A
Noviose	L-*lyxo*-hexose	5	6	3-*O*-carbamoyl, 4-*O*-Me	A
Olivomycose	L-*arabino*-hexose	3	2,6		A
Vancosamine	L-*lyxo*-hexose	3	2,3,6	3-NH_2	E
Vinelose	L-talose	3	6	2-*O*-Me	F

References: (A) H. Grisebach and R. Schmid, Angew Chem.internat.Edn., 1972, 11, 159; (B) P.J. Garegg and T. Norberg, Acta chem.Scand., 1975, B29, 507; (C) I. Dyong and D. Glittenberg, Ber., 1977, 110, 2721; (D) A.K. Ganguly *et al.*, Chem.Comm., 1977, 313; (E) A.W. Johnson, R.M. Smith and R.D. Guthrie, J.chem.Soc.Perkin I, 1972, 2153; (F) J.S. Brimacombe *et al.*, *ibid.*, 1975, 1292.

group and an amino-group at the branching points, while garosamine has a methylamino group adjacent to the branch point. The majority of the compounds noted in Table 9, which have branch-groups other than methyl, are also derived from microbial sources, but the well known apiose and hamamelose are found in plants, and 3-hydroxymethyl-D-riburonic acid has been tentatively identified as a component of a bilirubin conjugate (C.C. Kuenzle, Biochem.J., 1970, 119, 411), and as such would apparently be the only branched-sugar identified from mammalian sources.

Apiose (for a recent review see R.R. Watson and N.S. Orenstein, Adv.carbohydrate Chem.Biochem., 1975, 31, 135) occurs widely within the plant kingdom both as a constituent of low molecular weight substances (often flavonoid glycosides) and polysaccharides of both land and water plants. Several syntheses have been described (Grisebach and Schmid *loc.cit.*), a particularly simple one utilising the base-catalysed reaction of 2,3:5,6-di-*O*-isopropylidene-D-mannose with formaldehyde to introduce the branch-group (Scheme 116) (P.-T. Ho, Canad.J.Chem., 1979, 57, 381).

SCHEME 116

Introduction of the hydroxymethyl group into D-ribose to give D-hamamelose has a profound effect on the isomeric composition of the sugar in solution. While the unbranched pentose equilibrium contains 80% of the pyranoses (Table 2, p.32), the branched-sugar contains only 28%. The α:β-ratios of both furanoses and pyranoses are near unity (G. Schilling and A. Keller, Ann., 1977, 1475). Hamamelose has been synthesised in both enantiomeric forms (Grisebach and Schmid, *loc.cit.*), and also from methyl 3,4-*O*-isopropylidene-β-D-*erythro*-pentopyranosid-2-ulose by use of 2-lithio-1,3-dithiane (H. Paulsen, V. Sinnwell and P. Stadler, Ber., 1972, 105, 1978), and this reagent has also afforded a route to L-streptose. Glycosides of this sugar, including compounds

Table 9

Some naturally occurring sugars with branch-groups other than methyl

Trivial name	Parent sugar	Branch group and position	Other modifications	Reference
Apiose	D-*glycero*-tetrose	3-hydroxymethyl		A, B
Aldgarose	D-*ribo*-hexose	3-(1'-hydroxyethyl)	4,6-dideoxy 3,1'-cyclic carbonate	C
Hamamelose	D-ribose	2-hydroxymethyl		A, D
Dihydrostreptose	L-lyxose	3-hydroxymethyl	5-deoxy	A
Hydroxystreptose	L-lyxose	3-formyl		A
"γ-Octose"	L-*xylo*-hexose	4-(1'-hydroxyethyl)	2,6-dideoxy	E, F
-	L-*xylo*-hexose	4-acetyl	2,6-dideoxy	F
Pillarose	L-*threo*-hexose	4-glycolyl	2,3,6-trideoxy	E, G
Streptose	L-lyxose	3-formyl	5-deoxy	A, D
-	D-riburonic acid	3-hydroxymethyl		H

References: (A) Second Ed.; (B) R.R. Watson and N.S. Orenstein, Adv.carbohydrate Chem. Biochem., 1975, 31, 135; (C) H. Paulsen *et al.*, Ber., 1976, 109, 1362; (D) *idem.*, *ibid.*, 1972, 105, 1978; (E) H. Paulsen and V. Sinnwell, Angew Chem.internat.Edn., 1976, 15, 438, 439; (F) H. Grisebach *et al.*, Eur.J.Biochem., 1972, 29, 1; (G) B. Fraser-Reid and D.L. Walker, Canad.J.Chem., 1980, 58, 2694; (H) H. Paulsen and W. Stenzel, Ber., 1974, 107, 3020.

related to streptomycin, are obtainable as shown in Scheme 117. (H. Paulsen, *et al.*, *ibid.*, 1977, 110, 1896, 1916).

(R[1] = H, R[2] = Ph) ROH ← ; (R[2]= R[1]=O) ROH →

SCHEME 117

(b) Determination of structure (especially the configuration at branch-points)

General structural features are determined by the standard chemical and spectroscopic methods, mass spectroscopy being particularly useful for locating the branch (D.C. DeJongh and S. Hanessian, J.Amer.chem.Soc., 1966, 88, 3114; B. Giola and A. Vigevani, Org.mass Spectrometry, 1976, 11, 71). In the case of "Type B" compounds, which have a proton at the branching point, the normal methods of ^{1}H N.M.R. spectroscopy usually allow the assignment of configuration, but, as with the natural products of Tables 8 and 9, when a hydroxyl group replaces the proton ("Type A"), alternative procedures are necessary (Grisebach and Schmid *loc.cit.*). Degradations to recognisable fragments can be effected, but frequently the required information can be obtained more readily by interrelating the hydroxyl group at the tertiary centre to some intramolecular feature - often another hydroxyl group. For example, in cyclic compounds, the *cis*- or *trans*-relationship of a tertiary hydroxyl group to a neighbouring group can frequently be determined by measurement of the rate of cleavage of the diol with the periodate ion, or by the formation (or lack of formation) of 5-membered cyclic derivatives. In such cases the methods are more reliable if *cis*, *trans* epimeric pairs can be compared, but other methods can be applied dependably to individual epimers. These include electrophoretic and chromatographic methods which involve the complexing of the sugar with reagents which are configurationally selective.

For example, phenylboronic acid in chromatographic solvents forms specifically stable complexes with pyranoid compounds having contiguous *cis,cis*-triols (the 1,3-diaxial diols form cyclic esters which are stabilised by coordination from the central, equatorial hydroxyl groups and these compounds therefore have high mobilities). An alternative method depends on correlating axial tertiary hydroxyl groups on pyranoid rings with *cis*-related axial oxygen atoms by observation of the specific hydrogen bonding which occurs in such cases in dilute, non-hydrogen-bonding solvents. The O-H stretching frequency is characteristically low (*ca.* 3510 cm^{-1}) and this, for example, allows the assignment of configuration at $C_{(2)}$ of methyl 2-*C*-methyl-β-L-ribopyranoside (50) (W.G. Overend *et al.*, J.chem.Soc.(C), 1968, 1091).

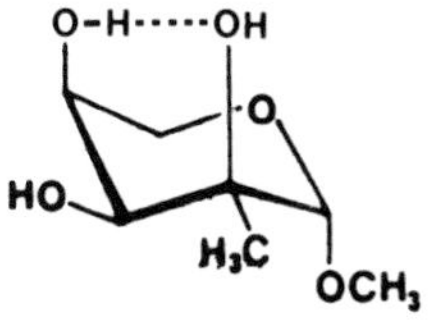

(50)

Optical rotatory dispersion and circular dichroism can be used in suitable cases, as for example, when a nitro-group is bonded to the branching-point.

N.M.R. methods can be applied even in the absence of protons at the critical centres; thus, comparison of the ^{1}H spectral changes of branched-chain alcohols and appropriate unbranched model compounds following the addition of $[Eu(fod)_3]$ (p. 51) is a powerful method (S.D. Gero *et al.*, J.org.Chem., 1975, 40, 1061). ^{13}C Spectroscopy is invaluable particularly when epimeric pairs of branched compounds are available and, consistent with observations on substituted cyclohexanes, pyranoid compounds with axial branched methyl groups (and, for example, 1,3-dithian-2-yl groups) give resonances for the exocyclic carbon atoms which are *ca.* 6 p.p.m. upfield from those resonances of the equatorial epimers. The resonances for the quaternary carbon atoms are another useful parameter (P.M. Collins and V.R.N. Munasinghe, Carbohydrate Res., 1978, 62, 19). The chemical shifts of tertiary acetoxy protons have also been used, but these and other ^{1}H methods can be unreliable (J. Yoshimura *et al.*, Bull.chem.Soc.Japan, 1980, 53, 189).

(c) Synthesis

Very considerable effort has gone into the synthesis of branched-chain compounds and extensive series of papers have been published by the following and their co-authors: H. Paulsen (Ber., 1980, 113, 2616), J.S. Brimacombe (J.chem. Soc.Perkin I, 1980, 995), J. Yoshimura (Bull.chem.Soc.Japan, 1980, 53, 189), A. Rosenthal (Carbohydrate Res., 1978, 60, 193), B. Fraser-Reid (Canad.J.Chem., 1980, 58, 2818), and J.M.J. Tronchet (Helv., 1979, 62, 2022). Brimacombe (Angew. Chem.internat.Edn., 1971, 10, 236) has reviewed the synthesis of cladinose, garosamine, mycarose and streptose, and general methods have also been surveyed (Grisebach and Schmid, *loc. cit.*; W.A. Szarek in M.T.P. International Review of Science, Organic Chemistry, Ser.1, Vol.7, 1973, p.71; Ser.2, Vol.7, 1976, p.89). Branched-chain amino-sugars have been separately treated (F.W. Lichtenthaler, Fortsch.Chem.Forsch., 1970, 14, 556).

(i) Synthesis of type A compounds (C-C-O)

Most commonly, methyl branches are introduced by treatment of ulose derivatives with methylmagnesium halides, methyllithium or diazomethane, and there are interesting stereoselectivities exhibited by these different reagents. Thus, methyl 2,3-di-*O*-methyl-6-*O*-trityl-α-D-*xylo*-hexopyranosid-4-ulose reacts with methyllithium to give the *gluco*-product stereospecifically, whereas the *galacto*-epimer is produced with methylmagnesium iodide. Conceivably this is because, in the former case the nucleophilic carbon is impeded during equatorial attack by electrostatic interaction with $O_{(3)}$, while in the latter case, coordination of the magnesium with this atom directs the nucleophile (Scheme 118) (M. Miljković *et al.*, J.org.Chem., 1974, 39, 1379). With methyl 3-*O*-benzoyl-

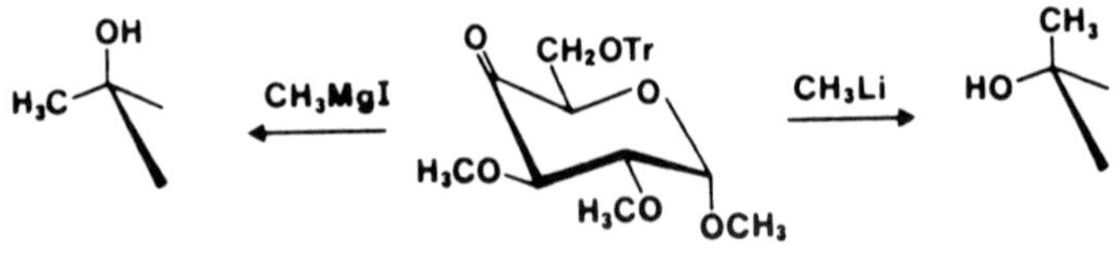

SCHEME 118

4,6-*O*-benzylidene-α-D-*arabino*-hexopyranosid-2-ulose the Grignard reagent gives mainly the *gluco*-compound while diazomethane gives an epoxide which affords the *manno*-epimer on reduction (Scheme 119) (J. Yoshimura *et al.*, Bull.chem.Soc.Japan, 1976, 49, 1686).

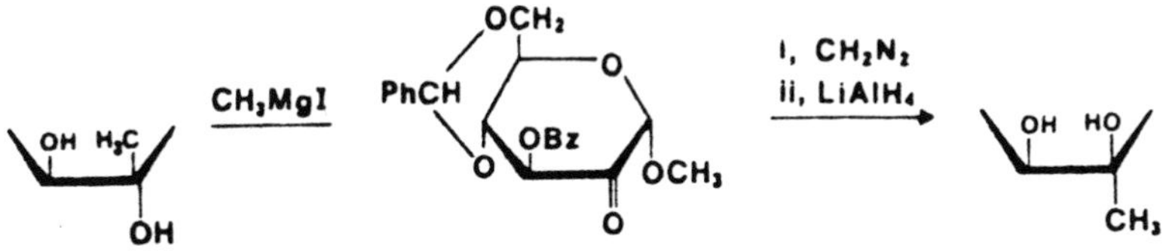

SCHEME 119

A feature of the reaction of diazomethane with ulose derivatives is that, as well as epoxides, ring expanded products can be produced (Scheme 120) (J.P. Horwitz, N. Mody and R. Gasser, J.org.Chem., 1979, 35, 2335). Spiro-epoxides can otherwise be

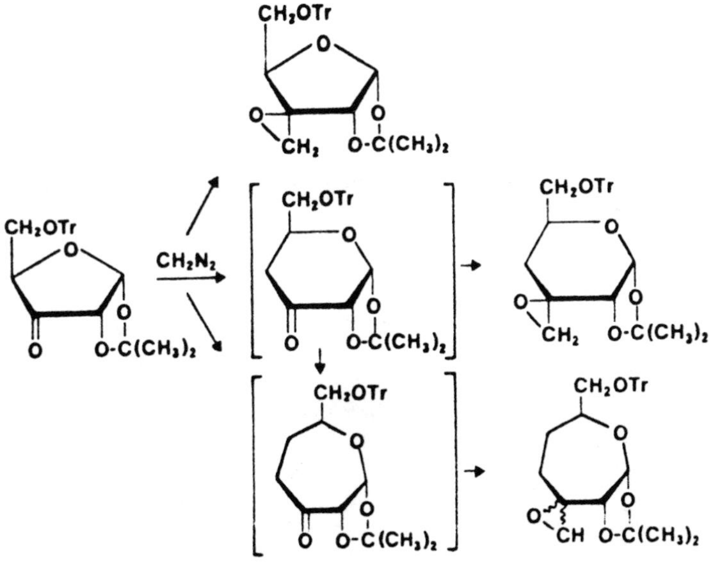

SCHEME 120

obtained by use of the methylene ylide derived from DMSO (Szarek, *loc.cit.*).

The application of sulphur-stabilised carbanions has been of major significance, and can permit the introduction of a range of branching groups which are found in natural products (Scheme 121).

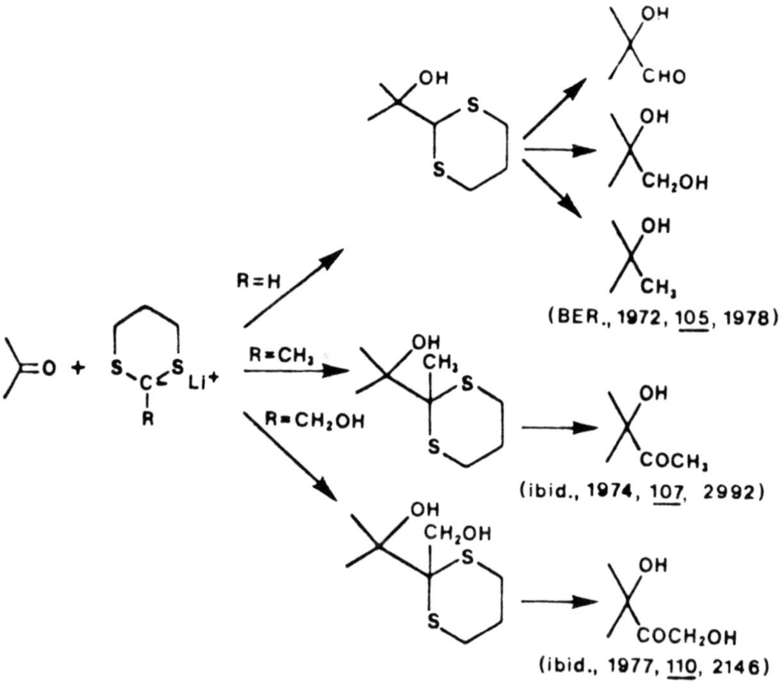
OH
CHO
OH
CH₂OH
OH
CH₃
(BER., 1972, 105, 1978)
R=H
R=CH₃
R=CH₂OH
Li+
R
S
OH
COCH₃
(ibid., 1974, 107, 2992)
OH
COCH₂OH
(ibid., 1977, 110, 2146)

SCHEME 121

Other carbanions which can be used with ulose derivatives are those derived from nitromethane (nitromethyl groups offer additional means of obtaining the formyl group, J.J. Nieuwenhuis and J.H. Jordaan, Carbohydrate Res., 1980, 86, 185), acetonitrile (A. Rosenthal and G. Schöllnhammer, Canad. J.Chem., 1974, 52, 51), ethyl bromoacetate (J. Yoshimura *et al.*, Bull.chem.Soc.Japan, 1972, 45, 1806), hydrogen cyanide, methyl nitroacetate (A. Rosenthal and B.L. Cliff, Canad.J. Chem., 1976, 54, 543), and ethyl isocyanoacetate (A.J. Brink and A. Jordaan, Carbohydrate Res., 1974, 34, 1).

An intramolecular radical reaction related to the above occurs on photolysis of methyl 4,6-*O*-benzylidene-2-*O*-methyl-α-D-*ribo*-hexopyranosid-3-ulose, and the product gives a 2-*C*-methyl-D-allose derivative with lithium aluminium hydride (Scheme 122) (P.M. Collins, P. Gupta and R. Iyer, J.chem.Soc. Perkin I, 1972, 1670). A further photochemical reaction con-

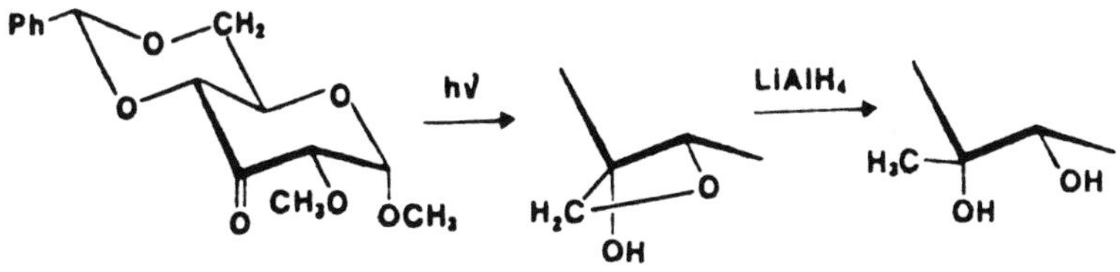

SCHEME 122

verts 4,6-*O*-ethylidene-1,2-*O*-isopropylidene-α-D-*ribo*-hexopyranos-3-ulose, in methanol solution, into 4,6-*O*-ethylidene-1,2-*O*-isopropylidene-3-*C*-hydroxymethyl-α-D-glucose and -allose (20 and 22% isolated). (P.M. Collins, V.R.N. Munasinghe and N.N. Oparaeche, *ibid.*, 1977, 2423).

Exocyclic alkenes derived from ulose compounds by use of the Wittig reaction, on appropriate hydration or hydroxylation, afford a further route to this class of branched-chain sugars (J.M.J. Tronchet and J. Tronchet, Helv., 1977, 60, 1984; A. Rosenthal and C.M. Richards, Carbohydrate Res., 1973, 29, 413), and photochemically activated addition of formamide to endocyclic enol esters also results in compounds of this class (Scheme 123) (A. Rosenthal and M. Ratcliff, *ibid.*, 1977, 54, 61).

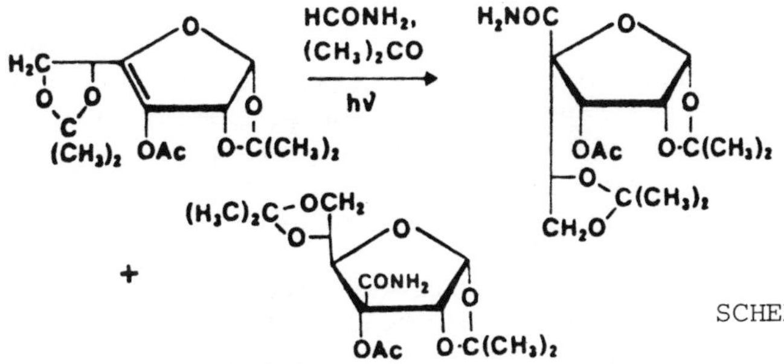

SCHEME 123

As well as permitting the generation of branching at carbonyl carbon atoms, carbonyl groups allow the introduction of carbon-bonded substituents at the α-positions. Apiose has been synthesised in this way (p.176), and hydroxymethyl groups can be introduced at $C_{(4)}$ of dialdose furanoid compounds by use of pentofuranose derivatives which have a free formyl group at this site (J.G. Moffatt *et al.*, Tetrahedron Letters, 1977, 435). Base-catalysed reactions of this type can, however, result in the formation of dimeric, complex branched-chain sugar derivatives (Scheme 124) (D. Horton and E.K. Just, Carbohydrate Res., 1971, 18, 81).

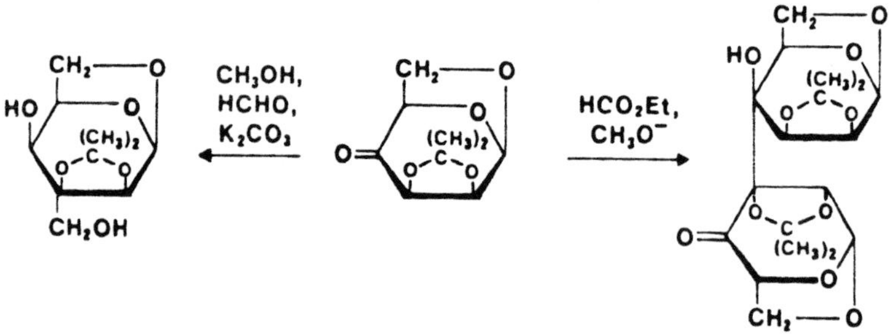

SCHEME 124

Saccharinic acids (Second Edition, p.251) are branched-chain compounds of this class which are formed by alkaline degradation of sugars. 2-*C*-Methyl-D-ribonic acid is available in practicable yields from D-fructose and its 1-*O*-substituted derivatives, and can be reduced, by way of the 1,4-lactone, to the free sugar (A.A.J. Feast, B. Lindberg and O. Theander, Acta chem.Scand., 1965, 19, 1127), and can also be used in synthesis of, for example, branched chain nucleosides (S.R. Jenkins, B. Arison and E. Walton, J.org.Chem., 1968, 33, 2490; J.J. Novák, I. Šmejkal and F. Šorm, Tetrahedron Letters, 1969, 1627). It is formed by way of the 1-deoxy-2,3-diulose which, like other α-dicarbonyl compounds, undergoes the benzylic acid rearrangement (A. Ishizu, B. Lindberg and O. Theander, Carbohydrate Res., 1967, 5, 329).

(ii) Synthesis of type B compounds (C-C-H)

This type of compound is seldom prepared by deoxygenation of corresponding tertiary alcohols, but in the special case of compounds having an acyl branching group (and presumably other radical stabilising groups) deoxygenation can be effected by *O*-acylation and reduction with tributytin hydride in the presence of a radical initiator (H. Redlich, H.-J. Neumann and H. Paulsen, Ber., 1977, 110, 2911).

While carbanion attack at carbonyl centres leads to type A compounds, similar reaction of conjugated enones affords type B products by the Michael process (M.B. Yunker, D.E. Plaumann and B. Fraser-Reid, Canad.J.Chem., 1977, 55, 4002; H. Paulsen and W. Koebernick, Carbohydrate Res., 1977, 56, 53; Ber., 1977, 110, 2127). Related additions of alcohols possessing α-hydrogen atoms, (and acetals and aldehydes) occur similarly on photochemical activation (Scheme 125) (S. Fraser-Reid, *et al.*, Canad.J.Chem., 1977, 55, 3986).

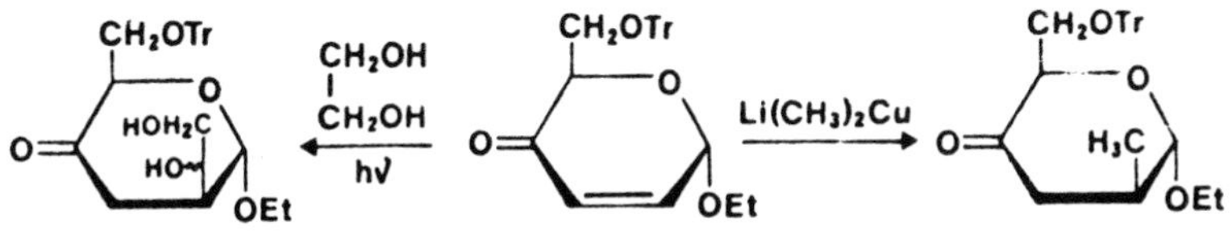

SCHEME 125

Conjugate additions also occur with other activated alkenes (Scheme 126) (T. Sakakibara, M. Yamada and R. Sudoh, J.org. Chem., 1976, 41, 736).

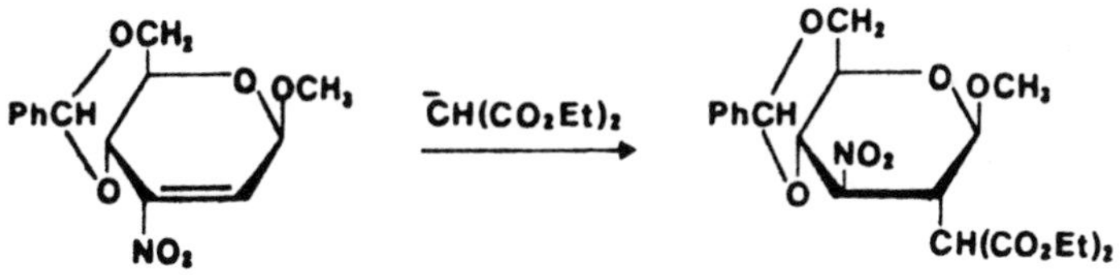

SCHEME 126

Sugar derivatives containing carbonyl-groups with adjacent deoxy-groups undergo α-alkylation to give compounds of this class. Thus, for example, methyl 4,6-*O*-benzylidene-3-deoxy-α-D-*erythro*-hexopyranosid-2-ulose can be alkylated at $C_{(3)}$ either directly or by way of a derived enamine (R.F. Butterworth, W.G. Overend and N.R. Williams, Tetrahedron Letters, 1968, 3239).

Exocyclic alkenes derived from uloses have played an important part in the synthesis of type B branched-chain sugars, and while the simplest members, i.e. methylene compounds, are usually produced by application of methylene Wittig reagents (J.K.N. Jones *et al.*, Canad.J.Chem., 1969, 47, 4473), they can also be made by use of methylene dimagnesium bromide (J. Yoshimura *et al.*, Bull.chem.Soc.Japan, 1976, 49, 1169) or by nitrous acid treatment of spiro-epimines. These are readily available from ulose cyanohydrins (J.-M. Bourgeois, Helv., 1974, 57, 2553). Methyl branch groups are obtained by reduction, and more complex branches can be introduced either by effecting other additions to the methylene alkenes (e.g. of nitryl iodide, J. Yoshimura *et al.*, Bull.chem.Soc.Japan, 1973, 46, 3207), or by initial preparation of more complex alkenes by use of appropriate Wittig reagents (Scheme 127) (A. Rosenthal and M. Ratcliffe, Canad.J.Chem., 1976, 54, 91).

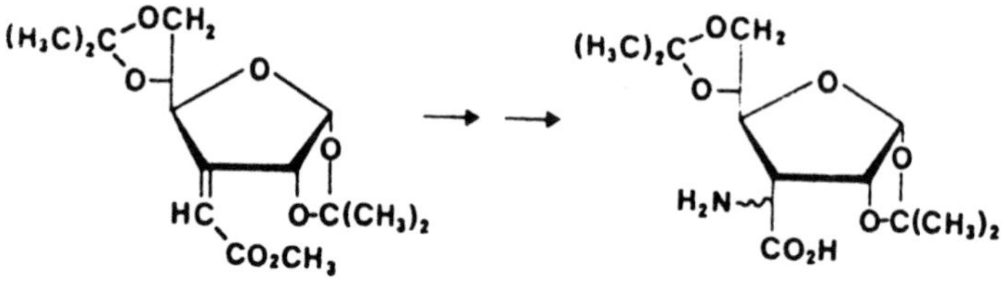

$(H_3C)_2C$ OCH_2 O O O HC $O\text{-}C(CH_3)_2$ CO_2CH_3 → → $(H_3C)_2C$ OCH_2 O O O H_2N CO_2H $O\text{-}C(CH_3)_2$

SCHEME 127

Various additions to endocyclic alkenes produce compounds of this category. For example, alcohol groups can be added photochemically (Scheme 128) (A. Rosenthal and K. Shudo, J.org. Chem., 1972, 37, 1608), and the oxo reaction gives products

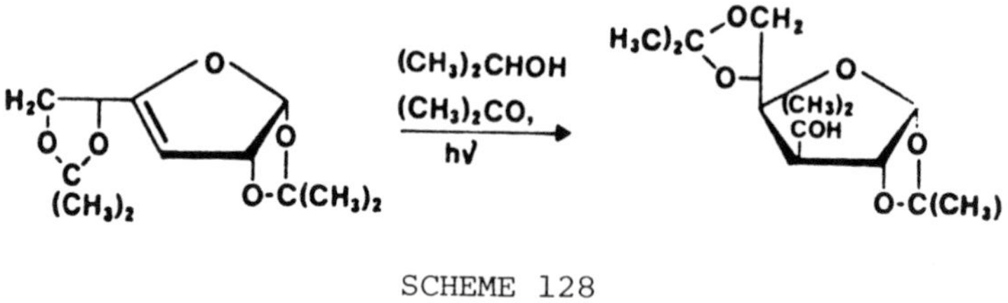

H_2C O O C $(CH_3)_2$ O O $O\text{-}C(CH_3)_2$ $(CH_3)_2CHOH$ $(CH_3)_2CO$, $h\nu$ → $H_3C)_2C$ OCH_2 O O $(CH_3)_2$ COH O $O\text{-}C(CH_3)$

SCHEME 128

with hydroxymethyl branches (A. Rosenthal, Adv.carbohydrate Chem., 1968, 23, 59). Glycals which cannot undergo allylic isomerisation (p.216) react with ortho-formates in the presence of boron trifluoride to give glycosidic products with acetalated formyl branches at $C_{(2)}$ (K. Heyns, *et al.*, Tetrahedron Letters, 1976, 1481).

Epoxides are another class of starting materials and have been used with a range of types of carbon nucleophiles. Alkyl magnesium chlorides react with pyranoid and furanoid epoxides to give branched-chain hydroxy compounds (T.D. Inch and G.J. Lewis, Carbohydrate Res., 1970, 15, 1; S.R. Jenkins and E. Walton, *ibid.*, 1973, 26, 71). Sometimes, however, chloride acts as a competitive nucleophile, and in the cases of the corresponding iodides and bromides halogenated products may predominate. (+)Blastmycinone (51), a component of blastmycin, has been made following the treatment of methyl 2,3-anhydro-4,6-*O*-benzylidene-α-D-mannopyranoside with butylmagnesium chloride which gives the 3-butyl-D-*altro*-product (Scheme 129) (S. Aburaki, N. Konishi, M. Kinoshita, Bull.chem.Soc.Japan, 1975, 48, 1254).

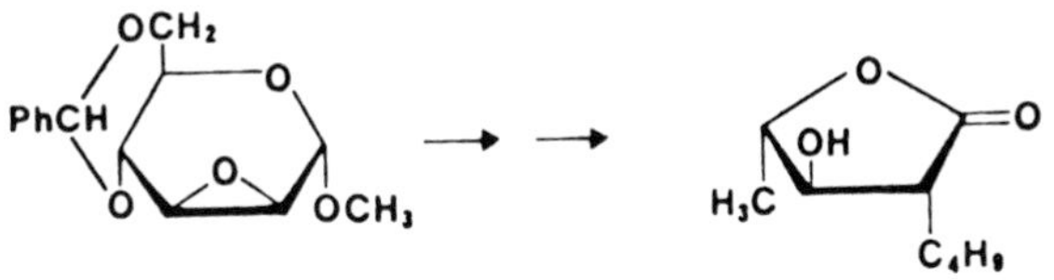

(51)

SCHEME 129

Other carbanions which have been used to open epoxides are those derived from 1,3-dithiane (A. Yamashita and A.Rosowsky, J.org.Chem., 1976, 41, 3422), lithium dimethyl cuprate (D.R. Hicks, R. Ambrose and B. Fraser-Reid, Tetrahedron Letters, 1973, 2507), diethyl malonate (S. Hanessian, P. Dextraze and R. Masse, Carbohydrate Res., 1973, 26, 264) and hydrogen cyanide (used with triethylaluminium, B.E. Davison and R.D. Guthrie, J.chem.Soc.Perkin I, 1972, 658).

Two general types of rearrangement processes can also be used to obtain type-B products. One involves the transfer of vinyl ether substituent groups via [3,3]-sigmatropic (Claisen) rearrangements (p.215), and the other, ring contraction of pyranoid derivatives via carbonium ion intermediates. The latter is observed during the deamination of some amino-sugars (Scheme 130) (J.M. Williams, Adv.carbohydrate Chem.Biochem., 1975, 31, 9), and during the displacement of sulphonyloxy

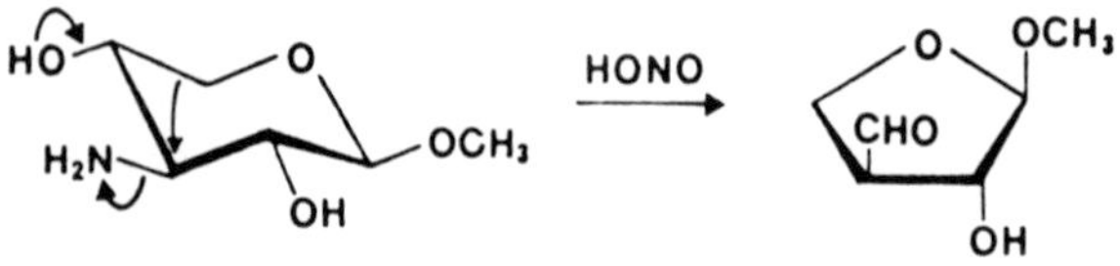

SCHEME 130

groups from related pyranosyl compounds (J.S. Brimacombe, Fortsch.chem.Forsch., 1970, 14, 367).

(iii) Synthesis of compounds with two branch-points

Two branches may be introduced successively as, for example, in the synthesis of 3-deoxy-3-*C*-ethyl-2-*C*-methylhexose derivatives from methyl 4,6-*O*-benzylidene-3-deoxy-3-*C*-ethyl-α-D-*ribo*-hexopyranosid-2-ulose (T.D. Inch, G.J. Lewis and N.E. Williams, Carbohydrate Res., 1971, 19, 17), or they may be introduced concurrently by application, for example, of the Diels Alder reaction (Scheme 131) (J.L. Primeau, R.C.Anderson and B. Fraser-Reid, Chem.Comm., 1980, 6). Applied to unsatu-

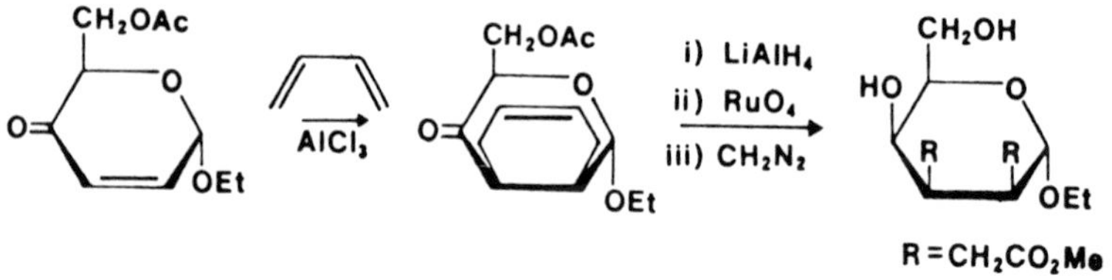

$R = CH_2CO_2Me$

SCHEME 131

rated sugar derivatives, the Simmons-Smith reaction and photochemical [2+2]cycloaddition of alkenes afford related methods of preparing cyclopropyl and cyclobutyl analogues, respectively. (B. Fraser-Reid, Acc.chem.Res., 1975, 8, 192). Alternative methods of introducing fused cyclopropyl groups in non-concerted fashion have been developed (B.J. Fitzsimmons and B. Fraser-Reid, J.Amer.chem.Soc., 1979, 101, 6123).

Doubly branched compounds with both branches at the same centre have been prepared by, for example, Michael addition of nitromethane to compounds with exocyclic nitromethylene groups (R.H. Hall, A. Jordaan and M. Malherbe, J.chem.Soc.

Perkin I, 1980, 126).

(iv) Synthesis of branched-chain amino-sugars (C-C-NH$_2$)

For a review see F.W. Lichtenthaler, Fortsch.Chem.Forsch, 1970, 14, 556. Base catalysed condensation of dialdehydes with nitroethane affords products with methyl- and nitro-groups bonded to the new ring carbon atom (c.f. p.160). (Scheme 132) (M.M. Abuaan, J.S. Brimacombe and J.N. Low, J.chem.Soc.Perkin I, 1980, 995). Reduction of the products leads to amino-

OHC, OHC, O, OCH$_3$ —— i) $CH_3CH_2NO_2$, $Et\bar{O}$; ii) Ac_2O ——→ O, OCH$_3$, NO$_2$, AcO, CH$_3$, OAc + 4 isomers

SCHEME 132

analogues, the nitro- and amino-compounds having the functionality at the branch points found in evernitrose and vancosamine, respectively. Analogous use of nitroethanol and ethyl nitroacetate gives nitro-products with hydroxymethyl and carboethoxy branch-groups, respectively.

Compounds of this series are also available by reduction of the products of addition of mercury(II) azide to exocyclic methylene alkenes, and from ulose cyanohydrins by tosylation, reduction with lithium aluminium hydride to give spiro-epimines, followed by catalytic hydrogenation. (J.S. Brimacombe, J.A. Miller and U. Zakir, Carbohydrate Res., 1976, 49, 233).

6. Thio-sugars

Because sulphur is the most versatile of the elements commonly found in carbohydrate derivatives, thio-sugars exhibit a wider range of reactions than other classes, and they are consequently of great value in synthesis. For a recent review see D. Horton and J.D. Wander in "The Carbohydrates", eds. W. Pigman and D. Horton, Vol.1B, Academic Press, New York, 1980, p.799. Their occurrence in natural products is not widespread, although further examples of "glucosinolates" i.e. "mustard-oil" glycosides and related 1-thioglucosides having isothiocyanate groups within the aglycones have been found in plants (A. Kjaer and A. Schuster, Acta chem.Scand., 1970, 24, 1631).

S-Methyl-5-thio-D-ribose, which previously has been detected as a component of modified adenosine (vitamin L2), has been isolated from *E. coli* (H.R. Schroeder *et al.*, Biochem.biophys. Acta, 1972, 273, 254), but the most notable sulphur-containing natural product is lincomycin (52) which has been examined extensively, particularly by the Upjohn Company, with a view to determining structure-function relationships. While modifications at $C_{(1)}$-$C_{(4)}$ of the carbohydrate ring drastically reduce antibacterial activity, changes at $C_{(7)}$ sometimes result in increased activity (B. Bannister, J.chem.Soc.Perkin I, 1974, 360).

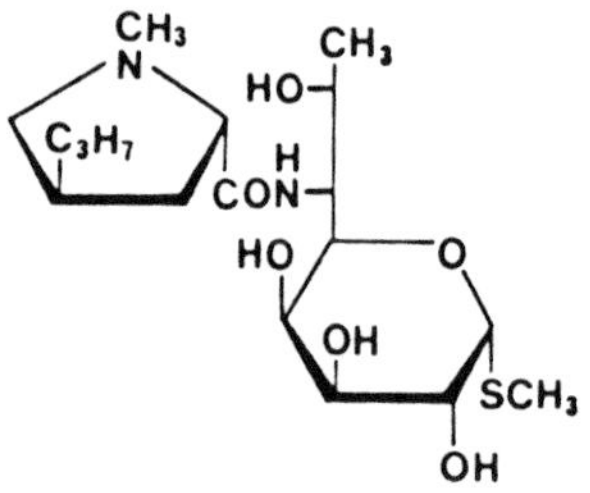

(52)

Some 1-thioglycosides act as enzyme inhibitors or inducers, and are of further biochemical interest in being useful in affinity chromatography, while some thio-sugars act as anti-metabolites. Of particular note are the biochemical properties of 5-thio-D-glucopyranose, i.e. the analogue with sulphur as the ring hetero-atom (p.200), and it has become commercially available. It inhibits transport and cellular uptake of D-glucose, it inhibits the growth of parasites with high requirements for glucose, and it can be used as a non-toxic agent to inhibit spermatogenesis and thus control male fertility. It is of consequence in cancer research since it is highly effective in killing hypoxic cells and protects oxic cells against radiation. Specific influences of the sugar on the actions of enzymes are known; for example, it inactivates phosphorylase *a* during the enzyme-catalysed degradation of glycogen (M. Chen and R.L. Whistler, Biochem.biophys.Res. Comm., 1977, 74, 1642).

(a) 1-Thio-derivatives

Whereas the acid-catalysed alcoholysis of aldoses gives

mixed glyco-furanosides and -pyranosides and only small proportions of dialkylacetals (p.108), analogous thiolysis often leads to good yields of the dithioacetals. To a large extent these products are favoured because of the relatively high nucleophilic character of thiols, and frequently their insolubility in the reaction media further facilitates their availability. Initially they were believed to be the dominant kinetically controlled reaction products and, under extended reaction conditions, were largely converted into 1-thioglycosides. However, D-xylose in ethanethiol-DMF (4:1) in the presence of hydrochloric acid acid first gives mainly the *S*-ethyl 1-thio-furanosides (with the α-anomer predominating) which then react to give the dithioacetal. The pyranosyl products are minor components throughout the reaction, and the dithioacetal reacts further by loss of water to give a furan derivative with an ethylthio-group at $C_{(3)}$ (Scheme 133) (R.J. Ferrier, L.R. Hatton and W.G. Overend, Carbohydrate Res., 1968, 6, 87). This study was carried out using radioactively

D-Xylose EtSH, H+ DMF → CH₂OH O OH OH SEt → CH₂OH OH SEt SEt OH OH → O CH₂SEt SEt

SCHEME 133

labelled sugar and chromatographic separation of the reaction components, and is one of the few to have used sensitive analytical techniques and a homogenous system. The presence of the DMF may have influenced the reaction significantly - in particular the final elimination step.

The intimate relationship between dithioacetals and 1-thioglycosides and the propensity for sulphur to enter at positions other than $C_{(1)}$ in acid-catalysed reactions is illustrated by the products of methanethiolysis of tetra-*O*-acetyl-β-D-ribofuranose and -pyranose in the presence of zinc chloride. Initially, the former gives (after subsequent deacetylation) the corresponding 1-thioglycoside and the dithioacetal, but then the main products become 4- and 5-thioderivatives. These are also produced from β-D-ribopyranose tetra-acetate indicating that a common intermediate is involved (Scheme 134) (N.A. Hughes, R. Robson and S.A. Saeed, Chem.Comm., 1968, 1381).

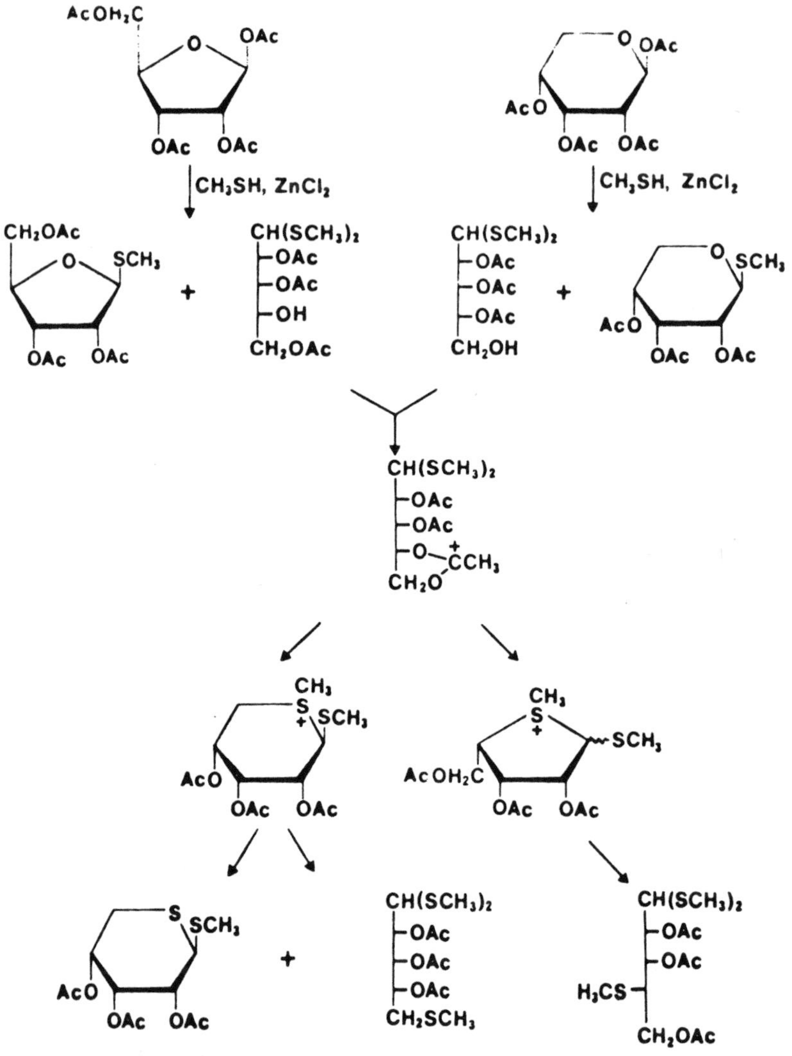

SCHEME 134

Analogous thiolysis of acyclic 1,3,4,5,6-penta-*O*-acetylhex-uloses gives the dialkyl dithioacetals and, as major secondary products, 1-alkylthio-analogues formed via 1,2-episulphonium species (G.S. Bethell and R.J. Ferrier, Carbohydrate Res., 1974, 34, 194). Whenever such intermediates can be formed, sulphur migration can occur (Scheme 135) (D. Horton, Pure. appl.Chem., 1975, 42, 301).

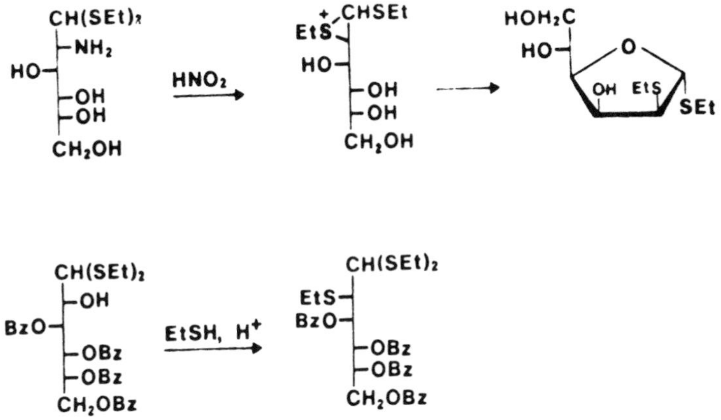

SCHEME 135

In more complex cases further sulphur can be introduced, so that ethanethiolysis of 3,5,6-tri-*O*-benzoyl-1,2-*O*-isopropylidene-α-D-glucofuranose gives 4,5,6-tri-*O*-benzoyl-2,3-di-*S*-ethyl-2,3-dithio-D-allose diethyl dithioacetal (G.S. Bethell and R.J. Ferrier, J.chem.Soc.Perkin I, 1972, 1033, 2873) in good yield. Analogous treatment of 3-*O*-benzoyl-1,2:5,6-di-*O*-isopropylidene-α-D-glucofuranose gives mainly a glycosidic product with thio-groups at $C_{(1)},_{(2)},_{(3)}$ and $C_{(6)}$ each having initially been introduced at $C_{(1)}$ and the ester group having migrated successively down the chain (Scheme 136) (*idem. ibid.*, 1973, 1400).

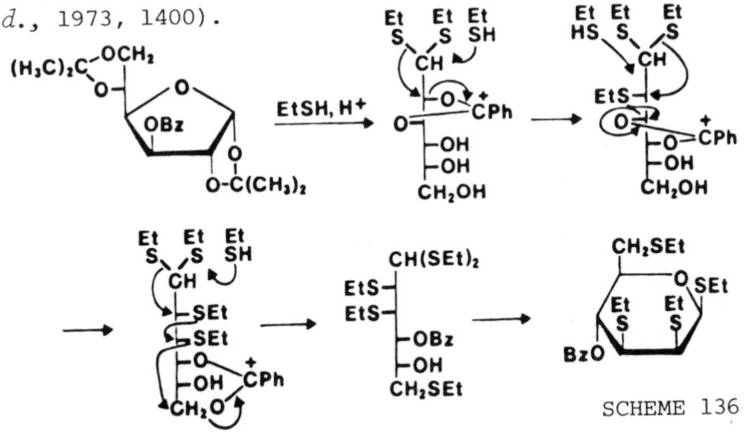

SCHEME 136

The close interrelationship between cyclic and acyclic 1-thio-compounds is illustrated by the above examples, but many features are specific to each category of derivative.

(i) Cyclic compounds

(1) *Synthesis*. Methods of preparing sugar derivatives with thiol groups at the anomeric centre have been reviewed by D. Horton and D.H. Hutson (Adv.carbohydrate Chem., 1963, 18, 123). A newer convenient method uses, for example, 1-thio-β-D-glucopyranose penta-acetate (easily made from the corresponding penta-*O*-acetyl compound using thioacetic acid with boron trifluoride as catalyst). Treated with phenylmercury(II) acetate in refluxing ethanol it gives a thiomercury compound which readily undergoes demercuration on treatment with hydrogen sulphide (Scheme 137) (R.J. Ferrier and R.H. Furneaux, Carbohydrate Res., 1977, 57, 73).

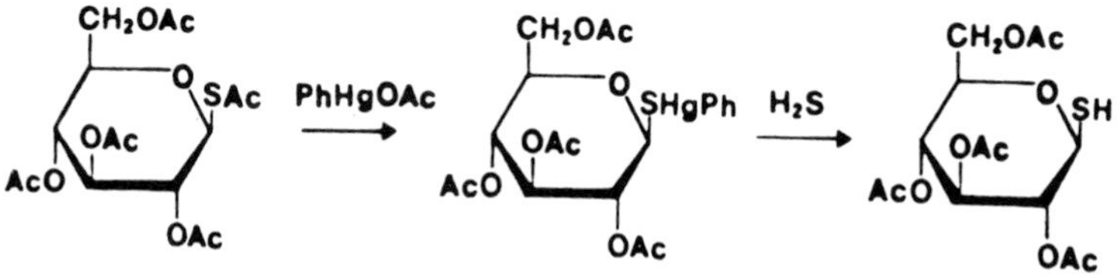

SCHEME 137

The thio-groups of such compounds offer potential for the preparation of other metal salts, and the gold derivative of 1-thio-β-D-glucopyranose is commonly used in the treatment of rheumatoid arthritis (D.H. Brown and W.E. Smith, Chem.Soc. Rev., 1980, 9, 217).

S-Alkylation of 1-thiosugars represents a common method for preparing 1-thioglycosides which are otherwise usually obtained by reactions involving displacements of anomeric leaving groups. A particularly convenient procedure uses the reaction undergone between 1,2-*trans*-related peresters and thiols in the presence of boron trifluoride etherate as catalyst (R.J. Ferrier and R.H. Furneaux, Carbohydrate Res., 1976, 52, 63), and because of participation by the ester group at $C_{(2)}$ this process gives 1,2-*trans*-thioglycosides as does the use of acylated glycosyl halides. A modification of this latter reaction involves the synthesis of 1,2-thio-orthoesters followed by their treatment with deactivated Raney nickel in the presence of the appropriate thiol, and this too gives 1,2-*trans*-related products (Scheme 138) (G. Magnusson, J.org.Chem., 1977, 42, 913).

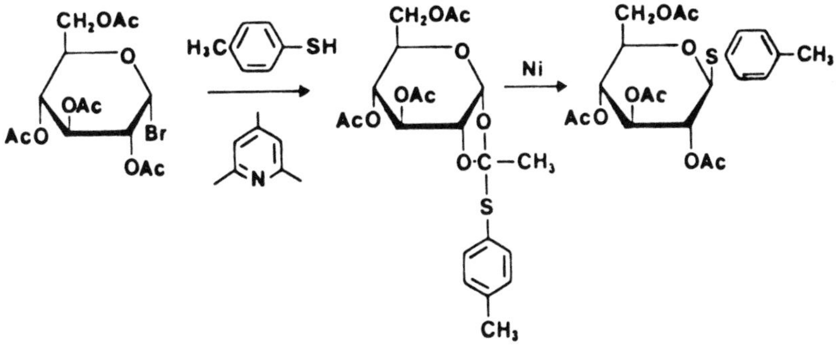

SCHEME 138

Use of alkyl or aryl tributylstannyl sulphides with acylated glycosyl halides in the presence of tin(IV) chloride also gives 1-thioglycosides in good yield, but often with similar proportions of α- and β-anomers (T. Ogawa and M. Matsui, Carbohydrate Res., 1977, 54, C17). Alternatively, 1-thio-α-D-glucopyranosides predominate as products following treatment of the per-*O*-trimethylsilyl ether of methyl α-D-glucopyranoside with trimethyl(alkyl- or aryl-thio)silanes in the presence of zinc iodide and tetrabutylammonium iodide (Scheme 139) (S. Hanessian and Y. Guindon, *ibid.*, 1980, 86, C3).

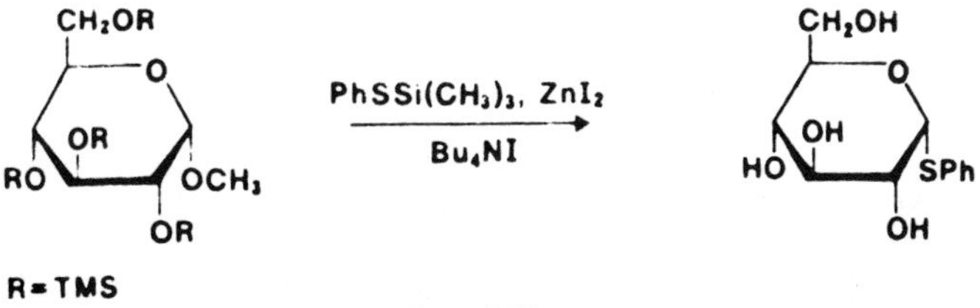

SCHEME 139

Otherwise, β-isomers can be anomerised to give 1-thio-α-D-glucopyranosides with boron trifluoride as catalyst (B.Erbing and B. Lindberg, Acta chem.Scand., 1976, B30, 611).

(2) *Reactions*. Acid-catalysed cleavage of the glycosidic bonds of 1-thioglycosides, as well as being slower than that of the corresponding *O*-glycosides, can be more complex, and the first products of aqueous acid treatment of methyl 1-thio-α-D-ribopyranoside are the anomer and the two 1-thiofuranosides. Protonation therefore must have occurred at the ring oxygen atom to give an acyclic $C_{(1)}$ carbonium ion (C.J. Clayton, N.A. Hughes and S.A. Saeed, J.chem.Soc.C, 1967, 644). The mechanisms of hydrolysis reactions are discussed by B. Capon (Chem.Rev., 1969, 69, 407).

1-Thioglycosides are subject to reaction with other electrophiles to give reactive sulphonium species which can undergo attack by nucleophiles at the anomeric carbon atoms. Thus, acylated glycosyl halides are obtainable by use of halogens, and 1-thiofuranosides therefore represent useful starting materials for making substituted furanosyl halides. However, further reaction can occur, as when a thio-group is also present at $C_{(2)}$ (Scheme 140) (D. Horton and M. Sakata, Carbohydrate Res., 1975, 39, 67).

SCHEME 140

Acylated alkyl and aryl 1-thio-β-D-glucopyranosides with dichloromethyl methyl ether and zinc chloride give the thermodynamically favoured α-glycosyl chloride esters (R. Bognar *et al.*, Acta chim.Acad.Sci.Hung., 1975, 84, 325).

Coordination between sulphur and metal ions [mercury(II) in particular] generates sulphonium species which are stabilised in the case of arylthio-derivatives, and new approaches to the synthesis of *O*-glycosides are based on this (p.116). 1-Thioglycosides are, therefore, subject to hydrolysis in the presence of metal ions, but mercury(II) chloride gives incomplete cleavage of glycosides bonded through sulphur atoms to polystyrene resins. However, these compounds are released satisfactorily when the glycosylated resins are treated in benzene with methyl iodide and benzyl alcohol, the former giving methylsulphonium complexes which are solvolysed to afford benzyl glycosides (S.-H. Lee Chui and L. Anderson, Carbohydrate Res., 1976, 50, 227).

In compounds containing suitably disposed leaving groups the sulphur atoms of 1-thioglycosides can effect intramolecular displacements to give sulphonium species which undergo nucleophilic attack at $C_{(1)}$ and lead to products with migrated thio-groups (Scheme 141).

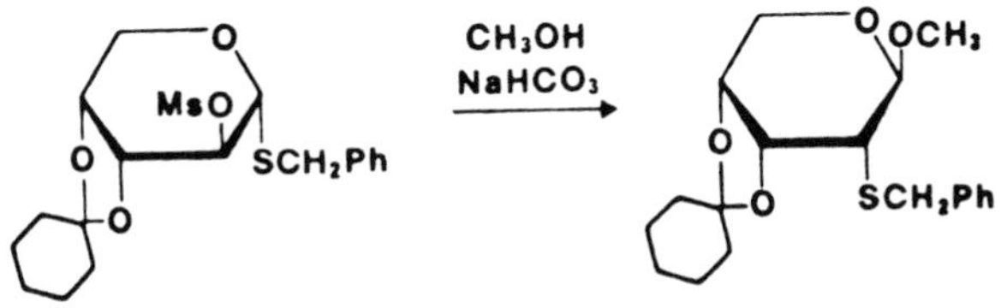

(K.J. Ryan, E.M. Acton, and L. Goodman, J.org.Chem., 1971, 36, 2646).

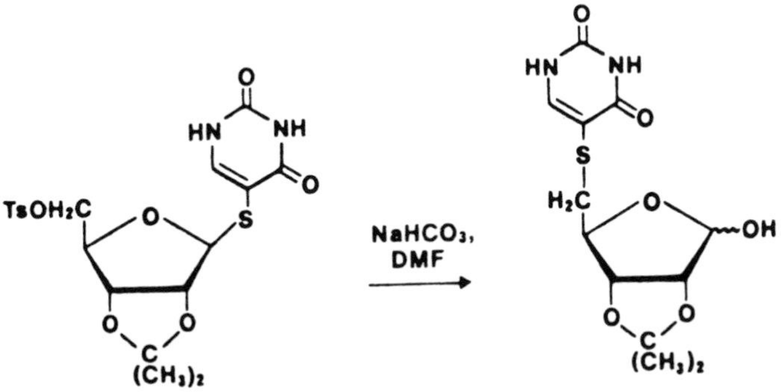

(E.J. Eustace *et al.*, J.heterocyclic Chem., 1974, 11, 1113).

SCHEME 141

An unusual reaction occurs with esters of phenyl 1-thioglycopyranosides on treatment with photochemically produced bromine radicals, and is presumably dependent upon initial hydrogen abstraction from $C_{(1)}$ to give sulphur-stabilised free radicals. While the reaction is specific in the case of benzoyl esters, with acetates some bromination also occurs within the ester groups at $C_{(2)}$ (Scheme 142) (R.J. Ferrier and R.H. Furneaux, J.chem.Soc.Perkin I, 1977, 1993).

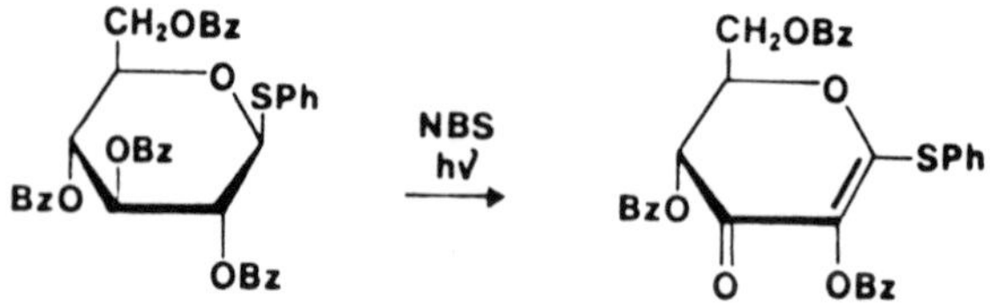

SCHEME 142

Oxidation of 1-thioglycosides with *m*-chloroperbenzoic acid gives diastereoisomeric sulphoxides, and with stronger reagents the corresponding sulphones. These latter substances are not suitable as glycosylating agents (R.J. Ferrier, R.H. Furneaux and P.C. Tyler, Carbohydrate Res., 1977, 58, 397). The phenylsulphones, on ultraviolet irradiation in benzene, undergo complex reactions and give a range of products derived from glycosylsulphonyl radicals which initially lose sulphur dioxide to give glycosyl radicals, and these then dimerise and react in various ways with the solvent (P.M. Collins and B.R. Whitton, J.chem.Soc.Perkin I, 1974, 1069).

(ii) Acyclic compounds

The chemistry of aldose dithioacetals has been reviewed (J.D. Wander and D. Horton, Adv.carbohydrate Chem.Biochem., 1976, 32, 15).

In similar manner to the way in which 1-thioglycosides react with bromine, acylated aldose dithioacetals give products with one thio-group replaced by bromine, and such compounds, when treated with appropriate derivatives of heterocyclic bases, give acyclic analogues of nucleosides (Scheme 143) (M.L. Wolfrom *et al.*, Carbohydrate Res., 1972, 23, 296).

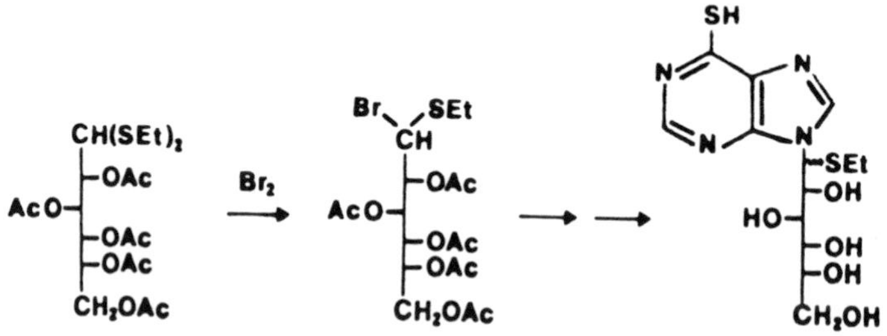

SCHEME 143

The acetal protons of dithioacetals are acidified because of the stabilisation by the sulphur atoms of the derived carbanions, and this can lead to specific elimination reactions in compounds with base-stable groups at $C_{(2)}$ (Scheme 144) (B. Berrang, D. Horton and J.D. Wander, J.org.Chem., 1973, 38, 187).

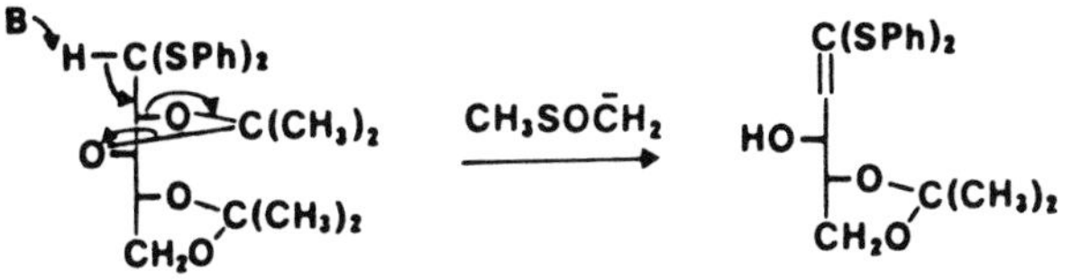

SCHEME 144

Sulphonyl groups stabilise carbanions at $C_{(1)}$ more effectively than do sulphur atoms, so that oxidation of aldose dithioacetals to the disulphonyl analogues is accompanied by loss of water to give 1,2-alkenes which are subject to cyclisations induced by attack of $O_{(5)}$ or $O_{(6)}$ at $C_{(2)}$ (A. Farrington and L. Hough, Carbohydrate Res., 1971, 16, 59). 2,5-Anhydrides are also produced during selective sulphonylation of some pentose dialkyl dithioacetals (p.138).

(b) Other thio-derivatives

(i) Synthesis

Newer methods for introducing sulphur into carbohydrate derivatives include treatment of primary tosyl esters with sodium ethyl xanthate which gives ethoxythiocarbonyl esters of thio-sugars; these cleave under basic conditions to give the thiols (W.M. Doane *et al.*, J.org.Chem., 1975, 40, 1337). Episulphides can be prepared by treatment of epoxides with potassium thiocyanate or thiourea and the products have the opposite configurations. Stereochemical factors inhibit the reaction in some cases, and alkenes are formed (M.V.Jesudason and L.N. Owen, J.chem.Soc.Perkin I, 1974, 2019). Such episulphides offer access to *trans*-related vicinal dithiols, and *cis*-isomers are obtainable from epoxides following their opening with sodium *N,N*-dimethyldithiocarbamate and tosylation of the resultant alcohols (Scheme 145) (S. Ishiguro and S. Tejima, Chem.pharm.Bull.Japan, 1968, 16, 1567).

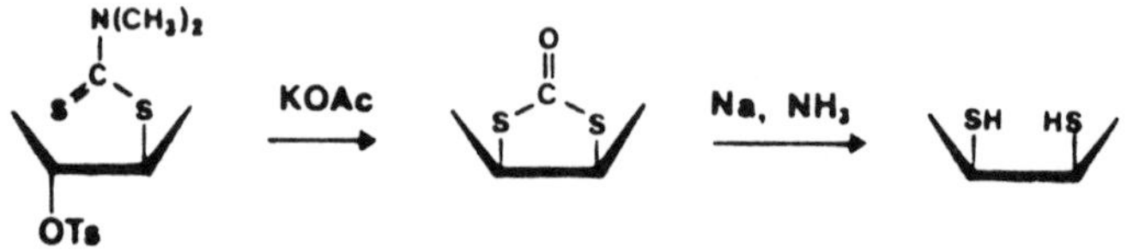

SCHEME 145

Sugars with sulphur as the ring hetero-atom have been examined thoroughly, and their synthesis has been reviewed by H. Paulsen and K. Todt (Adv.carbohydrate Chem., 1968, 23, 115). Their preparation requires the introduction of thiol groups in the 1,4-, 1,5- or 1,6-relationships to the anomeric centres, and furanoid, pyranoid and septanoid derivatives of aldoses, and furanoid and pyranoid ketose compounds with sulphur in the ring are known. Usually ring-opening reactions of appropriate epoxides or displacements of sulphonyloxy groups with such nucleophiles as thiocyanate, thioacetate or benzylthiolate ions are involved. For example, 5-thio-D-fructose is obtainable from an L-sorbose 5-sulphonate (Scheme 146) (M. Chmielewski and R.L. Whistler, J.org.Chem., 1975, 40, 639).

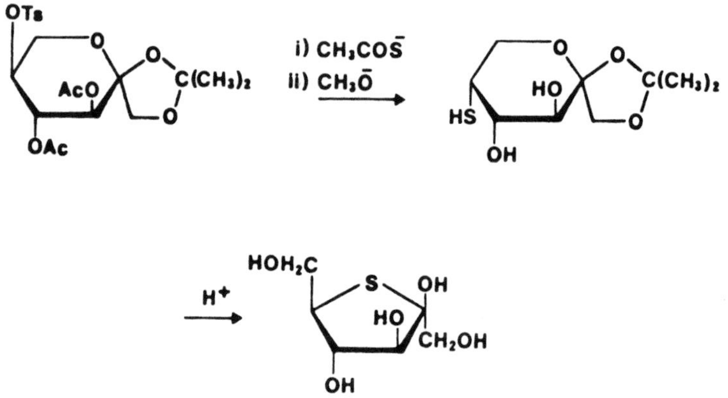

SCHEME 146

Earlier synthetic work is referred to in the Second Edition.

6-Thio-D-fructose can be made enzymically from the D-glucose analogue by use of D-glucose isomerase. Fortuitously, the product of the isomerisation is not a substrate for the enzyme, and the reaction is therefore irreversible leading to very efficient conversion (M. Chmielewski, M.-S. Chen and R.L. Whistler, Carbohydrate Res., 1976, 49, 479).

Improved methods for preparing 5-thio-D-glucose which is of considerable biochemical interest (p.190) are based on the use of 5,6-anhydro-3-*O*-benzyl-1,2-*O*-isopropylidene-β-L-idofuranose (U.G. Nayak and R.L. Whistler, J.org.Chem., 1969, 34, 97; see also C.-W. Chiu and R.L. Whistler, *ibid.*, 1973, 38, 832).

The finding of derivatives of sugars with sulphur in the rings amongst the products of thiolysis of aldose peresters is noted on p.192, and extension of the principles of using cyclic sulphonium ions gives an alternative synthetic method (Scheme 147) (J. Harness and N.A. Hughes, Chem.Comm., 1971, 811).

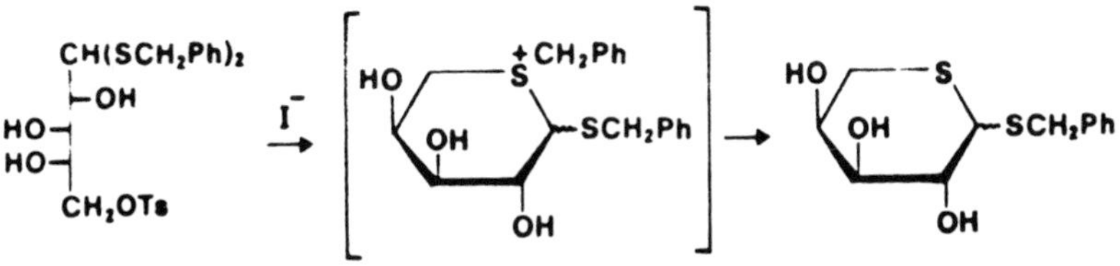

CH(SCH₂Ph)₂ / OH / HO / HO / CH₂OTs → I^- → [HO, +CH₂Ph S, OH, SCH₂Ph, OH] → HO, S, OH, SCH₂Ph, OH

SCHEME 147

An entirely different approach which results in racemic products uses the Pummerer rearrangement reactions (Scheme 148) (J.E. McCormick and R.S. McElhinney, J.chem.Soc.Perkin I, 1976, 2533).

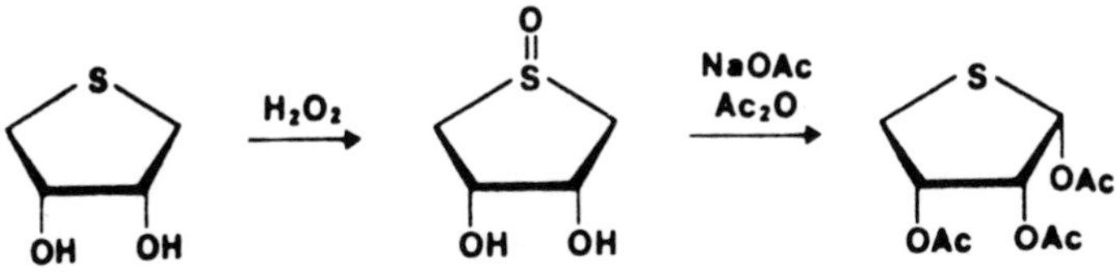

S, OH, OH → H_2O_2 → O=S, OH, OH → NaOAc, Ac_2O → S, OAc, OAc, OAc

SCHEME 148

(ii) Reactions

Thio-sugars which can form 5- and 6-membered sulphur-containing rings show a marked tendency to do so as evidenced, for example, by the simple mutarotation curves of both anomers of 5-thio-D-ribopyranose which suggest the absence of furanoid forms. This is in marked distinction to the behaviour of D-ribose itself. However, compounds such as 6-deoxy-4-thio-D-gulose which have sterically strained forms when ring closed through sulphur show little tendency to exist in these forms (R.-A. Boigegrain and B.E. Gross, Carbohydrate Res., 1975, 41, 135). Methanolysis of 5-thio-D-ribose gives initially the pyranosyl compounds which again distinguishes it from ribose (C.J. Clayton and N.A. Hughes, *ibid.*, 1967, 4, 32). Mutarotation of 5-thio-α-D-glucopyranose is faster than that of the non-thio analogue (M. Suzuki and R.L. Whistler, *ibid.*, 1972, 22, 473), and the rate of the acid-catalysed hydrolysis of glycosides is enhanced by factors of 10-20 by replacement of the ring oxygen atom by sulphur.

Whereas acetylation of 5-thio-D-glucose with cold acetic anhydride in pyridine gives just the pyranose peresters, treatment of the 4-thio-isomer with hot acetic anhydride and sodium acetate affords products containing 30% of the 4-*S*-acetyl pyranoid esters showing that the nucleophilicity of the thiol group renders it subject to substitution as well as inclined to take part in ring formation (L. Vegh and E. Hardeggar, Helv., 1973, 56, 2020).

5-Thio-D-ribose and -D-xylose react with acetone and 2,2-dimethoxypropane, respectively, under acidic conditions to give acetals of the 6-membered forms (53) and (54) (N.A.Hughes and C.J. Wood, Carbohydrate Res., 1976, 49, 225) showing again

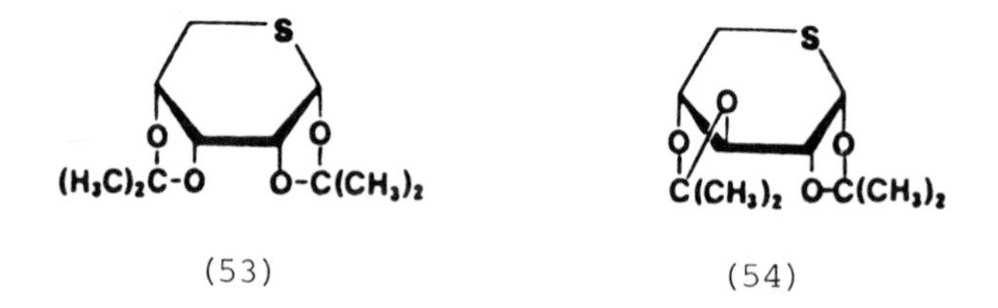

(53) (54)

the propensity of such compounds to react by way of the isomers with sulphur in the rings. D-Ribose and -xylose give large proportions of furanoid products under these conditions.

Diastereoisomeric sulphoxides, which can be configurationally characterised by N.M.R. methods, are obtained by controlled oxidation of sugars with sulphur in the ring, and these react further to afford common sulphones (Scheme 149) (C.J. Clayton, N.A. Hughes and T.D. Inch, *ibid.*, 1975, 45, 55).

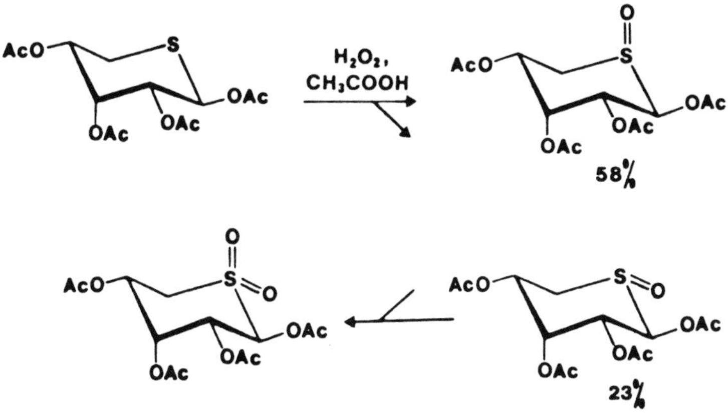

SCHEME 149

7. Unsaturated sugars

Unsaturated derivatives have represented a major area of development in monosaccharide chemistry because of their reactivity and consequent convertibility into a wide range of functionalised products, and also because several further natural products containing sugars of this group have been found. Examples from outside the nucleoside field are sisomycin (55) (H. Reimann *et al.*, J.org.Chem., 1974, 39, 1451) and virescenoside D (56) (N. Cagnoli-Bellavita *et al.*, J. chem.Soc.Perkin I, 1977, 351).

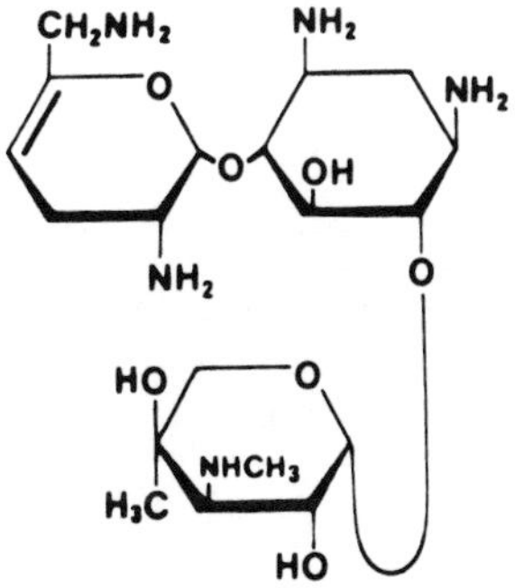

(55) 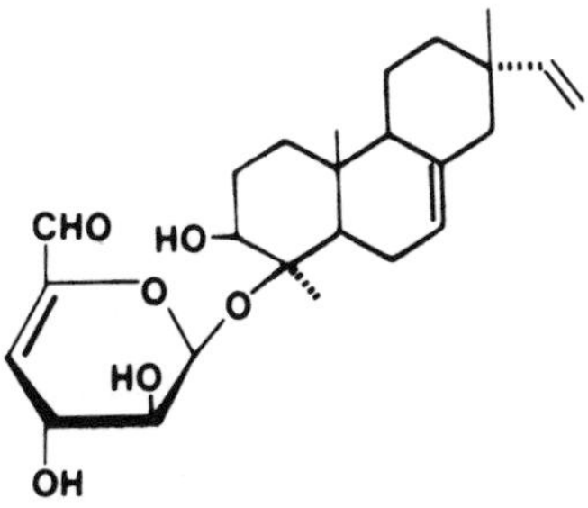 (56)

An updated comprehensive review has appeared (R.J. Ferrier, Adv.carbohydrate Chem.Biochem., 1969, 24, 199; see also *idem.*, in "The Carbohydrates", 2nd Edn., Vo.1B, eds. W. Pigman and D. Horton, Academic Press, New York, 1980, p.843). Other surveys have dealt with allylic displacement and rearrangement reactions (R.J. Ferrier, International Review of Science, Organic Chemistry, Ser. 2, Vol.7, ed. G.O. Aspinall, Butterworths, London, 1976, p.35), the substantial contributions of the Waterloo University group (B. Fraser-Reid, Acc.chem. Res., 1975, 8, 192) and the chemistry of enolones (F.W. Lichtenthaler, Pure and appl.Chem., 1978, 50, 1343).

(a) Glycals or 1,2-dideoxy-1-enoses and derivatives

(i) Preparation

The primary method of making glycals remains the treatment of acylated glycosyl halides with zinc in acetic acid (W.Roth and W. Pigman, in "Methods in Carbohydrate Chemistry", Vol.2, eds. M.L. Wolfrom and R.L. Whistler, Academic Press, New York, 1963, p.405), and it has been applied to the synthesis of tri-*O*-acetyl-D-allal (M. Haga and S. Tejima, Carbohydrate Res., 1974, 34, 214) amongst more common members of the class. While not generally applicable in furanoid systems, the reaction has been used with modest success with compounds having relatively poor leaving groups at $C_{(3)}$ which are less subject to changes involving allylic rearrangement (Scheme 150) (K.Bischofberger and R.H. Hall, *ibid.*, 1976, 52, 223).

SCHEME 150

Otherwise, furanoid or pyranoid glycals may be made by treating glycosyl halides having base-stable protecting groups with sodium naphthalide in tetrahydrofuran (S.J. Eitelman, R.H. Hall and A. Jordaan, J.chem.Soc.Perkin I, 1978, 595) or lithium in liquid ammonia (Scheme 151) (R.E. Ireland, C.S. Wilcox and S. Thaisrivongs, J.org.Chem., 1978, 43, 786; 1980, 45, 48).

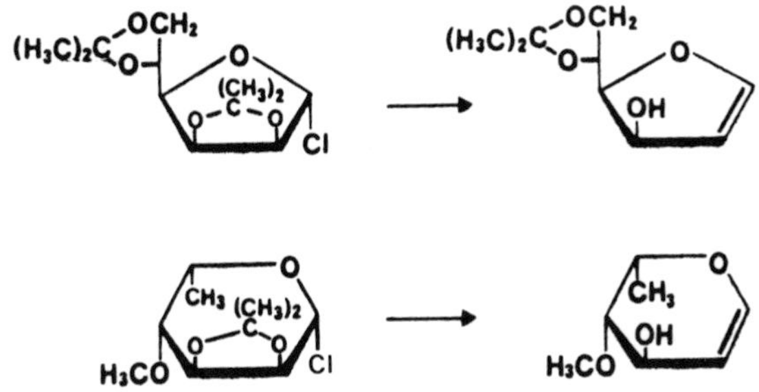

SCHEME 151

Methyl 2,3-anhydro-4,6-*O*-benzylidine-α-D-allopyranoside, treated with methyllithium containing lithium iodide, gives the 2-deoxy-2-iodo-D-altroside and then 4,6-*O*-benzylidine-D-allal and, likewise, 4,6-*O*-benzylidene-D-gulal (57) is obtainable from the anhydroguloside (58) (Scheme 152). Halide-free

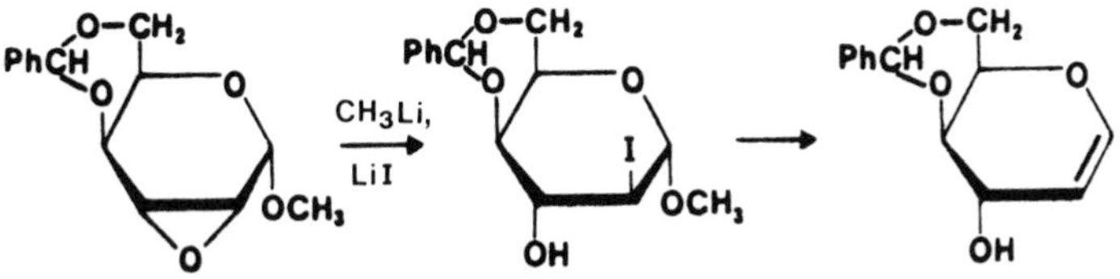

(58) SCHEME 152 (57)

methyllithium leads to epoxide ring opening by the carbon nucleophile and ultimate production of 2-methylglycal derivatives (Scheme 153) (R.U. Lemieux, E. Fraga and K.A. Watanabe, Canad.J.Chem., 1968, 46, 61; M. Sharma and R.K. Brown, *ibid.*, 1968, 46, 757).

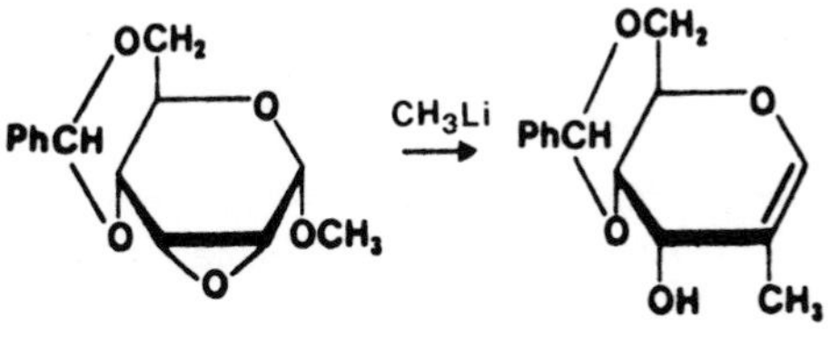

SCHEME 153

Treatment of acylated glycosyl halides with 1,5-diazabicyclo[5.4.0]undec-5-ene (DBU) in *N,N*-dimethylformamide offers an efficient means of removing hydrogen halide from acylated glycosyl halides to give esters of 2-hydroxyglycals (D.R. Rao and L.M. Lerner, Carbohydrate Res., 1972, 22, 345), but this reaction is confined to halides which are able to undergo *trans*-elimination of hydrogen halide. With halides in which these elements are *cis*-related elimination can be induced by the addition of halide ion to promote isomerisation to the more reactive anomer (Scheme 154) (N.A. Hughes, *ibid.*, 1972, 25, 242).

SCHEME 154

Applied to furanosyl analogues the above base gives furanosyl 2-hydroxyglycal esters which readily undergo isomerisation, elimination and nucleophilic displacement reactions (R.J. Ferrier and J.R. Hurford, *ibid.*, 1974, 38, 125), and in the ketose series an anomeric mixture of tetra-*O*-benzyl-D-fructofuranosyl chloride, on treatment with potassium hydroxide in *p*-dioxane, gives both the *endo*-cyclic and *exo*-cyclic products (Scheme 155) (E. Zissis, R.K. Ness and H.G. Fletcher, *ibid.*, 1971, 20, 9).

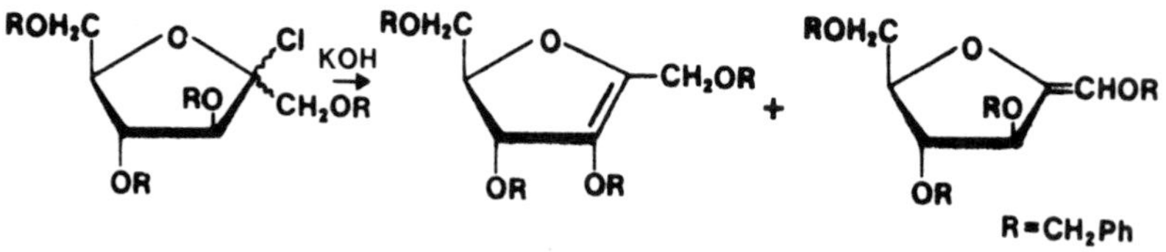

SCHEME 155

Dehydrohalogenation of 2-acetamido-3,4,6-tri-*O*-acetyl-2-deoxy-α-D-glucopyranosyl chloride with DBU gives the corresponding 2-acetamidoglycal derivative, and treatment with boiling isopropenyl acetate in the presence of catalytic amounts of toluene-*p*-sulphonic acid affords efficient means of obtaining the *N*-acetylacetamido analogue (Scheme 156) (N. Pravdić, I.Franjić-Mihalić and B.Danilov, *ibid.*, 1975, 45, 302).

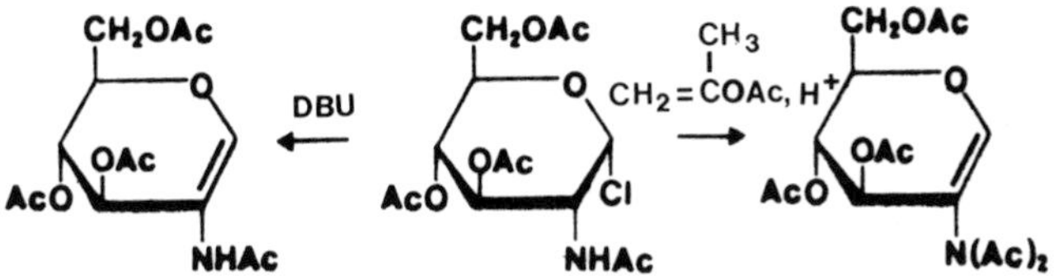

SCHEME 156

Analogous elimination of hydrogen bromide from acylated 2-bromo-2-deoxyglycosyl halides affords 2-bromoglycal esters (C. Pedersen *et al.*, Acta chem.Scand., 1977, B31, 768), and reaction of glycal derivatives not prone to undergo allylic rearrangements (i.e. those without ester groups at $C_{(3)}$ or with the *cis*-configuration at $C_{(3)}$, $C_{(4)}$) with chlorosulphonyl isocyanate, followed by triethylamine, gives 2-cyanoglycal esters (Scheme 157) (R.H. Hall and A. Jordaan, J.chem.Soc. Perkin I, 1973, 1059).

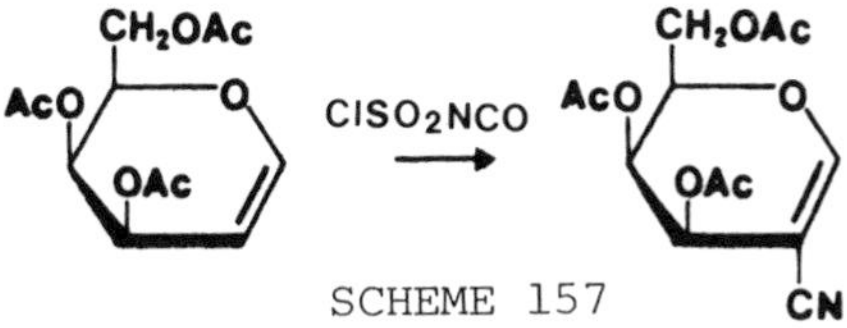

SCHEME 157

1-Substituted glycals are less common, but such a compound bearing a carbon substituent was obtained on treatment of 1,4, 5-tri-*O*-benzoyl-3-*O*-methanesulphonyl-β-D-fructopyranosyl bromide with sodium iodide in acetone (R.K. Ness and H.G.Fletcher, J.org.Chem., 1968, 33, 181), and nucleoside derivatives have been synthesised by base-catalysed elimination reactions applied to compounds with good leaving groups at $C_{(2)}$ (Scheme 158) (M.J. Robins and E.M. Trip, Tetrahedron Letters, 1974, 3369; M.J. Robins and R.A. Jones, J.org.Chem., 1974, 39, 113).

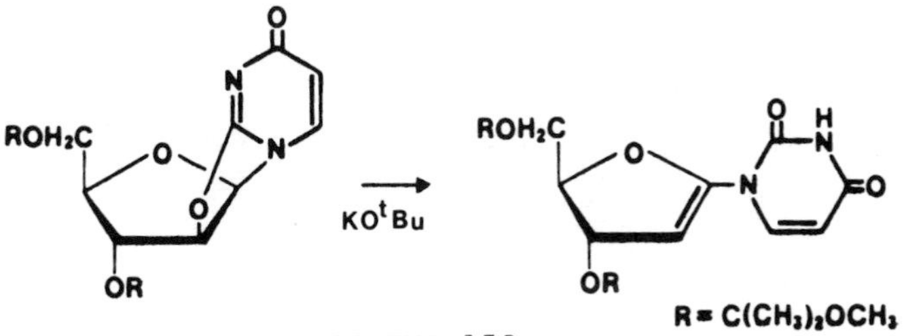

SCHEME 158

(ii) Properties and reactions

In the crystalline state tri-*O*-acetyl-D-glucal and its 2-acetoxy derivative exist in the $^4H_5$ conformation, (K. Vangehr, P. Luger and H. Paulsen, Carbohydrate Res., 1979, 70, 1), and the former also favours this ring shape in solution. The latter, however, which, in this form has a destabilising interaction between the vinylic substituent at $C_{(2)}$ and the *quasi*-equatorial ester group at $C_{(3)}$ (allylic effect) adopts the alternative half-chair conformation predominantly (59) (M. Rico and J. Santoro, Org.magn.Resonance, 1976, 8, 49). Similarly, di-*O*-acetyl-D-xylal preferentially adopts the $^5H_4$ ring form (60) in solution indicating that, in its case, the allylic effect is the determining factor (A.A. Chalmers and R.H. Hall, J.chem.Soc.Perkin II, 1974, 728).

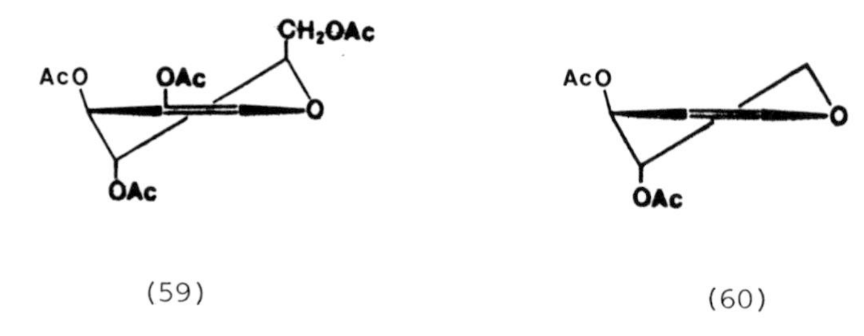

(59) (60)

In the mass spectrometer the main fragmentation pattern of glycal derivatives usually involves retro-Diels-Alder cleavage to give radical ions which abstract a hydrogen atom to afford resonance stabilised species (61) retaining the substituent at $C_{(3)}$ (G. Descotes *et al.*, Carbohydrate Res., 1979, 71, 305).

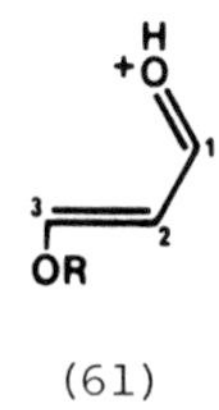

(61)

The enzymes β-D-glucosidase and β-D-galactosidase act on D-glucal and D-galactal, respectively, to give the corresponding 2-deoxyhexoses and, in the presence of glycerol, glyceryl 2-deoxy-β-D-*arabino*- and -*lyxo*-hexopyranoside (J. Lehmann and E. Schröter, *ibid.*, 1972, 23, 359). For sweet almond β-glucosidase and *Candida tropicalis* α-glucosidase, it was shown, by use of D_2O and high resolution ^1H N.M.R. techniques, that both enzymes cause *trans*-hydration of D-glucal to give the β- and α-adducts, respectively (E.J. Hehre *et al.*, Biochemistry, 1977, 16, 1780). In the case of the hydration of D-glucal catalysed by emulsin β-glucosidase, a dimeric compound (62) is also formed in substantial proportions. Rather

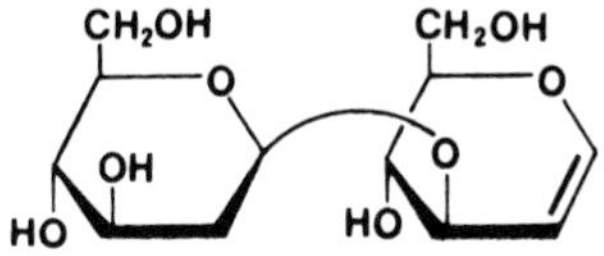

(62)

than being produced by a mechanism involving glycosylation of the initial substrate (i.e. addition of D-glucal across its double bond), the evidence suggests that the dimer is formed by specific substitution of 2-deoxy-β-D-*arabino*-hexose by the allylic cation formed from D-glucal (J. Lehmann and E. Schröter, Carbohydrate Res., 1977, 58, 65).

The stereochemistry of the chlorination of glycals is critically dependent on solvent, tri-*O*-acetyl-D-glucal giving almost exclusively the α-*gluco*- and β-*manno*-products of *cis*-addition in nonpolar solvents (with the former representing 85% of the products), while the main adducts in the polar solvent propylene carbonate are the *trans*-related β-*gluco*- and α-*manno*-dichlorides (formed in similar proportions) (K. Igarashi, T. Honma and T. Imagawa, J.org.Chem., 1970, 35, 610). For bromination the situation is different, and *trans*-addition occurs dominantly in non-polar solvents, while the main products formed in polar media have the α-configuration and result presumably following anomerisations. A further important variable is the substitution form of the glycal, tri-*O*-benzyl-D-glucal affording only small amounts of β-adducts with chlorine or bromine in all solvents. Chlorination in carbon tetrachloride gives the α-D-*gluco*-dichloro-adduct almost exclusively (P. Boullanger and G. Descotes, Carbohydrate Res., 1976, 51, 55).

The addition reaction undergone by glycal esters with nitrosyl chloride which affords dimeric *cis*-related products (p.112) has been extensively studied particularly by Lemieux and his associates (R.U. Lemieux, T.L. Nagabhushan and I.K. O'Neill, Canad.J.Chem., 1968, 46, 413), and the products have been used in important syntheses of α-D-glycosides, 2-amino-2-deoxyglycosides and the free amino sugars (R.U. Lemieux *et al.*, *ibid.*, 1968, 46, 397, 401, 405, 413; 1973, 51, 1, 7, 19, 27, 33, 42, 48, 53). The adducts dissociate in DMF to give monomers which rearrange to 2-oximo compounds, and these undergo dehydrochlorination to afford 2-nitrosoglycal intermediates, and it is these which react with alcohols in the critical step in the glycosidation processes. Dinitrogen tetroxide similarly adds to acetylated glycals to give corresponding dimeric 2-deoxy-2-nitroso-α-D-aldopyranosyl nitrates or acetylated 2-nitroglycals.

A modification of the method of preparation of 2-deoxy-2-halogenopyranosides involves treatment of glycal esters with *N*-halogenosuccinimide and equimolar proportions of alcohols, so that specific disaccharide derivatives and 2-deoxy-analogues can be synthesised in high yield by this method (K. Tatsuta *et al.*, Carbohydrate Res., 1977, 54, 85; J. Thiem, H. Karl and J. Schwentner, Synthesis, 1978, 696). This approach is preferable to the use of a modified alkoxymercuration procedure by which disaccharides with a 2-deoxyhexopyranosyl non-reducing moiety can be obtained, since the latter reaction gives poor stereoselectivity and the yields are not high (S. Honda *et al.*, Carbohydrate Res., 1973, 29, 477).

The vinyl ether system of tri-*O*-acetyl-D-glucal undergoes several photochemically induced addition reactions. Irradiation in acetone - propan-2-ol mixtures with a high-pressure mercury lamp gives good yields of an oxetane derivative (63) when the solvent contains high acetone proportions, but mainly a *C*-glycoside derivative (64) when the alcohol is the dominant component (Scheme 159) (K. Matsuura *et al.*, Carbohydrate Res., 1973, 29, 459). 1,3-Dioxolane also adds photochemically to tri-*O*-acetyl-D-glucal in the presence of acetone as sensitiser, but equal proportions of anomers are formed, and some of the monocyclic acetone adduct is also produced (Scheme 159) (*idem. et al.*, Bull.chem.Soc.Japan, 1973, 46, 2538). Radical induced addition of thiolacetic acid occurs to give products with acetylthio-groups at $C_{(2)}$, the major having the D-*manno*-configuration (K. Igarashi and T. Honma, J.org.Chem., 1970, 35, 606).

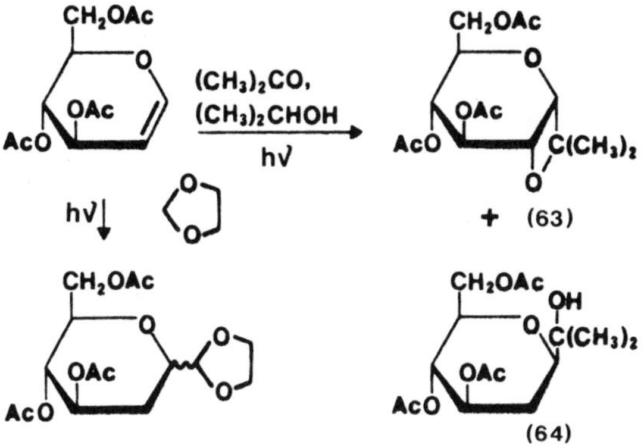

SCHEME 159

Addition of chlorine to tri-*O*-benzoyl-2-benzoyloxy-D-glucal occurs, when the reaction is conducted in toluene, in *cis*-fashion to give the β-*gluco*- and α-*manno*-adducts in the ratio 4:1, but with poorer selectivity in more polar solvents. Bromination, in this case, occurs similarly which suggests the intermediacy of 2-halo-1,2-benzoxonium ions (F.W. Lichtenthaler, T. Sakakibara and E. Oeser, Carbohydrate Res., 1977, 59, 47). When chlorination is carried out at -30°C, again the α-*manno*-dichloride is obtained, but the main product is the benzoxonium ion salt (65) which gives the β-*gluco*-chloride (66) on heating and readily and specifically hydrolyses to the α-2-ulose tetrabenzoate (67). This loses benzoic acid on treatment with mild base to give the corresponding enolone (68) which can be prepared, in this way, in 65% yield from the glycal derivative. With hydrogen bromide, the ulose ester gives the unsaturated α-bromide (69) and, from this, β-glycosides (70) can be obtained (Scheme 160) (F.W. Lichtenthaler and U. Kraska, *ibid.*, 1977, 58, 363).

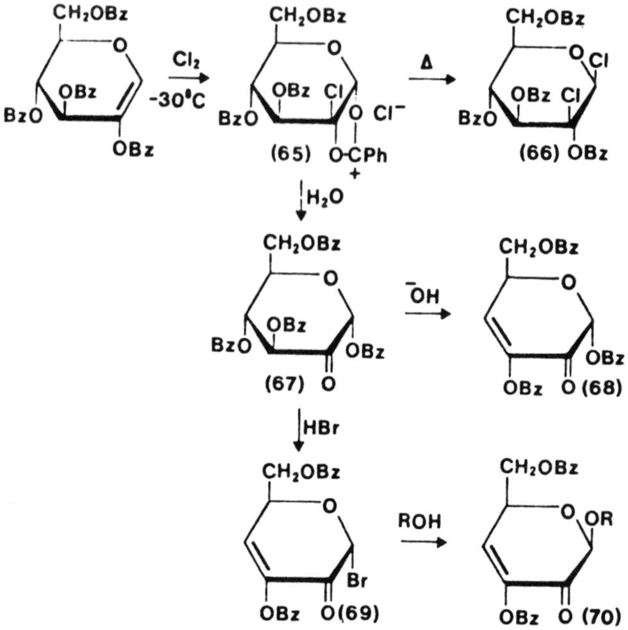

SCHEME 160

Lichtenthaler has reviewed the chemistry of sugar enolones (Pure appl.Chem., 1978, 50, 1343) pointing out that treatment of 2-acetoxy-tri-*O*-acetyl-D-glucal with *m*-chloroperbenzoic acid gives the β-2-ulose tetra-acetate which is a source of a β-enolone derivative (Scheme 161).

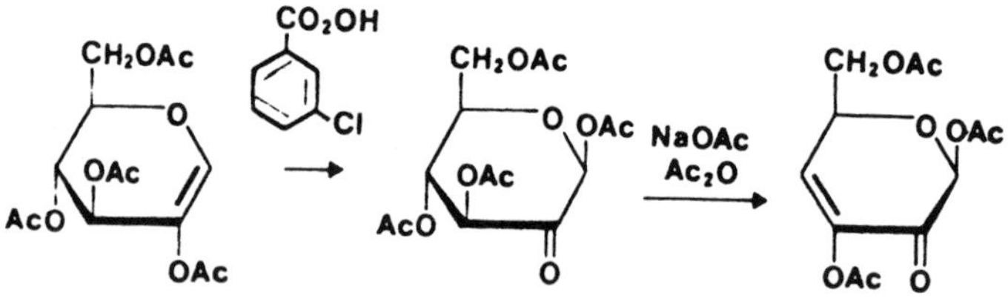

SCHEME 161

Photochemical addition of formamide to 2-acetoxy-tri-*O*-acetyl-D-glucal in the presence of acetone as sensitiser affords mainly the α-D-*gluco*-amide (but also some of the α-

D-*manno*-isomer and a *C*-glycosidic product of addition of acetone) (Scheme 162) (A. Rosenthal and M. Ratcliffe, Canad. J.Chem., 1976, 54, 91).

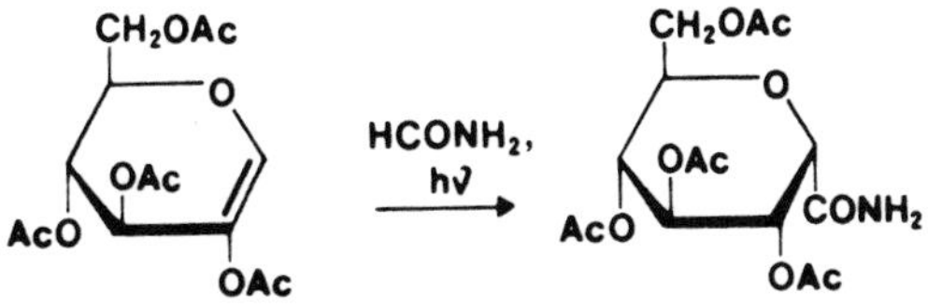

SCHEME 162

Oxidation of D-glucal by use of platinum oxide or, preferably, silver carbonate on Celite (p. 98) affords the corresponding conjugated enone amongst the products, and the 4,6-*O*-isopropylidene acetal, prepared by use of 2,2-dimethoxypropane in DMF in the presence of a little toluene-*p*-sulphonic acid, undergoes smooth allylic oxidation on treatment with manganese dioxide in chloroform solution (B. Fraser-Reid *et al.*, Canad. J.Chem., 1973, 51, 3950). However, this reagent, amongst several, is unsuitable for the oxidation of 4,6-*O*-benzylidene-D-allal from which the enone is obtainable satisfactorily by use of chromium trioxide in pyridine (P.M. Collins, Carbohydrate Res., 1969, 11, 125).

D-Glucal can be converted into the tribenzyl ether by use of benzyl chloride in DMSO in the presence of sodium hydride, and selective introduction of benzoyl and toluene-*p*-sulphonyl ester and t-butyldimethylsilyl ether groups at the primary centre can be achieved. Dibenzoylation affords the 3,6-diester in good yield. From these partially substituted derivatives a wide range of D-glucal based compounds [e.g. (71)-(73)] can be obtained (I.D. Blackburne, P.M. Fredericks and R.D.Guthrie, Austral.J.Chem., 1976, 29, 381).

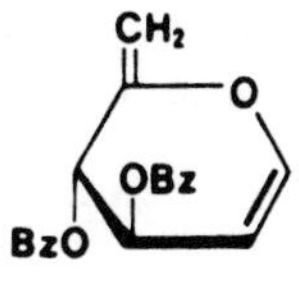

(71) 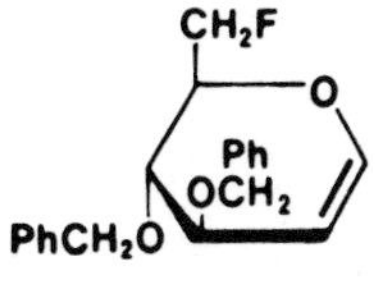 (72) 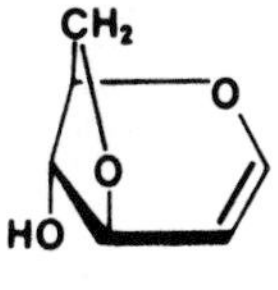 (73)

The allylic rearrangement reactions in which certain glycal derivatives take part to give 2,3-unsaturated products with nucleophile bonded to $C_{(1)}$ have been greatly extended, and an α-glycosylating procedure follows from the observation that tri-*O*-acetyl-D-glucal reacts with equimolar proportions of alcohols in inert solvents and in the presence of Lewis acid catalysts - particularly boron trifluoride - to give 2,3-unsaturated glycosides (R.J. Ferrier and N. Prasad, J.chem.Soc. (C), 1969, 570). The method has the advantage of being mild and of giving good yields of products with high α:β ratios (*ca.* 10:1). Many unsaturated disaccharides have been produced in this way, and a pseudotrisaccharide derivative (74) of the aminoglycoside antibiotic class has been synthesised directly in 73.5% yield by simultaneous double glycosylation of a 2,5-dideoxystreptamine compound (A. Canas-Rodriguez and A. Martinez-Tobed, Carbohydrate Res., 1979, 68, 43).

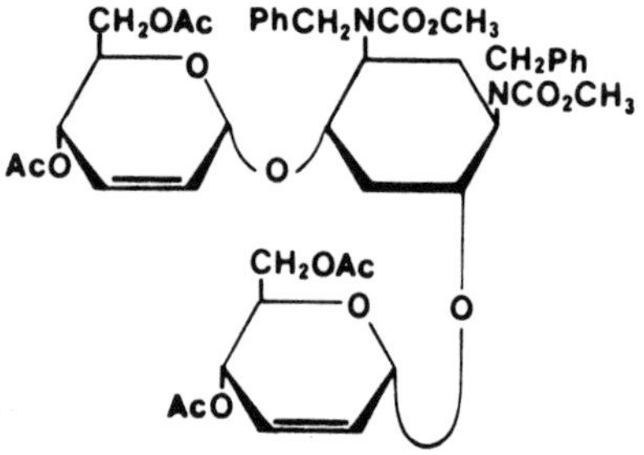

(74)

In the absence of alcohols, tri-*O*-acetyl-D-glucal, on treatment with Lewis acids undergoes isomerisation to give a 2,3-unsaturated glycosyl acetate, probably by way of a resonance stabilised allylic cation (p.209), and it is this species which reacts with nucleophiles such as alcohols. All nucleophiles, however, do not give 2,3-unsaturated products exclusively, thiols affording mainly 3-thioglycal derivatives, and carbanions also giving products of this general type, while nitrogen nucleophiles give mixtures. A generalisation has been proposed that soft bases lead to 3-substituted compounds, and that the harder oxygen, fluorine, chlorine, phosphorus (and, to some extent nitrogen) attack the harder $C_{(1)}$ centre (W. Priebe and A. Zamojski, Tetrahedron, 1980, 36, 287). Such generalisations are however hazardous because many of the initially formed products can, like tri-*O*-acetyl-D-glucal itself, undergo allylic rearrangement. Treated with the ambident nucleophiles sodium azide, potassium thiocyanate or potassium *O*-ethyl dithiocarbonate in acetonitrile in the presence of boron trifluoride etherate, tri-*O*-acetyl-D-glucal initially affords the products of allylic rearrangement, but these isomerise thermally (partially for azides, and completely for the sulphur-containing compounds) and revert to 3-substituted glycals (Scheme 163) (K. Heyns and R. Hohlweg, Ber., 1978, 111, 1632). Like azide ion, benzotriazole and purine derivatives

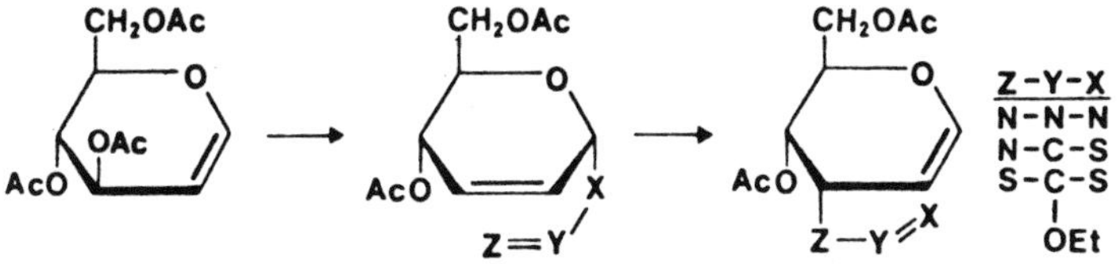

SCHEME 163

give mixed products with the bases bonded to $C_{(1)}$ or $C_{(3)}$, the former being formed under kinetic control and the latter by rearrangement processes. Acid also catalyses further additions to give saturated 2-deoxy-products with bases bonded at both these carbon atoms (F.G. de las Heras and M. Stud, Tetrahedron, 1977, 33, 1513).

The Claisen rearrangement of ethers is an analogous [3,3] sigmatropic isomerisation process, and has been effected to give glycal derivatives with branched-chains at $C_{(3)}$ and, in the reverse direction, 4,5-unsaturated 3,7-anhydro-octoses (Scheme 164) (Heyns and Hohlweg, *loc.cit.*).

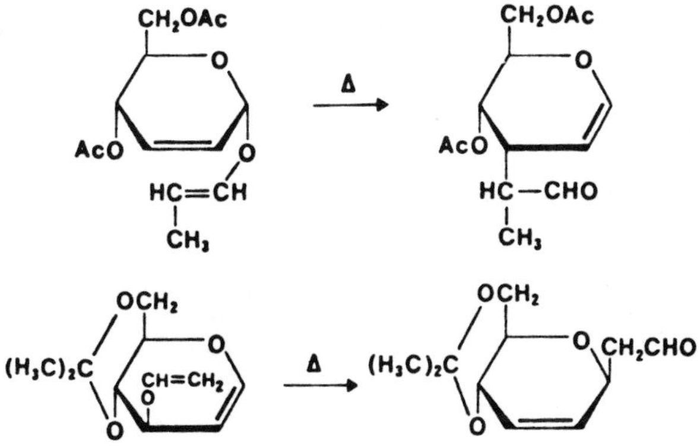

SCHEME 164

Although the above glycosylation reaction which is based on the rearrangement of glycal esters may only proceed when there is an allylic ester group at $C_{(3)}$ and also when there is a *trans*-related ester group at $C_{(4)}$, neither factor is absolutely required. Tri-*O*-benzyl-D-glucal isomerises to benzyl 4,6-di-*O*-benzyl-2,3-dideoxy-α,β-D-*erythro*-hex-2-enopyranoside in the presence of boron trifluoride and reacts with methanol to give the corresponding glycosides (G. Descotes and J.-C. Martin, Carbohydrate Res., 1977, 56, 168); tri-*O*-acetyl-D-galactal [*cis*-related groups at $C_{(3)}$-$C_{(4)}$] can be converted efficiently into 2,3-unsaturated α-glycoside derivatives by use of tin(IV) chloride as catalyst (G. Grynkiewicz, W.Priebe and A. Zamojski, *ibid.*, 1979, 68, 33).

After isomerisation, tri-*O*-acetyl-D-glucal undergoes dimerisation on treatment in inert solvents with boron trifluoride, and the *C,C*-linked product (75) is formed. Probably this arises by the *trans*-addition of one molecule of the first product (1,4,6-tri-*O*-acetyl-2,3-dideoxy-D-*erythro*-hex-2-enopyranose) to the double bond of a second molecule (R.J.Ferrier and N. Prasad, J.chem.Soc.(C), 1969, 581). This dimer and related compounds have therefore appeared amongst the products of several reactions which have involved the use of glycal derivatives under acidic conditions.

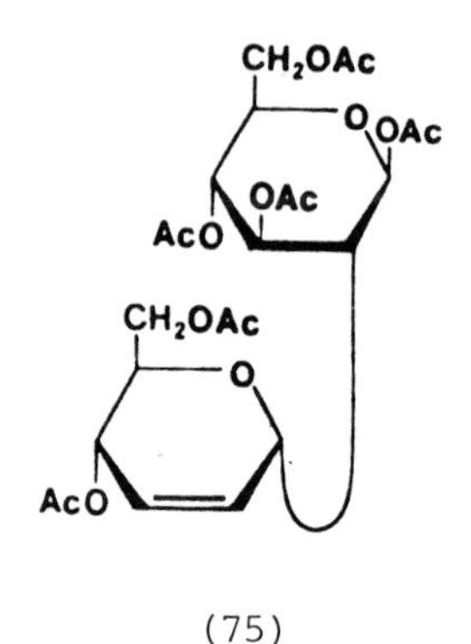

(75)

Heating of tri-*O*-acetyl-D-glucal under reflux in aqueous dioxane initially causes the expected allylic rearrangement to the 2,3-unsaturated pyranose which, however, then isomerises to give the acyclic *trans*-related aldehyde (76) as main product. This latter process is photochemically induced, and this aldehyde may have been responsible for the reducing properties in samples of "glucal" which led Emil Fischer to introduce this anomalous name. (B. Fraser-Reid and B. Radatus, J.Amer.chem.Soc., 1970, 92, 5288). This same product is also obtained quantitatively when tri-*O*-acetyl-D-glucal is dissolved in dioxane and dilute sulphuric acid in the presence of mercury(II) sulphate, the reaction occurring by an addition, elimination pathway (Scheme 165) (F. Gonzalez, S. Lesage and A.S. Perlin, Carbohydrate Res., 1975, 42, 267).

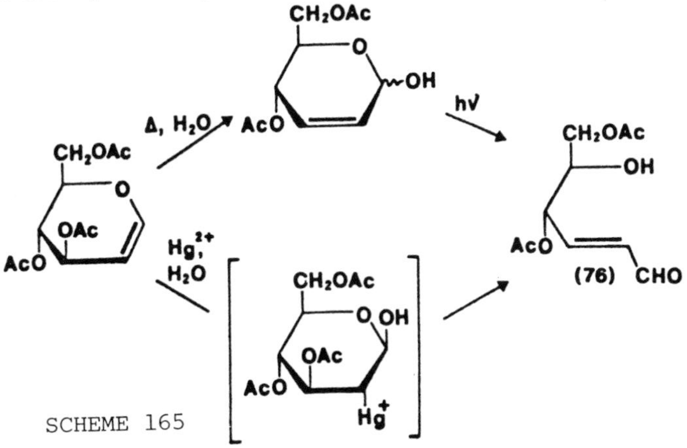

SCHEME 165

In the 2-hydroxyglucal series, esters rearrange thermally to give 2,3-unsaturated isomers with retained stereochemistry and, in the presence of boron trifluoride, to equilibrated anomeric mixtures containing mainly the α-anomers (Scheme 166) (R.J. Ferrier, N. Prasad and G.H. Sankey, J.chem.Soc.(C), 1968, 974).

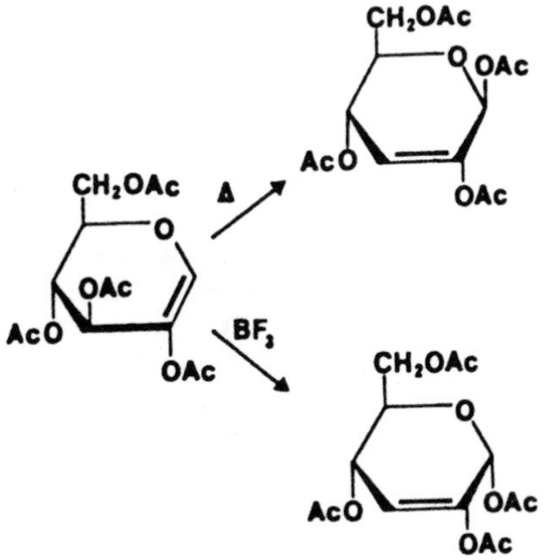

SCHEME 166

As with tri-*O*-acetyl-D-glucal, the 2-acetoxy analogue can be used in glycosylation reactions, and with equimolar proportions of alcohols and catalytic boron trifluoride, it gives mainly α-anomers. These products, which can be formed in high yield, on hydrogenation, give access to 3-deoxy-α-D-*ribo*-hexopyranosides [e.g. the aminoglycoside antibiotic analogue (77)] (S.D. Gero *et al.*, J.Amer.chem.Soc., 1980, 102, 857).

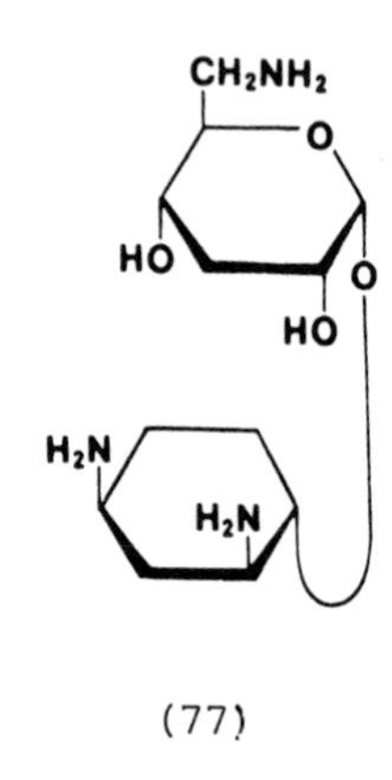

(77)

(b) Other unsaturated monosaccharides

(i) Preparation

The Tipson-Cohen reaction (Carbohydrate Res., 1965, 1, 338) by which vicinally related disulphonates are converted to corresponding alkenes by treatment with sodium iodide and zinc in DMF is strongly influenced by steric factors including, for 2,3-diesters of pyranosyl compounds, the configuration at the anomeric centre. Contrary to an earlier report, methyl 4,6-*O*-benzylidene-2,3-di-*O*-toluene-*p*-sulphonyl-β-D-glucoside is particularly readily and efficiently converted into the alkene by way of the 3-deoxy-3-iodo-D-*allo*-intermediate which undergoes metal-induced *cis*-elimination (Scheme 167)(T. Yamazaki *et al.*, J.chem.Soc.Perkin I, 1977, 1981).

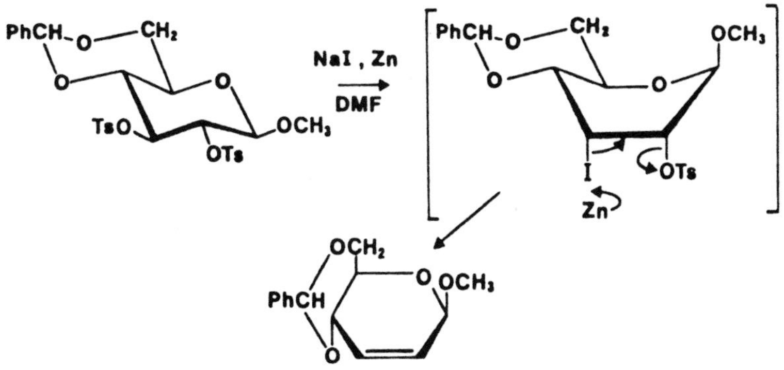

SCHEME 167

Methanesulphonates react less readily, and α-anomers, although they afford alkenes with moderate efficiency, also react more slowly because of the interaction between the iodide nucleophile and the aglycone, and because of unfavourable dipolar interaction between the sulphonyloxy group and the anomeric bond in the transition state for elimination. The method has found wide use, for example in the synthesis of 3',4'-dideoxykanamycin B, an antibiotic analogue, which is obtained by hydrogenation of a derivative containing a 3,4-unsaturated pyranoid ring prepared from a 3,4-disulphonylated precursor (S. Umezawa *et al.*, Bull.chem.Soc.Japan, 1972, 45, 3624). However, application of the Tipson-Cohen procedure to 3,4-benzylsulphonyl esters of certain 2-aminosugar derivatives causes additional reactions, and the required eliminations are effected only by repeating the sodium iodide treatment in the absence of zinc (T. Miyake *et al.*, Carbohydrate Res., 1976, 49, 141). Normally it is mandatory that the zinc be present, and the use of a zinc-copper couple can be advantageous (B. Fraser-Reid and B. Boctor, Canad.J.Chem., 1969, 47, 393). Both *cis*- and *trans*-vicinal sulphonates undergo elimination under standard conditions for the reaction (B. Fraser-Reid and N.L. Holder, Chem. Comm., 1972, 31).

A modified procedure for preparing alkenes involves the ring opening of epoxides by treatment with sodium iodide, sodium acetate and acetic acid in refluxing acetone to give iodohydrins which, on treatment with sulphonyl chlorides in refluxing pyridine, undergo the required eliminations (R.U. Lemieux, E. Fraga and K.A. Watanabe, Canad.J.Chem., 1968, 46,

61). In the furanose series 2,3-unsaturated compounds have been made by further modification of this procedure, intermediate iodotosylates being treated with tetraethylammonium chloride in pyridine in the presence of zinc dust (R.U. Lemieux, K.A. Watanabe and A.A. Pavia, *ibid.*, 1969, 47, 4413).

A convenient and efficient method applicable to both epoxides and vicinal disulphonates involves treatment with potasium selenocyanate (T. van Es, Carbohydrate Res., 1967, 5, 282).

Epimino-derivatives undergo elimination by way of *N*-nitroso analogues to give alkenes on treatment with sodium nitrite in aqueous acetic acid (R.D. Guthrie and D. King, Carbohydrate Res., 1966, 3, 128), *spiro*-compounds giving exocyclic alkenes (Scheme 168) (J.-M. Bourgeois, Helv., 1974, 57, 2553).

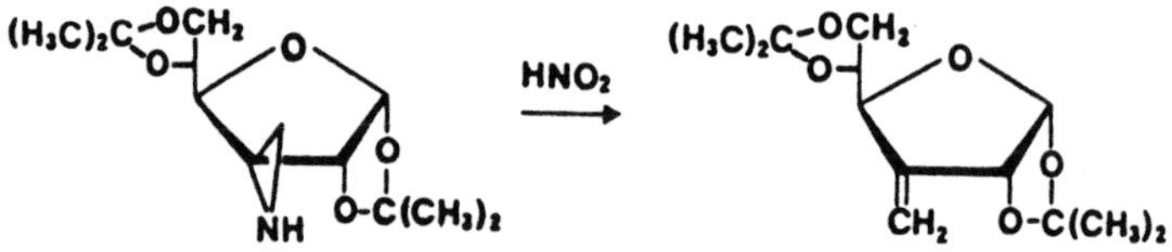

SCHEME 168

Oxidation of *N*-substituted epimines with *m*-chloroperbenzoic acid also results in alkene formation (H.W. Heine, J.D.Myers and E.T. Peltzer, Angew.Chem.internat.Edn., 1970, 9, 374).

An attractive method of preparing alkenes directly from vicinal diols involves refluxing in toluene in the presence of triphenylphosphine, tri-iodoimidazole and imidazole. Yields are good for both diequatorial and diaxial diols, but with *cis*-related systems excess of iodide as tetrabutylammonium iodide should be added (P.J. Garegg and B. Samuelsson, Synthesis, 1979, 813).

Treatment of bisdithiocarbonates of vicinal diols with tri-*n*-butylstannane in refluxing toluene gives a mild general method for obtaining alkenes (A.G.M. Barrett, D.H.R. Barton and R. Bielski, J.chem.Soc.Perkin I, 1979, 2378). A further approach involves the conversion of α-diols to cyclic 1-dimethylamino(methylene)acetals by treatment with excess of DMF dimethylacetal in dichloromethane (which restricts the method to *cis*-diols of cyclic systems or any acyclic diols), and reaction of the products with methyl iodide in toluene. This causes the precipitation of the corresponding trimethylammonium salts which, on heating under reflux, readily decompose to give the alkenes in near quantitative yield (S. Hanessian, A. Bargiotti and M. LaRue, Tetrahedron Letters, 1978, 737).

Other reactions to have been applied efficiently include the Cope and Hofmann eliminations of 4-aminosugar *N*-oxides and quaternary ammonium salts to afford 4-deoxypyranosyl-4-enes (C.L. Stevens and D. Chitharanjan, J.org.Chem., 1975, 40, 2474), and the base-catalysed (*n*-butyllithium) elimination of methanol from methyl 4,6-*O*-benzylidene-2,3-di-*O*-methyl-α-D-hexopyranosides to give the 3-deoxy-2-ene. Although the *gluco*, *altro*- and *allo*-isomers all gave this product, the third of these, which loses diaxially related hydrogen and methoxyl groups reacts most efficiently (Scheme 169) (A. Klemer and G. Rodemeyer, Ber., 1975, 108, 1896).

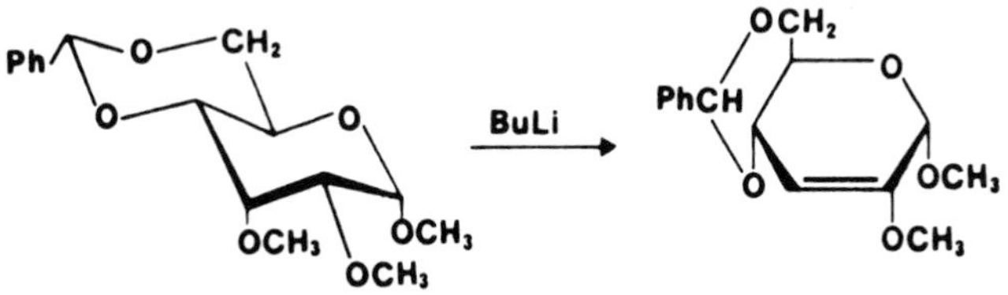

SCHEME 169

Eliminations occur very much more readily with compounds having leaving groups β-related to carbonyl functions, and J. Kiss has reviewed the many such eliminations which have been recorded for uronic acid derivatives (Adv.carbohydrate Chem.Biochem., 1974, 29, 229). Oxidations of methyl 2,3,4-tri-*O*-acetyl-α-D-glucopyranoside and 2,3,4,6-tetra-*O*-acetyl-D-glucose with sulphur trioxide, pyridine in DMSO and triethylamine give the 6-aldehyde and the lactone, respectively, both of which undergo elimination of acetic acid, but a more complex set of reactions occurs with 1,3,4,6-tetra-*O*-acetyl-α-D-glucose from which di-*O*-acetylkojic acid (78) is produced (Scheme 170) (D.M. Mackie and A.S. Perlin, Carbohydrate Res., 1972, 24, 67).

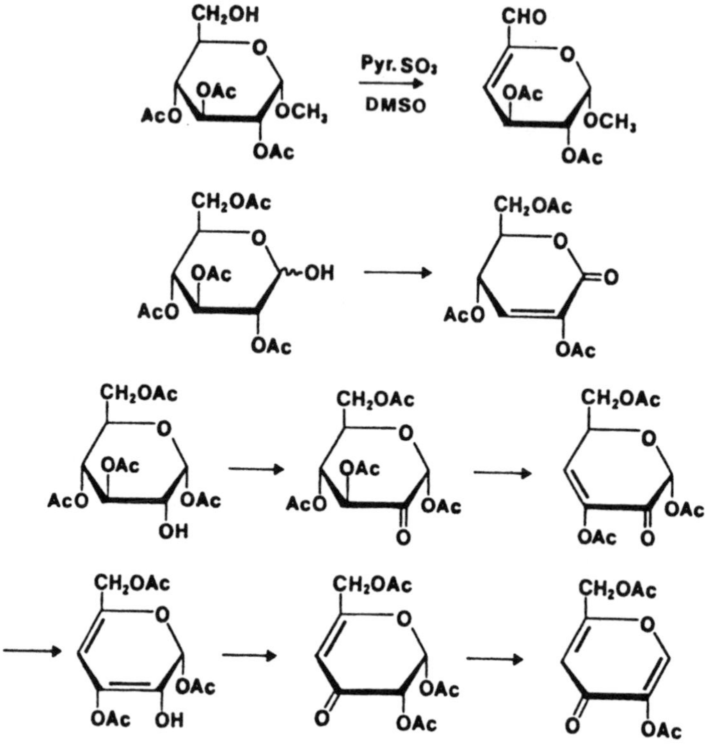

(78)

SCHEME 170

A wide range of syntheses based on the reaction of nucleophilic carbon reagents with carbonyl compounds have been applied to give both extended-chain and branched-chain unsaturated compounds (Scheme 171). The application of the Wittig reaction in carbohydrate chemistry has been reviewed (Yu. A. Zhdanov, Yu. E. Alexeev and V.G. Alexeeva, Adv. carbohydrate Chem.Biochem., 1972, 27, 227).

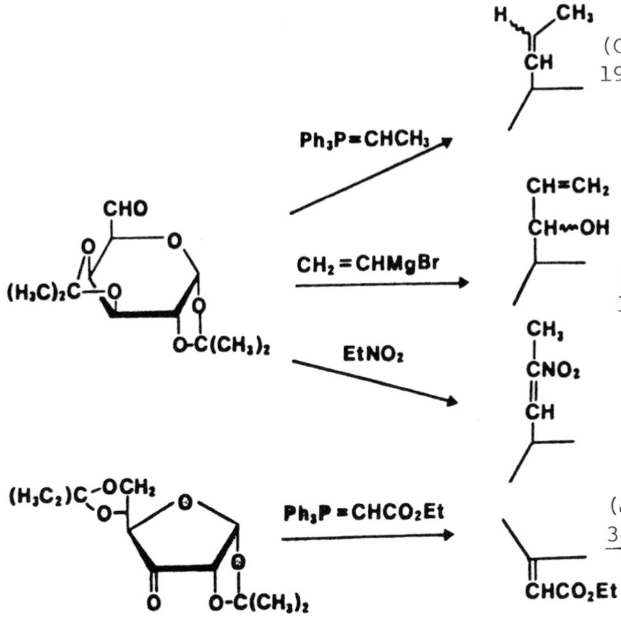

(Carbohydrate Res., 1969, 10, 306)

(Canad.J.Chem., 1969, 47, 75)

(J.org.Chem., 1969, 34, 1029)

SCHEME 171

(ii) Properties and reactions

Functionalisation by additions to alkenes has led to a wide range of products of value in synthetic work. Bromine, for example, adds to the double bond of methyl 4,6-*O*-benzylidene-2,3-dideoxy-α-D-*erythro*-hex-2-enopyranoside to give 70% of the crystalline D-*altro*-product and, likewise, acetyl hypobromite gives the 2-*O*-acetyl-3-bromo-3-deoxy-D-*altro*-adduct in high yield (together with small proportions of the 3-*O*-acetyl-2-bromo-2-deoxy-isomer), showing that the bromonium ion was formed on the α-face of the ring. Consistent with this, the alkene reacts to give the *allo*-cyclopropyl adduct when treated with methylene carbene (Simmons-Smith reaction), but in this case the yield is low (25%) (E.L. Albano, D. Horton, and J.H. Lauterbach, Carbohydrate Res., 1969, 9, 149).

cis-Hydroxylation of 2,3-dideoxy-α-D-*erythro*-hex-2-enopyranosides with hydrogen peroxide and osmium tetroxide or with neutral permanganate leads to attack from the upper face i.e. to D-*manno*-products (R.J. Ferrier and N. Prasad, J.chem.Soc.

(C), 1969, 575), and likewise, *cis*-oxyamination with Chloramine T and osmium tetroxide (Sharpless reagent) affords mainly D-*manno*-adducts, but in this case, two products are obtained (p.157). Since *N*-detosylation can be effected by treatment with sodium in liquid ammonia this addition reaction can be used in the synthesis of aminosugar derivatives.

Epoxidation of alkyl 2,3-dideoxy-α-D-*erythro*-hex-2-enopyranoside esters with hydrogen peroxide and benzonitrile similarly gives *manno*-products predominantly, but the *manno*-, *allo*-ratio is only about 3:2 with simple aglycones; in the case of the t-butyl glycoside diacetate the ratio is increased to *ca.*, 4:1. However, in keeping with expectations from alicyclic chemistry the hydroxyl group at C-4 of unesterified compounds directs the incoming oxygen atom to enter from the *cis*-direction; the products from methyl 2,3-dideoxy-α-D-*erythro*-hex-2-enopyranoside are the D-*manno*- and D-*allo*-epoxides in the ratio *ca.* 1:3 (R.J. Ferrier and N. Prasad, *loc.cit.*).

Compounds with mixed functionality and with heterocyclic rings can also be produced by addition processes (Scheme 172 a-c).

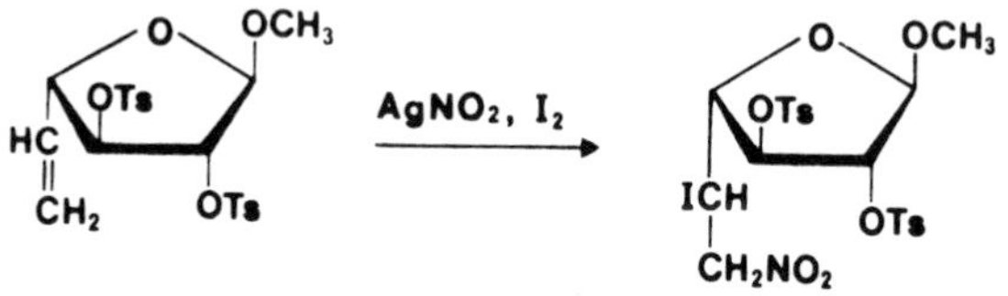

(J.K.N. Jones *et al.*, Carbohydrate Res., 1972, 22, 163)

SCHEME 172a

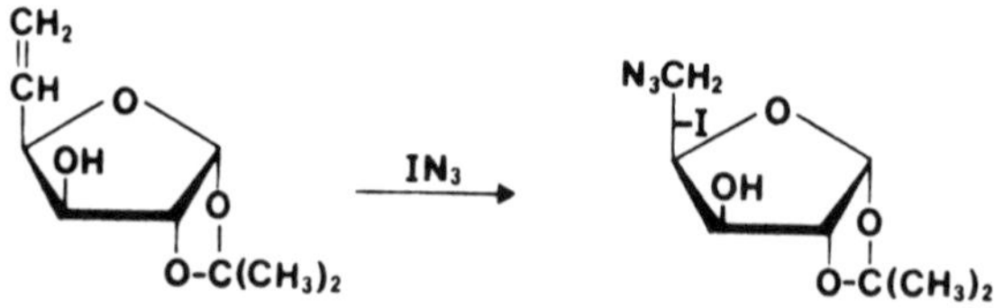

(J.S. Brimacombe *et al.*, Chem.Comm., 1968, 1401)

SCHEME 172b

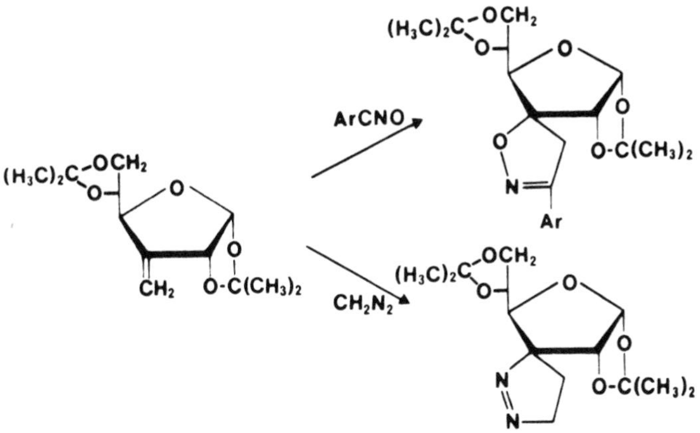

(J.M.J. Tronchet *et al.*, Helv., 1972, 55, 2813)

SCHEME 172c

Photochemical additions of thiols occurs with compounds containing vinyl ether groups to give saturated β-thioethers. Pyranosyl glycals therefore afford 1,5-anhydro-2-thioalditols, and compounds with terminal exocyclic alkene groups react in comparable fashions (Scheme 173) (K. Matsuura *et al.*, Tetrahedron Letters, 1970, 2869).

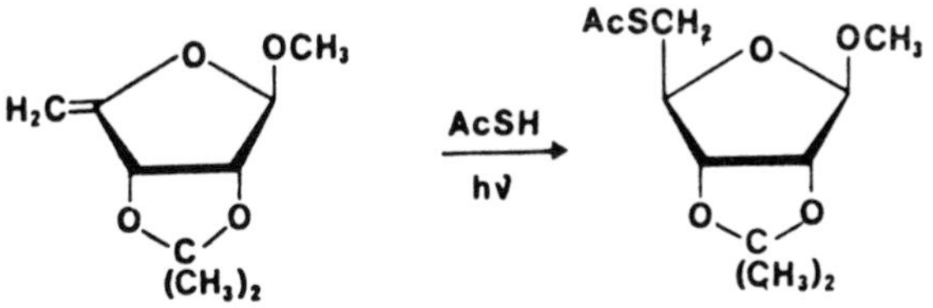

SCHEME 173

The orientations of such addition reactions are dependent on many factors - sometimes on steric features removed from reaction centres. For example, hydrogenation of 6-deoxy-5-enopyranose derivatives depends critically on the orientation of the aglycone (Scheme 174) (C. Monneret, C. Conreur and Q. Khuong-Huu, Carbohydrate Res., 1978, 65, 35).

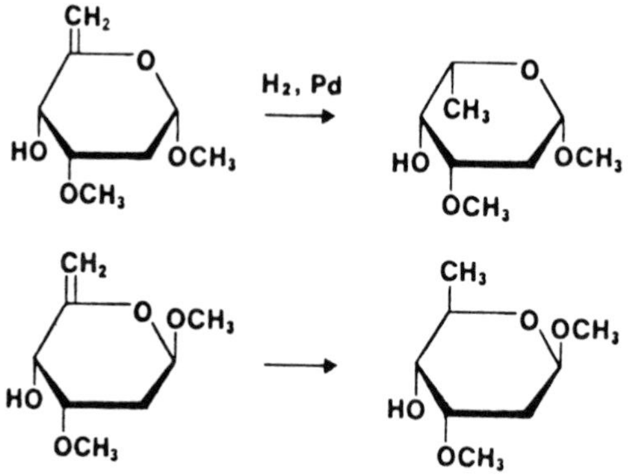

SCHEME 174

Oxidations of polyhydroxy unsaturated sugar derivatives can be effected specifically at allylic positions to give good yields of conjugated enones by use of manganese dioxide in chloroform (B. Fraser-Reid *et al.*, Canad.J.Chem., 1970, 48, 2877), and the enone (79) undergoes Simmons-Smith cyclopropanation to afford the *lyxo*-adduct (Scheme 175) (*idem. ibid.*, 1972, 50, 2928). Conjugate additions occur with carbanions or

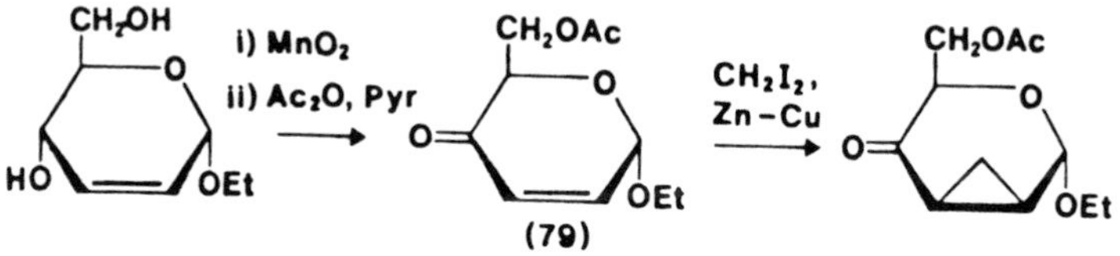

SCHEME 175

photochemically (p.185). With alkenes, (2+2) cycloadditions give cyclobutanes (mixed diastereoisomers), and, with diazomethane, conjugated pyrazolines are formed (B. Fraser-Reid, Acc.chem.Res., 1975, 8, 192). In the presence of aluminium trichloride, Diels Alder reaction of buta-1,3-diene and compound (79) gives a bicyclic product (p.188).

Many additions carried out under basic conditions to 2,3-dideoxy-3-nitro-D-*erythro*-hex-2-enopyranosides give mainly 2-substituted compounds with the thermodynamically preferred D-*gluco*-configuration. Alternatively, under neutral, or slightly acid conditions D-*manno*-adducts are obtained predominantly

(Scheme 176) (T. Sakakibara and R. Sudoh, Carbohydrate Res., 1976, 50, 191). As with enones, cyclic adducts can be produced, and with diazomethane, pyrazolines are obtained which on heating lose nitrogen to give 2-*C*-methyl derivatives of the original alkenes (H.H. Baer, F. Linhart and H.R. Hanna, Canad. J.Chem., 1978, 56, 2087).

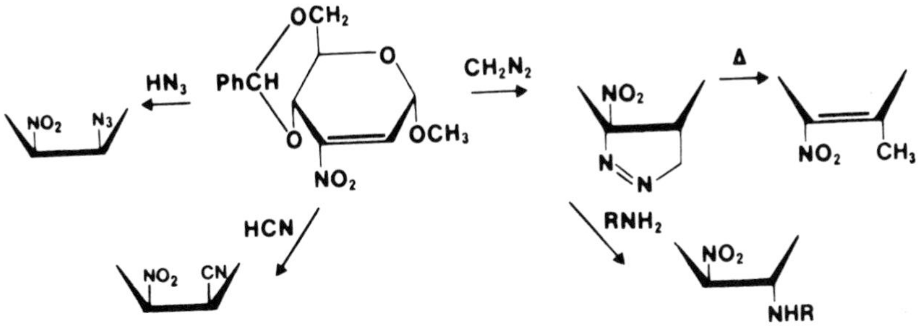

SCHEME 176

The nucleophilic displacement of leaving groups from allylic positions of unsaturated compounds has been observed many times, and most frequently direct S_N2 reaction occurs (R.J. Ferrier, M.T.P. International Rev. of Science, Organic Chemistry, Ser.2, Vol.7, 1976, p.35). Only occasionally are allylic rearrangements involved (*cf.* p.214), but reduction of methyl 6-deoxy-2,3-*O*-isopropylidene-4-*O*-mesyl-α-D-*lyxo*-hex-5-enopyranoside with lithium aluminium hydride gives a 6-deoxy-4-enose as product (J. Lehmann, Angew.Chem.internat.Edn., 1965, 4, 874) and this seems to be the main type of compound obtained from such 6-deoxy-5-enoses (Scheme 177) (M. Brockhaus and J. Lehmann, Ann., 1974, 1675).

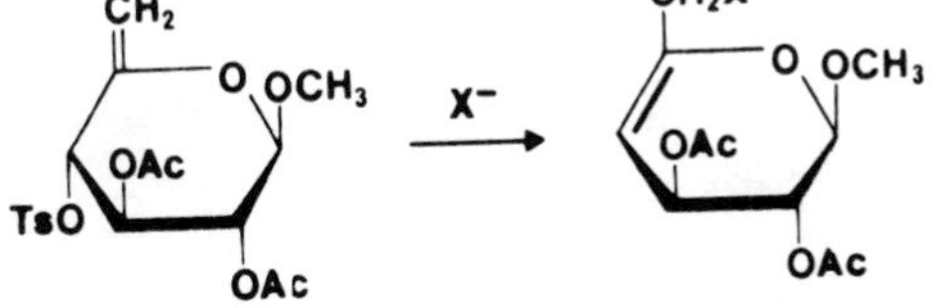

SCHEME 177

More commonly, major products are the unrearranged isomers, and introduction of a nucleophile which is itself a satisfactory leaving group provides means of obtaining products with either configuration at an allylic centre (Scheme 178) (J.S. Brimacombe *et al.*, J.chem.Soc.Perkin I, 1972, 2977; 1973, 1295).

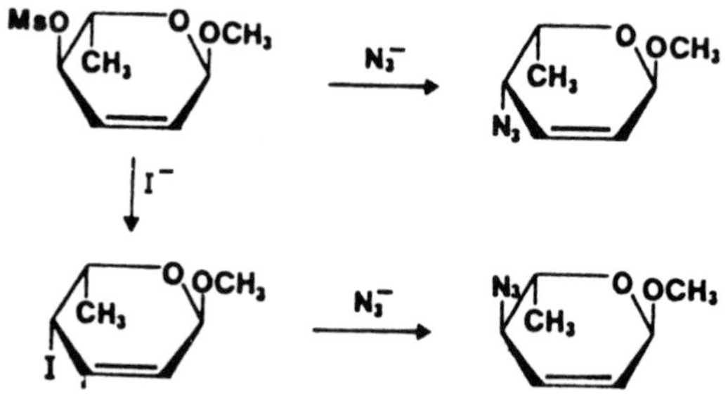

SCHEME 178

Such allylic 4-azides rearrange thermally by [3,3]sigmatropic processes and equilibrate with 2-azide isomers, analogous thiocyanates give 3,4-unsaturated isothiocyanates completely (R.J. Ferrier and N. Vethaviyasar, J.chem.Soc.(C), 1971, 1907), xanthate esters give 3,4-unsaturated 2-(methylthio)carbonylthio esters (*idem.*, Carbohydrate Res., 1977, 58, 481), and 4-vinyl ethers allow introduction of branched chains at C-2 by application of the Claisen rearrangement (*idem.*, J.chem.Soc.Perkin I, 1973, 1791) (Scheme 179).

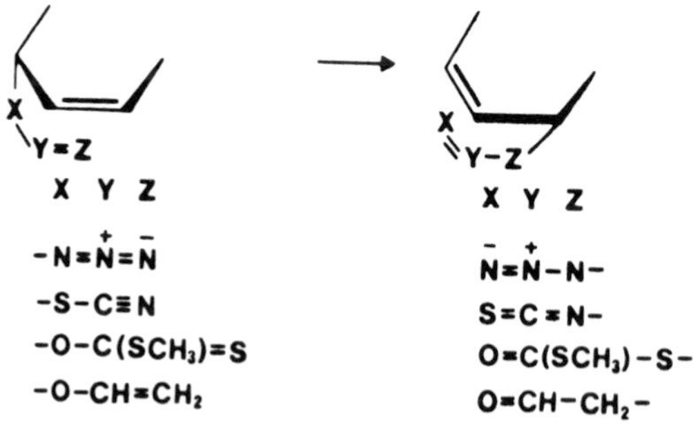

SCHEME 179

This last type of reaction is of great value in the synthesis of branched-chain compounds and was used in modified form by G. Stork and coworkers (J.Amer.chem.Soc., 1978, 100, 8272) to introduce an acetic acid side chain into an acyclic unsaturated compound in their synthesis of prostaglandin $F_{2\alpha}$ from D-glucose.

Special use can be made of 6-deoxy-hex-5-enose compounds in the efficient synthesis of either 5- or 6-membered carbocyclic compounds (Scheme 180).

(B. Bernet and A. Vasella, Helv., 1979, 62, 1990, 2400, 2411)

(R.J. Ferrier, *et al.*, J.chem.Soc.Perkin I, 1979, 1455; Chem. Comm., 1980, 944)

SCHEME 180

Acknowledgments

Dr. Richard Furneaux is thanked for reading the manuscript and for his helpful criticism. Ann Beattie's and Eric Stevens' excellent typing and drawing are also gratefully acknowledged.

Chapter 24

OLIGOSACCHARIDES, POLYSACCHARIDES AND RELATED COMPOUNDS

R. KHAN, J. K. WOLD, AND B.S. PAULSEN

Introduction

Since the subject was last discussed in this series (G.O. Aspinall, E. Percival, D.A. Rees and M. Rennie, in Rodd's Chemistry of Carbon Compounds, ed., S. Coffey, Elsevier Publishing Company, 1967, vol. 1F, p. 596), the progress in the field of stereocontrolled synthesis of 1,2-*cis*-and 1,2-*trans*-glycosides has been rapid. It is, therefore, considered appropriate to include a discussion on the subject of oligosaccharide synthesis. The development and application of physcial methods such as ^{1}H-n.m.r., ^{13}C-n.m.r. and mass spectrometry has played a major role in oligosaccharide chemistry. Some of these methods are briefly reviewed.

The polysaccharide section includes plant, microbial and animal polysaccharides, and glycoconjugates as glycoproteins, proteoglycans and glycolipids; thus they represent a large and varied group among the natural products. In view of the increasing understanding of the many important biological functions in which complex carbohydrates are involved, chemical and biochemical studies of these polymers have proceeded at a great pace during recent years. The considerable chemical complexity of most of these compounds necessitates the use of physical, chemical, biochemical and immunological methodologies for their detailed structural examination.

The general arrangement of this chapter follows closely that of the corresponding chapter of the second edition.

1. *Oligosaccharides*

(a) Synthesis

General methods of synthesis of oligosaccharides has been dealt with by several reviews and monographs (J. Staněk, M. Černý, and J. Pacak, "The Oligosaccharides", Czechoslovak Academy of Science, Prague, 1965; R.W. Bailey, "Oligosaccharides", Pergamon Press, Oxford, 1965 ; J.H. Pazur, "The Carbohydrates; Chemistry and Biochemistry", Vol. II, A, eds, W. Pigman, D.Horton, and assistant ed. A. Herp, Academic Press, New York and London, 1970, p.69; N.K. Kochetkov and A.F. Bochkov, in "Recent Development in the Chemistry of Natural Carbon Compounds", Vol.4, Akadémia Kiado, Budapest, 1971, p.77; N.K. Kochetkov, O.S. Chizhov, and A.F. Bochkov, in "MTP International Review of Science, Organic Chemistry", Ser.1, Vol. 7, ed. G.O. Aspinall, Butterworths, London and University Park Press, Baltimore, 1973, p.147; S.E. Zurabyan and A.Ya. Khorlin, Uspekhi Khimii, 1974, 43, 1865; A.F. Bochkov and G.E. Zaikov, "Chemistry of the *O*-Glycosidic Bond : Formation and Cleavage", Pergamon Press, Oxford,1979). Some of the methods of oligosaccharide synthesis will be discussed.

(i) Classical Koenigs-Knorr Reaction

The condensation reaction is carried out between an acylglycosyl halide and a suitably protected carbohydrate derivative with an isolated hydroxyl group in the presence of silver oxide or silver carbonate (as hydrogen halide acceptor) in dichloromethane or chloroform at room temperature with addition of Drierite (as a desiccant) and iodine to supress side reactions (W.L. Evans, D.D. Reynolds and E.A. Talley, Adv. carbohydrate Chem., 1951, 6, 27; L.J. Haynes and F.H. Newth, *ibid.*, 1955, 10, 207; K. Igarashi, Adv. carbohydrate Chem. Biochem., 1977, 34, 243).

The condensation reaction is stereospecific and results usually into 1,2-*trans*-glycosides. The actual mechanism involved in the reaction is not well understood. However, it has been suggested that the reaction proceeds either by an S_N2 process (scheme 1, route 1) or by heterolysis of C-1 halogen bond leading to glucosyl cation which is then stabilised immediately by intramolecular nucleophilic attack by an ester group at C-2, followed by ring opening of acyloxonium ion by an alcohol (route 2). Alternatively, the initial heterolysis leads to the formation of an intimate ion pair, which retains the stereochemistry of the starting substrate and provides inversion as a total result of condensation wth an alcohol (route 3). (A.F. Bochkov, and G.E. Zaikov, *loc.cit.*, N.K. Kochetkov, O.S. Chizov, and A.F. Bochkov, *loc.cit.*)

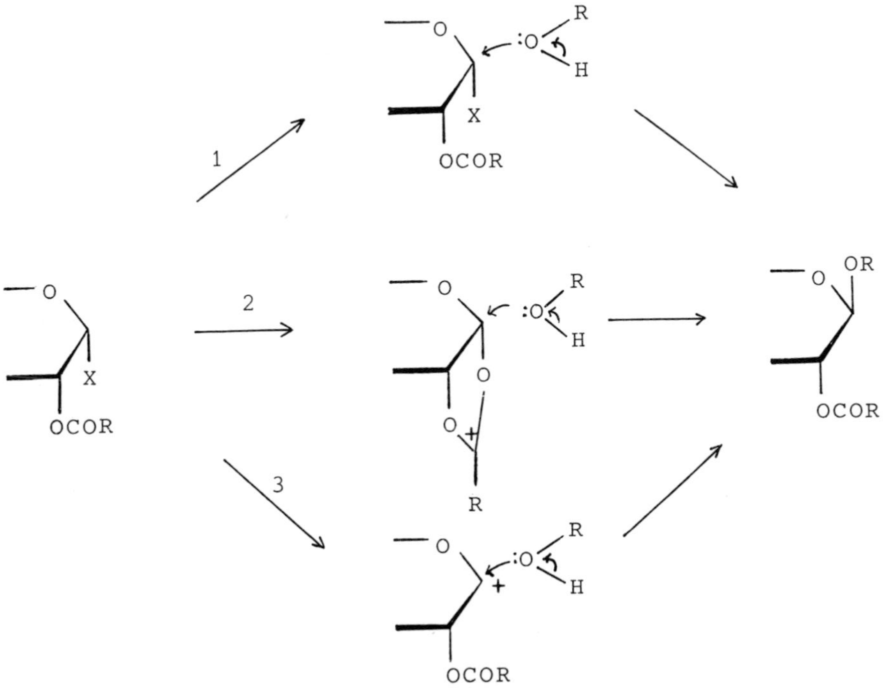

O
X
OCOR
1
2
3
O
:O
R
H
X
OCOR
O
:O
R
H
O
O
+
R
O
:O
R
H
+
OCOR
O
OR
OCOR

Scheme 1

It is of interest to note that in the L-fucose series under the classical Koenigs-Knorr reaction condition, 1,2-*cis*-glycosidic linkage was formed (M.A.E. Shaban and R.W. Jeanloz, Carbohydrate Res., 1971, 20, 399).

(ii) Modified Koenigs-Knorr Reaction

The reaction in modified form is extensively used for the synthesis of sugar glycosides and oligosaccharides. In the Helferich modification the reaction is performed by condensation of an alcohol with 1,2-*cis*-or 1,2-*trans*-acylglycosyl halide in equimolar amount or with some excess of halide in nitromethane or acetonitrile admixed with a small amount of benzene at 20-50° in the presence of mercury (II) cyanide. The reaction is non-stereospecific and leads, in most cases, to a mixture containing 1,2-*trans*-anomer as the major product (G.M. Bebault, G.G.S.Dutton and N.A. Funnell, Canad. J. Chem., 1974, 52, 3844; G.G.S.Dutton and coworkers, Carbohydrate Res., 1974, 34, 174; C. Augé and A.Veyrières, *ibid.*, 1976, 46, 293; B.R. Bhattacharyya, K. Ramaswamy, and R.K.Crane, *ibid.*, 1976, 47, 167; P.J.Garegg and T. Norberg, *ibid.*, 1976, 52, 235; 1957, K. Schwabe, A. Graffi and B. Tschiersch, *ibid.*, 1976, 48, 277; A. Lipták and P. Nánási, Tetrahedron Letters, 1977, 921), and 1,2-*cis*-anomer as the minor product (K. Matsuda, Nature, 1957, 180, 985; M.A.E.Shaban and R.W.Jeanloz, Carbohydrate Res., *loc. cit.*; K. Eklind, P.J.Garegg and B. Gotthammer, Acta. chem. Scand., 1976, B30, 300, 305).

With few exceptions (J.E. Wallace and L.R. Schroeder, J. chem. Soc.Perkin I, 1976, 1938, J. chem. Soc. Perkin II, 1977, 795; K. Igarashi, J. Irisawa and T. Honma, Carbohydrate Res., 1975, 39, 213; G. Wulff, U.Schröder and J.Wichelhaus, *ibid.*, 1979 72, 280) the mechanism of the glycosylation reaction has not been studied systematically. Earlier speculations, including catalysed heterolysis of a C-1 bromine bond followed by a non-stereospecific attack of an alcohol on the glycosyl cation form, seem unsatisfactory (R.J. Ferrier, Topics in current Chemistry, 1970, 14, 389; M. Shaban and R.W.Jeanloz, Carbohydrate Res., 1971, 17, 193). In order to select reaction conditions which would lead to stereoselective synthesis of 1,2-*cis*- or 1,2-*trans*-glycosides an understanding of the reaction mechanism is essential.

In the glycosylation reaction factors such as the stereochemistry of the blocking groups in the halide, the polarity of the solvent (K. Igarashi, J.Irisawa and T.Honma, Carbohydrate Res., 1975, 39, 341; S.Wulff,U.Schröder and J. Wichelhaus, *loc.cit.*), the concentration of the nucelophile (J.F.Kronzer and C.Schuerch,Carbohydrate Res.,1973,27, 379; J.E.Wallace andL.R.Schroeder,*loc.cit.*), the nature of the promotor (J.E.Wallace and L.R.Schroeder, *loc.cit.*), and to some extent the nature of the substituents in the neighbourhood of the hydroxyl group in the nucleophile (M.Dejter-Juszynski andH.M.Flowers, Carbohydrate Res.,1975, 37, 75, 41, 308; T.J.Lucas and C.Schuerch, *ibid.*, 1975,39, 39) are important in determining the steric outcome.

In the synthesis of 1,2-*trans*-glycosides advantage can be taken of the fact that glycosyl halide having an *O*-acyl group at C-2 *cis* to the halogen at C-1 react normally under the conditions of a Koenigs-Knorr reaction with inversion of configuration at C-1. 1,2-*cis*-Glycosides have been successfully synthesised from glycosyl halides having a non-participating group at C-2 using the Helferich reaction. Most studies on 1,2-*cis*-glycoside synthesis involve benzyl ether as non-participating group, used both as uniform protection and in combination with other substituents (A.F. Bochkov and G.E. Zaikov, *loc.cit.*; P.A.Gent, *et al.*, J.chem.Soc. Perkin I, 1976, 1395; G. Wulff, U. Schröder, and J. Wichelhaus, *loc.cit.*; K.Igarashi, J.Irisawa, and T.Honma, *loc.cit.*; G.Excoffier, D.Y.Gagnaire, and M.R.Vignon Carbohydrate Res., 1976,46, 215; J.Hirsch and P.Kovác, J.carbohydrate Nucleosides Nucleotides, 1978, 5, 373). The use of 2-*O*-benzyl glycosyl halide in glycoside synthesis has advantage in that the hydroxyl group at C-2 in the product can be regenerated under neutral conditions by catalytic hydrogenolysis. The free 2-hydroxyl group can then be used for further chemical transformations, for example, converting into 1 → 2 linked di- and/or higher branched oligosaccharides.

β-D-Mannopyranosides, however, are not formed with high selectivity from simple derivatives of this type (A.S.Perlin, Pure applied Chem., 1978, 50, 1401). The value of such derivatives as 2,3-*O*-carbonyl-α-D-mannopyranosyl bromide (P.A.J.Gorin and A.S.Perlin, Canad.J.Chem., 1961, 39, 2474; G.M.Bebault and G.G.S.Dutton, Carbohydrate Res., 1974, 37,

309) and 2,3-*O*-cyclohexylidene-α-D-mannopyranosyl chloride (P.J.Garegg and T.Iversen, *ibid.*, 1979, 70, C13) as glycosylating reagents for the synthesis of 1,2-*cis*-mannopyranosides has been demonstrated. A highly stereoselective route to the synthesis of β-D-mannopyranoside using a glycosylating reagent having a mesyl as non-participating group at C-2 and a tosyl group at C-1 as a leaving group has been described (V.K.Srivastava and C.Schuerch, *ibid.*, 1980, 79, C13).

(iii) The Orthoester Method

Sugar 1,2-orthoesters react with alcohols leading stereospecifically to 1,2-*trans*-glycosides (N.K.Kochetkov, and coworkers, *ibid.*, 1971, 16, 17; 1980, 85, 209; N.K.Kochetkov, O.S. Chizhov, and A.F.Bochkov, *loc.cit.*; A.F.Bochkov and G.E. Zaikov, *loc.cit.*; N.K.Kochetkov and A.F.Bochkov, Methods in carbohydrate Chem., 1973, 6, 480; N.K.Kochetkov and A.F.Bochkov in "Recent Developments in the Chemistry of Natural Carbon Compounds", Akademia Kiado, Budapest, vol.4, 1972, p.77; W.G.Overend, in "The Carbohydrates", eds.W.Pigman and D. Horton, Academic Press, New York, 1A, 1972, p.279). The glycosylation reaction by this method can be carried out either directly or in two stages. In the former case the reaction is catalysed either by mercury (II) bromide in boiling nitromethane with removal by azeotropic distillation of the lower alcohol or by 2,6-dimethylpyridinium perchlorate in boiling chlorobenzene also accompanied by azeotropic distillation. The latter two-stage glycosylation reaction is usually performed in solvents of low polarity, preferably 1,2-dichloroethane or chlorobenzene, in the presence of toluene-*p*-sulphonic acid and pyridinium perchlorate (N.K. Kochetkov, A.F.Bochkov, T.A. Sokolovskaya, and V.J.Snyatkova *loc.cit*).

The influence of reaction conditions on the composition of the products has been studied in detail (N.K.Kochetkov, A.Ya.Khorlin and A.F.Bochkov, Tetrahedron, 1967, 23, 693; A.F. Bochkov, V.I.Betanely and N.K.Kochetkov, Bioorg.Khim., 1976, 2, 927; 1977, 3, 39). In spite of some minor side reactions (A.F.Bochkov, V.I.Betanely and N.K.Kochetkov, Carbohydrate Res., 1973, 30, 418; R.U.Lemieux, Chem.Canad., 1964, 14), the versatility and the reproducibility of the 1,2-orthoester method to oligosaccharide synthesis is well recognised.

Carbohydrate 1,2-orthoesters containing a variety of alkoxy groups can be prepared from both 1,2-*cis*-and 1,2-*trans*-acylglycosyl halides by well established methods (N.K Kochetkov,A.Ya.Khorlin and A.F.Bochkov,*loc.cit.*;R.U.Lemieux and A.R.Morgan, Canad.J.Chem.,1965,43,2199;G.Wulff and W. Krüger,Carbohydrate Res.,1971,19,139; N.K.Kochetkov,A.F. Bochkov,T.A.Sokolovskaya and V.J. Sanyatkov, *loc.cit.*; S.Hanessian and J.Banoub, *ibid.*, 1975,44,C14). *t*-Butyl-orthoesters have been found to be much more effective as glycosylating reagents than their lower homologues, for example, methyl, ethyl and benzoate. The application of the 1,2-orthoester approach can be exemplified by the synthesis of gentiobiose octaacetate (4) from 1,2,3,4-tetra-*O*-acetyl-β-D-glucopyranose (1) and 3,4,6-tri-*O*-acetyl-α-D-glucopyranose 1,2-(ethyl-orthoacetate) (2) (N.K.Kochetkov,A.F. Bochkov, T.A.Sokolovskaya and V.J.Snyatkov, *loc.cit.*) The condensation of (1) with (2) in dichloroethane in the presence of toluene-*p*-sulphonic

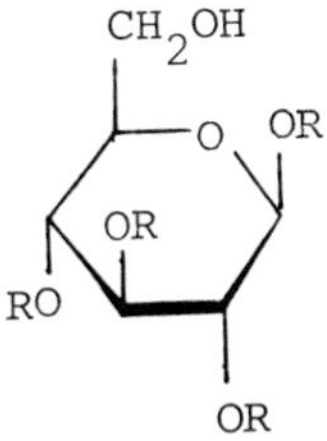

(1)

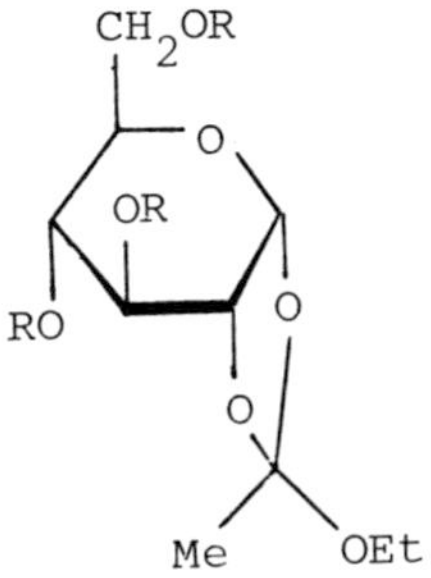

(2)

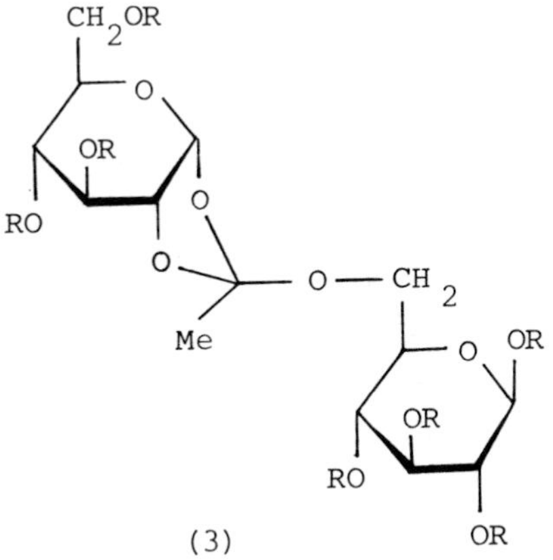

(3)

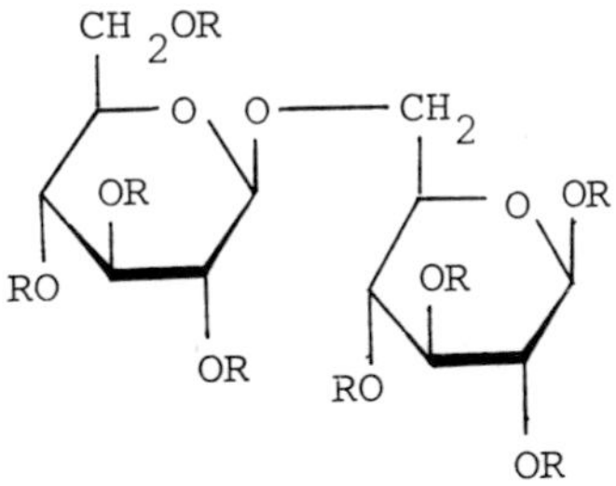

(4)

R = Ac

acid gives the orthoester (3) in 75% yield. Treatment of (3) in boiling dichloroethane in the presence of pyridinium perchlorate and toluene-*p*-sulphonic acid affords (4) in 59% yield. In the absence of toluene-*p*-sulphonic acid practically no conversion of (3) into (4) occurs. From a preparative point of view, it is convenient to carry out the two-stage glycosylation reaction without isolation of the intermediate ester. The yield of (4) under these conditions is 60.5%. The direct (one-stage) condensation of (1) with (2) under comparable conditions gives (4) in lower (38%) yield.

(iv) The Oxazoline Method

This method has been extensively used in recent years for the synthesis of 1,2-*trans*-glycosides of 2-acylamido-2-deoxyhexopyranose. Mechanistically, it is related to the orthoester method. The reactive intermediate which undergoes nucleophilic attack at the anomeric carbon atom is presumed to be the protonated 1,2-oxazolinium ion (scheme 2) (N.K. Kochetkov, O.S.Chizhov and A.F.Bochkov, in MTP Internatioal Review of Sciences, ed.G.O.Aspinall, Butterworth,London,1973, p.147; A.F.Bochkov,and G.E.Zaikov, *loc.cit.*)

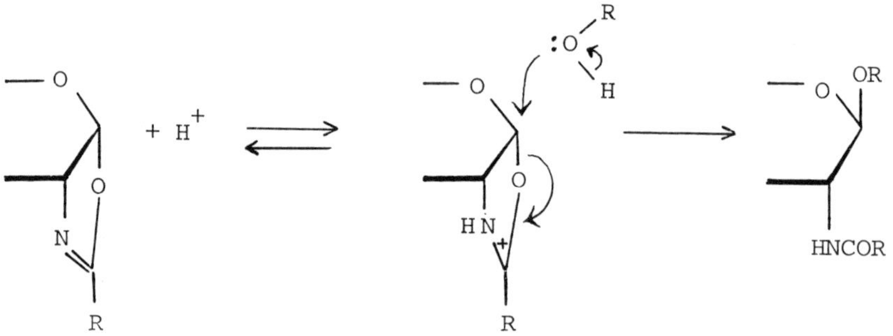

Scheme 2

Sugar oxazolines are formed more easily than the analogous orthoesters. A well developed method for the synthesis of oxazolines includes treatment of an *N*-acetylated aminosugar with hydrogen chloride in acetyl chloride followed by treatment with a silver salt and 2,4,6-collidine (scheme 3) (G.E. Zurabyan, I.M. Privalova, and Yu. L. Kopaevitch, Izv. Akad. Nauk. SSSR, Ser. Khim., 1968, 2094).

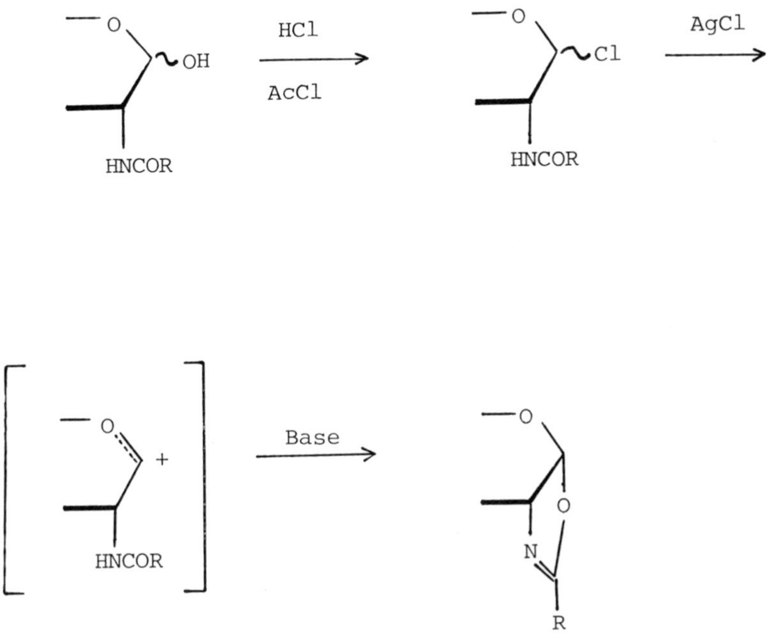
HCl
AcCl
AgCl
OH
Cl
HNCOR
HNCOR
HNCOR
Base
O
N
R

Scheme 3

The fully acylated aminosugar glycosyl halides can be directly converted into the oxazoline in near quantitative yield by treatment with tetraethylammonium chloride as catalyst, acetonitrile as solvent, and sodium carbonate as base (R.U.Lemieux and H.Driguez,J.Amer.chem.Soc.,1975,97,4063; R.W.Jeanloz and coworkers,Carbohydrate Res.,1979,71, C5; C.D.Warren, *et al.*, *ibid.*,1980,82, 71). Alternatively, a fully acetylated amino sugar may be treated with anhydrous iron (III) chloride to give an oxazoline in high yields (K.L.Matta and O.P.Bahl, *ibid.*, 1972,21,460; K.L.Matta,E.A.Johnson and J.J.Barlow, *ibid.*, 1973,26,215; C.Augé and A.Veyrières, *ibid.*, 1976, 46, 293). Recently disaccharide oxazolines have been prepared in high yield by acid catalysed acetolysis of the corresponding methyl glycoside derivatives (K.L.Matta and J.J.Barlow, *ibid.*, 1977, 53, 47).

The reaction of oxazolines with alcohols is usually carried out in boiling nitromethane or toluene in the presence of anhydrous toluene-*p*-sulphonic acid (S.E.Zurabyan, T.P.Volosyuk and A.Ya.Khorlin,*loc.cit.*, S.David and A. Veyrières, Carbohydrate Res., 1975, 40, 23; R.Kaifu,T.Osawa and R.W.Jeanloz, *ibid.*, 1975,40,111; K.L.Matta and J.J.Barlow *loc.cit.*) The reaction of 2-methyl-(2-acetamido-3,4,6-tri-*O*-acetyl-1,2-dideoxy-α-D-glycopyrano)-[2,1-*d*]-2-oxazoline with 2,2,2-trichloroethanol has been reported to give superior yields when catalysed by trifluoromethanesulphonic acid rather than by toluene-*p*-sulphonic acid (R.U.Lemieux and H. Driguez,*loc.cit.*) Other workers, however, could not support this improvement (C.D.Warren and R.W.Jeanloz, *loc.cit.*) The poor solubility of oxazoline tosylate formed during the reaction in toluene-nitromethane prompted Warren and Jeanloz (*loc.cit.*) to change the solvent to 1,2-dichloroethane in which the solubility of the toluene-*p*-sulphonium salt is higher (C.Augé, C.D.Warren and R.W.Jeanloz, Carbohydrate Res. 1980,82, 85). It has been observed (M.Kiso and L.Anderson, *ibid.*,1979,72,C12,C15) that when alcohol was included in the preparation of the oxazoline from 2-acetamido-3,4,6-tri-*O*-acetyl-2-deoxy-β-D-glucopyranose by treatment with iron (III) chloride in dichloromethane (K.L.Matta and O.P.Bahl,*loc.cit.*) the product obtained is the corresponding β-glycoside. The potential of this "direct" glycosylation reaction has been demonstrated by the synthesis of several (1→3)- and (1→6)-β-linked disaccharides of 2-amino-2-deoxy-D-glucopyranose.

(v) 1-O-Sulphonyl Derivatives for the Synthesis of 1,2-cis-glycosides

The steric outcome of reactions between equal amounts of a glycosyl derivative having a C-2 non-participating group and an alcohol can be controlled by a careful choice of the C-1 leaving group, solvent, and substituents of C-4 and C-6 (E.R. Rachaman, R. Eby and C. Schuerch, Carbohydrate Res., 1978, 67, 147 and references therein; R. Eby and C. Schuerch, 1979, 77, 61; V.K. Srivastava and C. Schuerch, *ibid*, 1980, 79, C 13; J. Leroux and A.S. Perlin, *ibid*., 1976, 47, C 8; 1978, 67, 163).

The glucopyranosyl trifluoromethanesulphonates (triflates) have been found to be extremely reactive giving high yields of glucosides at low concentration of alcohols and at low temperature

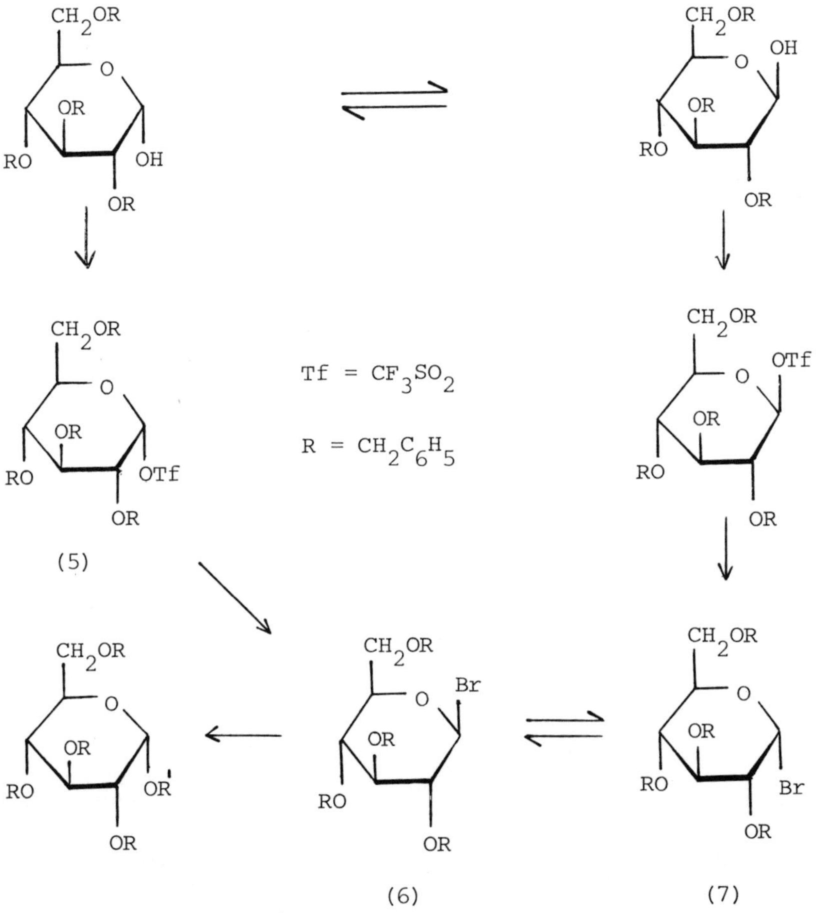

$Tf = CF_3SO_2$

$R = CH_2C_6H_5$

(5)

(6)

(7)

Scheme 4

(-78°), in 30 min or less. At higher temperatures, the glucopyranosyl trifluoromethanesulphonates either reacted with the solvent or decomposed and gave a mixture of undesired products (F.J.Kronzer and C.Schuerch, *ibid.*, 1973,27,379). Leroux and Perlin (*loc.cit.*) have shown that α-D-glucopyranosides, including disaccharides, can be prepared in high yield from 2,3,4,6-tetra-*O*-benzyl-D-glucose by direct addition of the appropriate alcohol, bromide ion, *s*-collidine, and trifluoromethanesulphonic anhydride. It has been suggested that the 1,2-*cis*-glycosides are formed through the successive intermediary of a 1-triflate (5) and a glycosyl bromide (6,7) (scheme 4). Due to the presence of excess bromide ion in the medium, (6) and (7) are in equilibrium, permitting glycoside formation to take place under kinetic control, that is, (6) reacts more rapidly with alcohol, with inversion, favouring formation of the α-D-glucosides.

The glycoside-forming reactions of D-glucose and D-mannose having a 1-*O*-tosyl group and either a non-participating benzyl ether or a participating ester group at C-2 have been studied by Schuerch and coworkers (*loc.cit.*) 1-*O*-Tosyl-D-glucopyranose derivatives having a non-participating group at C-2 have been shown to react rapidly in various solvents with low concentration of alcohols, either methanol or methyl 2,3,4-tri-*O*-benzyl-α-D-glucopyranoside. The stereospecificity of the reaction can be varied from 80% of β to 100% of α-anomer by changing the solvent or modifying the substituents on the 1-*O*-tosyl-α-D-glucopyranose derivatives. 2,3,4-Tri-*O*-benzyl-6-*O*-(*N*-phenylcarbomyl)-1-*O*-tosyl-α-D-glucopyranose in ether gave a high yield of α-D-glucoside (R.Eby and C.Schuerch, *loc.cit.*) The 1-*O*-sulphonyl derivatives are prepared from the corresponding glycosyl halides (bromide or chloride) by treatment with the appropriate silver salt in acetonitrile. Since the 1-*O*-tosyl derivatives are hydrolysed very rapidly, high vacuum techniques are used to exclude moisture from the system. All additions and transfers are carried out *in vacuo* or under an atmosphere of dry nitrogen.

The stereoselective synthesis of β-D-mannopyranoside from the reaction of 3,4,6-tri-*O*-benzyl-2-*O*-mesyl-1-*O*-tosyl-α-D-mannopyranose with methanol at room temperature gives the β-glycoside in 90% yield (V.K.Srivastava and C.Schuerch, Carbohydrate Res.,1980,79,C13). For the synthesis of methyl 2,3,4-tri-*O*-benzyl-6-*O*-(3,4,6-tri-*O*-benzyl-2-*O*-methylsulphonyl- β-D-mannopyranosyl)- α-D-mannopyranoside, the glycosylating reagent 3,4,6-tri-*O*-benzyl-2-*O*-methylsulphonyl-1-*O*-(2,2,2-trifluoroethylsulphonyl)- α-D-mannopyranose is found to be more selective and to react faster than the 1-*O*-tosyl derivative.

(vi) Halide Ion Catalysed Synthesis of 1,2-cis-glycosides

Rhind-Tutt and Vernon (J. chem. Soc., 1960, 4637) observed that methanolysis of 2,3,4,6-tetra-*O*-methyl-α-D-glucopyranosyl chloride in the presence of a soluble chloride (lithium chloride) increased considerably the proportion of the methyl-α-D-glucoside. This effect was also recognised by Ishikawa and Fletcher (J. org. Chem., 1969, 34, 563) who found that the reaction of 2,3,4,6-tetra-*O*-benzyl- α-D-glucopyranosyl bromide with excess of methanol in the presence of tetra-*n*-butylammonium chloride gave an anomeric mixture of the methyl glycoside with a higher content of 1,2-*cis* anomer than in the absence of the salt. The effect of added tetra-alkyl-ammonium halide to favour retention of configuration at C-1 was further supported by Frechét and Schuerch (J. Amer. chem. Soc., 1972, 94, 604). This idea has been developed and applied to stereospecific synthesis of α-linked oligosaccharides (R.U. Lemieux and coworkers, Canad. J. Chem., 1965, 43, 2199; J. Amer. chem. Soc., 1975, 97, 4056, 4063, 4069).

Glycosylation by this method usually involves the condensation of an equimolar amount or excess of totally benzylated glycosyl bromide and an alcohol in dichloromethane with an excess of tetra-ethylammonium bromide in the presence of diisopropylethylamine and/or molecular sieves (Lemieux and coworkers, *loc.cit.*, J.-C. Jacquinet and P. Sinaÿ, Tetrahedron, 1976, 32, 1693, J. org. Chem., 1977, 42, 720; G. Excoffier, D.Y. Gagnaire, and M.R. Vignon, Carbohydrate Res., 1976, 46, 215; P.J. Garegg and T. Norberg, *ibid.*, 1976, 52, 235).

Aminosugar α-glycosides have been synthesised from glycosylating reagents having azidodeoxy group as non-participating group at C-2 and halide at C-1 as a leaving group (H. Paulsen and W. Stenzel, Angew. Chem. internat. Edn., 1975, 14, 558; H. Paulsen, Pure and appl. Chem., 1977, 49, 1169). The reaction of 6-*O*-acetyl-3,4-di-*O*-benzyl- α -D-glucopyranosyl bromide with tetra-ethylammonium chloride in acetonitrile gives the corresponding β-chloride in greater than 90% yield. After separation of the ammonium salt the β-chloride on condensation with 2-

azido-1,3,4-tri-*O*-benzyl-2-deoxy- β -D-glucopyranoside in dichloromethane in the presence of silver carbonate and catalytic amount of silver perchlorate or collidine and silver perchlorate at -5° for 15 min gives the α-linked 1 ⟶ 6 disaccharide (85% purity) in 85% yield (H. Paulsen and W. Stenzel, *loc. cit.*), (Scheme 5).

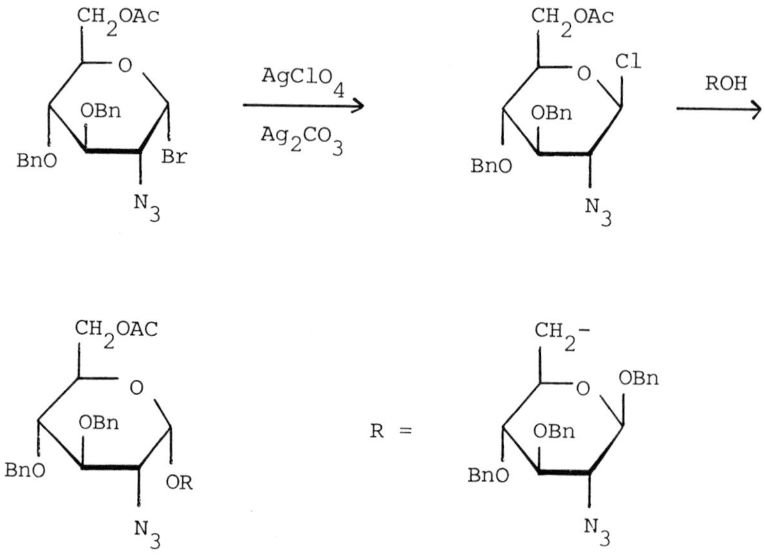

Scheme 5

(vii) *Use of Electropositive Substituents at C-1 for 1,2-cis-glycosides*

Use of various onium salts derived from fully benzylated α-D-*gluco*- and α-D-*galacto*-pyranosyl bromide as intermediates in the synthesis of α-glycosides has been proposed (A.C. West and C. Schuerch, J. Amer. chem. Soc., 1973, 95, 1333; F.J. Kronzer and C. Schuerch, Carbohydrate Res., 1974, 33, 273; R. Eby and C. Schuerch, *ibid.*, 1975, 39 33). 2,3,4,6-Tetra-*O*-benzyl- α-D-glucopyranosyl bromide is readily converted into the corresponding quaternary ammonium salt by treatment with a tertiary amine. It has been shown to exist in the β-configuration and to react with alcohols in solvents of low di-electric constant with predominant inversion at C-1 to give good yields of 1,2-*cis*-glycosides (Scheme 6). Complete stereo-selectivity is not observed (R.Eby and C.Schuerch, *loc.cit.)*

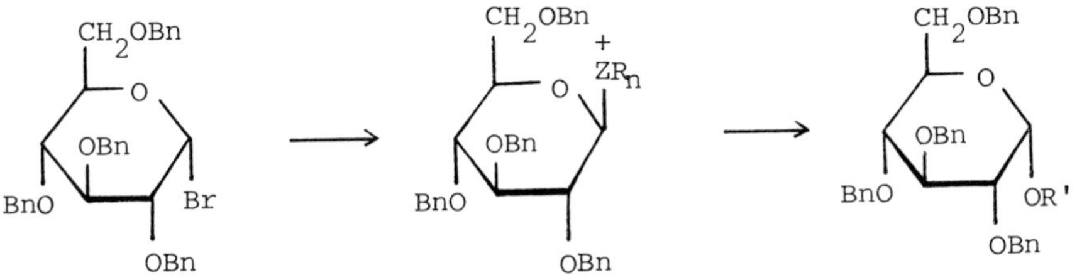

$Z = N, S, P$

$Bn = CH_2C_6H_5$

Scheme 6

Synthesis of an α-linked disaccharide from 2,3,4-tri-*O*-benzyl-6-*O*-acyl-α-D-glucopyranosyl bromide in the presence of pyridine can be regarded as proceeding by way of a glycosylpyridinium intermediate (M.Petitou and P.Sinaÿ,Carbohydrate Res.,1975,40,13). A similar reaction pathway can be suggested for some of the reactions of glycosyl halides in the presence of triethylamine (P.A.Gent and R.Gigg,J.chem. Soc.Perkin I,1974, 1446 and 1835; 1975,361; P.A.Gent, R. Gigg and A.A.E.Penglis, *ibid.*, 1976, 1395).

(viii) Configurational Inversion at C-2 leading to 1,2-cis-glycosides

The approaches described in this section involve the creation of an asymmetric centre at C-2 with 1,2-*cis*-configuration when a glycosidic linkage with the desired absolute configuration pre-exists in the molecule.

Synthesis of 1,2-*cis*-glycosides has been accomplished by using 2-*C*-nitroso-glycosyl chloride as the glycosylating reagent (R. U. Lemieux and T.L.Nagabhushan, in "Methods in Carbohydrate Chemistry", eds. R.L. Whistler and J.N. BeMiller, Academic Press, Vol. 6, 1972, p.487; R.U. Lemieux and co-workers, Canad. J. Chem., 1968, 46, 405, 1040; 1973, 51, 42, 48; T.L. Nagabhushan and C.G. Chin, *ibid.*, 1970, 48, 3097). The key stage of the process utilises the highly stereoselective formation of tri-*O*-acetyl-2-Oximino- α -D-*arabino*-hexopyranosides (8) on treatment of tri-*O*-acetyl-2-deoxy-nitroso- α -D-glucopyranosyl chloride dimer (9) with alcohols. The reaction is considered to proceed by way of the intermediate tri-*O*-acetyl-1,2-dideoxy-2-nitroso-D-*arabino*-hex-1- enopyranose (10). Mild hydrolysis of the oximino group in (8) to a keto group followed by reduction with boron hydride gives the α -D-glucopyranosides. However, it

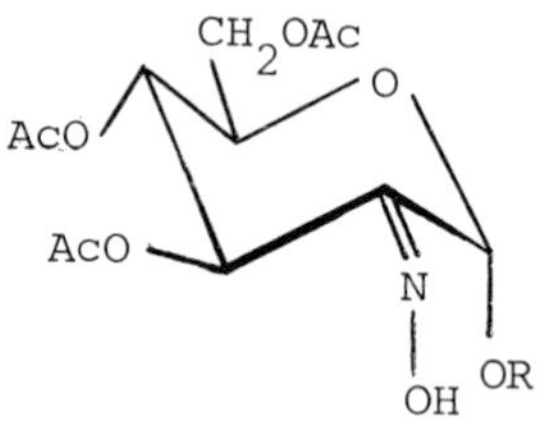

(8)

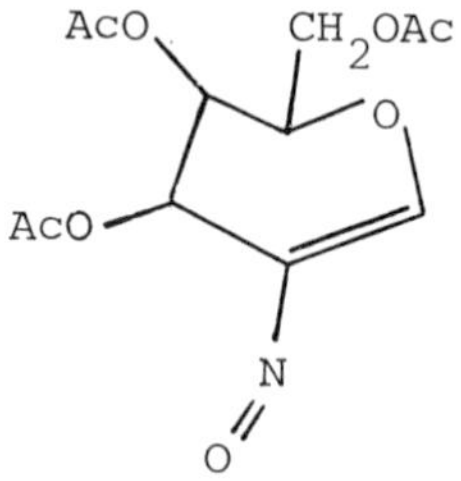

(10)

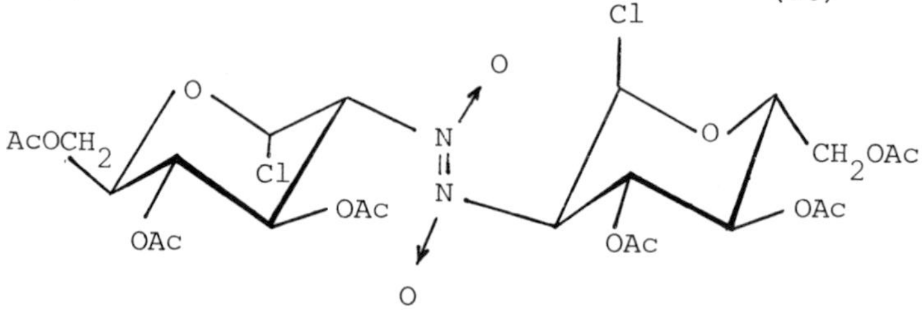

(9)

has been shown that in certain cases it can lead to an anomeric mixture (K. Miyai and R.W.Jeanloz, Carbohydrate Res., 1972, 21, 45).

The oxime (8) can be reduced to give a mixture of epimeric 2-amino-2-deoxy-glycosides. Thus providing a route to the synthesis of complex 2-amino-2-deoxy- α -D-glucosides and - mannosides (R.U. Lemieux *et al.*, Canad. J. Chem., 1968,

46,405; 1973,51,42, 48; K. Miyai and R.W.Jeanloz, Carbohydrate Res., 1972, 21,45; W.A.R.van Heeswijk, P.de Hann and J.F.G.Vliegenthart, *ibid*., 1976, 48, 187), or 2-amino-deoxy-α-D-galactosides and -talosides, if 4-epimers of the starting materials are used (Lemieux, *et al*., Canad. J. Chem., 1973, 51, 48).

The principle of configurational inversion at C-2 has been invoked to prepare α-D-glucopyranosides, β-D-mannopyranosides and 2-amino-2-deoxy-α-D-glucopyranosides (N.K. Kochetkov, *et al*., Carbohydrate Res., 1974,33, C5; 1975,45, 283; G.Ekborg and coworkers, Acta. Chem.Scand., 1972, 26, 3287; 1973,27,2639; 1975, B 29, 1031; "Abstracts of the VI International Symposium on Carbohydrate Chemistry", August 1974, Bratislava, p.30).

Koenigs-Knorr reaction of 2-*O*-benzoyl-3,4,6-tri-*O*-benzyl-α-D-glucopyranosyl bromide with 1,2:3,4-di-*O*-isopropylidene-α-D-galactopyranose gave 6-*O*-(2-*O*-benzoyl-3,4,6-tri-*O*-benzyl-β-D-glucopyranosyl)-1,2:3,4-di-*O*-isopropylidene-α-D-galactopyranose, which on sequential reactions involving debenzoylation, oxidation, reduction and removal of the protecting groups afforded 6-*O*-β-D-mannopyranosyl-D-galactose (G. Ekborg, B. Lindberg and J. Lönngren, Acta.Chem.Scand., 1972, 26, 3287) (Scheme 7).

The Koenigs-Knorr reaction in the above scheme has been replaced by the orthoester condensation method because of the latter's high yield and stereoselectivity (G.Ekborg, *et al*.,Acta.Chem.Scand.,1975, B 29, 1031, 1085).Treatment of 3,4,6-tri-*O*-benzyl-1,2-*O*-(methoxyethylidene)-β-D-mannopyranoside with 1,2:3,4-di-*O*-isopropylidene-α-D-galactopyranose in nitromethane in the presence of mercuric bromide under the general reaction conditions described by Kochetkov and co-workers (Tetrahedron, 1967,23,693) gave 6-*O*-(2-*O*-acetyl-3,4,6-tri-*O*-benzyl-α-D-mannopyranosyl)-1,2:3,4-di-*O*-isopropylidene-α-D-galactopyranose. Removal of the acetyl group at C-2, followed by oxidation, reduction and removal of the protecting groups gave 6-*O*-α-D-glucopyranosyl-D-galactose. The keto disaccharide, 6-*O*-(3,4,6-tri-*O*-benzyl-α-D-*arabino*-2-hexulopyranosyl)-1,2:3,4-di-*O*-isopropylidene-α-D-galactopyranose, has also been converted into 6-*O*-(2-amino-3,4,6-tri-*O*-benzyl-2-deoxy-α-D-glucopyranosyl)-1,2:3,4-di-*O*-isopropylidene-α-D-galactopyranose by oximation and reduction

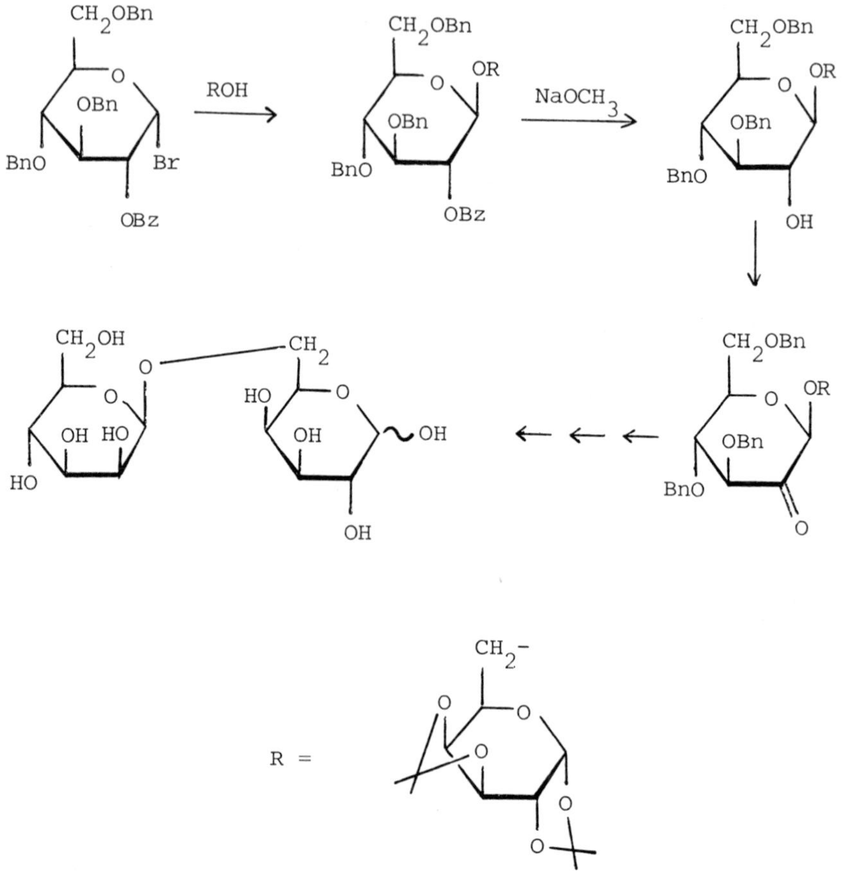

CH$_2$OBn, OBn, BnO, Br, OBz, ROH, CH$_2$OBn, OR, OBn, BnO, OBz, NaOCH$_3$, CH$_2$OBn, OR, OBn, BnO, OH, CH$_2$OH, CH$_2$, OH, HO, HO, OH, OH, OH, CH$_2$OBn, OR, OBn, BnO, O

R = CH$_2$-

Scheme 7

reactions (H.B. Borén, G. Ekborg and J.Lönngren, *ibid.*, 1975, B29, 1085).

(ix) Miscellaneous Glycosylation Reactions

Use of 1-*O*-(*N*-methyl) acetimidyl derivatives of D-glucose, D-galactose, L-fucose, and 2-deoxy-D-*arabino*-hexopyranose as glycosylating reagent has led to a variety of 1,2-*cis*-disaccharides (P. Sinaÿ and coworkers, J. Amer. chem. Soc., 1977, 99, 6762; Pure appl. Chem., 1978, 50, 1437; Nouveau Journal de Chemie, 1978, 2, 389; Tetrahedron Letters, 1979, 545; Carbohydrate Res., 1979, 77, 99). The imidate (11) for example is prepared by the reaction of 2,3,4,6-tetra-*O*-benzyl- α -D-glucopyranosyl chloride with *N*-methylacetamide in benzene in the presence of silver oxide, diisopropylethylamine and powdered 4A° molecular sieves. Treatment of (11) with methyl 2,3,4-tri-*O*-benzyl- α -D-glucopyranoside in anhydrous benzene

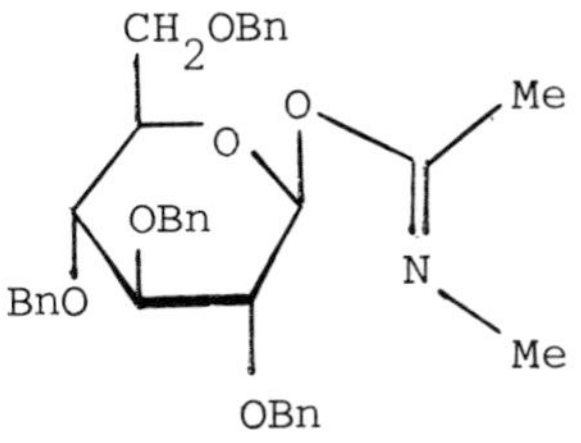

CH_2OBn O Me OBn N BnO Me OBn

(11)

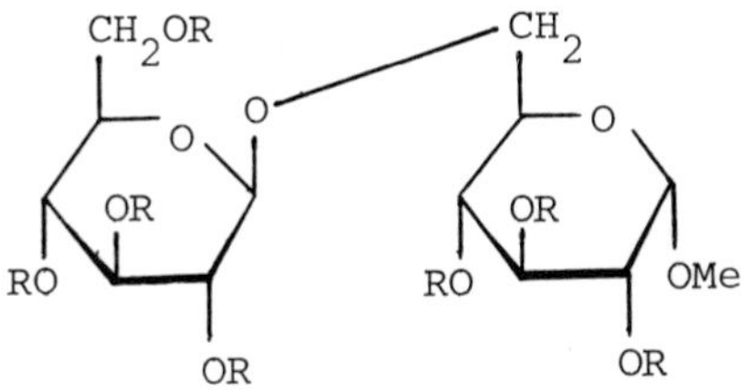

CH_2OR O CH_2 O OR RO OR OMe OR OR

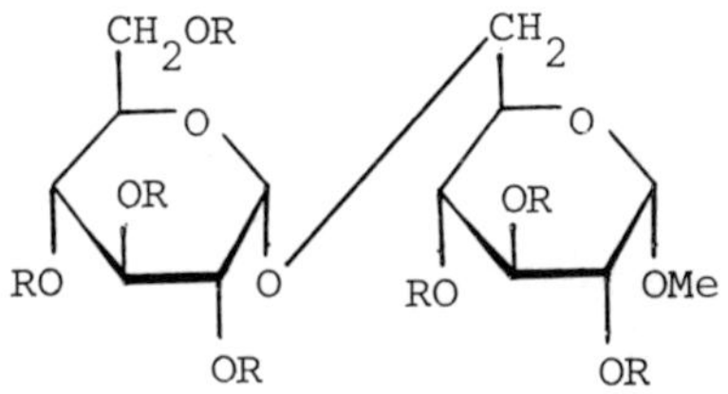

CH_2OR O CH_2 O OR RO O OR OMe OR OR

(12) R = $CH_2C_6H_5$ (13)

(diethyl ether or nitromethane gives similar results) with one equivalent of anhydrous toluene-*p*-sulphonic acid gives an excellent yield (90%) of disaccharides (12) and (13) (α : β ratio, 84 : 16, g.l.c.). These results compare well with the halide ion-catalysed reaction (75%, α : β ratio, 9 : 1), (P. Sinaÿ, *loc. cit.*). The presence of a non-participating group at C-2 in the imidate intermediate is necessary for the synthesis of the 1,2-*cis*-disaccharide. When the imidate intermediate contains a participatory group at C-2, it gives an orthoester as the main product.

The potential of 3,4,6-tri-*O*-acetyl-1,2-*O*-(1-cyanoethylidene)- α -D-glucopyranose as a glycosylating reagent for 1,2-*trans*-glycoside synthesis has been demonstrated (Y.V. Wozney, L.V. Backinowsky , and N.K. Kochetkov, Carbohydrate Res., 1979, 73, 282). Suitably protected trityl ethers of sugars are used as aglycons in the condensation reaction, which is usually performed in dichloromethane in the presence of triphenylmethylium perchlorate.

Phenyl 1-thio-D-glucopyranosides in the presence of mercury (II) salts are readily solubilised to give alkyl D-glucopyranosides (R.J. Ferrier, R.W. Hay, and N. Vethaviyasar, *ibid.*, 1973, 27, 55). Condensation of phenyl tetra-*O*-benzyl-1-thio- β -D-glucopyranoside with 1,2:3,4-di-*O*-isopropylidene- α -D-galactopyranose in refluxing tetrahydrofuran in the presence of mercury (II) sulphate has been described to give 6-*O*-(2,3,4,6-tetra-*O*-benzyl- α -D-glucopyranosyl)-1,2:3,4-di-*O*-isopropylidene- α -D-galactopyranose.

Formation of glycosides by boron trifluoride etherate catalysed addition of alcohols to glucal has been described (R.J. Ferrier and N. Prasad, Chem. Comm., 1968, 476). The addition reaction is followed by an elimination reaction to give unsaturated anomeric glycosides, which can be converted by means of hydroxylation or epoxidation reactions into the desired glycoside (Scheme 8).

The 2,3-diphenyl-2-cylcopropen-1-yl-(Ph_2cp) group has been used for the activation of secondary as well as primary hydroxyl groups of sugar aglycons. Sugar Ph_2cp ethers are easily obtained by the action of 2,3-diphenyl-2-cylcopropen-1-ylium perchlorate on suitably protected sugars in the presence of 2,4,6-trimethylpyridine. Condensation of Ph_2cp ethers with a

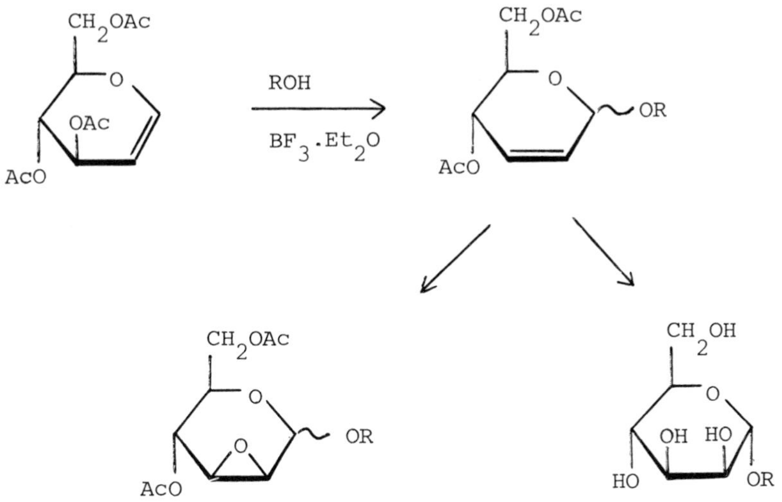

CH_2OAc O OAc AcO

ROH

$BF_3.Et_2O$

CH_2OAc O OR AcO

CH_2OAc O O OR AcO

CH_2OH O OH HO HO OR

Scheme 8

slight excess of glycosyl bromide and silver perchlorate has been used to prepare several 1 → 3, 1 → 4 and 1 → 6 β-linked di- and trisaccharides (A. Ya. Khorlin, V.A. Nesmeyanov and S.E.Zurabyan, Carbohydrate Res., 1975, 43, 69).

The Lewis acid catalysed syntehsis of 1,2-*trans*-glycosides from sugar lactim ethers and amide acetals has been reviewed (S.Hanessian and J.Banoub, in "Synthetic Methods for Carbohydrates", ed., H.S.El Khadem, ACS Symposium Series 1976, 39, 36).

Synthesis of oligosaccharides on a polymeric support provides a new approach to the problem of protecting groups in this field

(J.M. Frechét and C. Schuerch, J. Amer. chem.Soc., 1971, 93, 492; R.D. Guthrie, A.D.Jenkins and J. Stechlicek, J.chem. Soc. C., 1971, 2690; R.D. Guthrie, A.D. Jenkins, and G.A.F. Roberts, J. chem. Soc.Perkin I, 1973, 2414; U. Zehavi and A. Patchornik, J. Amer. chem. Soc., 1973, 95, 5673; G.Excoffier, D.Y. Gagnaire and M.R. Vignon, Carbohydrate Res. , 1976, 46, 215; S.-H.L. Chiu and L. Anderson, *ibid.*, 1976, 50, 227). Several methods of linking the first sugar to the support and its coupling to the next sugar have been studied. In most cases a hydroxyl group of the bound sugar serves as the glycosyl acceptor, but in one instance (R.D. Guthrie,A.D.Jenkins and G.A.F. Roberts, *loc.cit.*) the anomeric centre is activated for coupling by converting it into the orthoester.

A fundamentally different approach to the synthesis of disaccharides involves the preparation of a dienyl ether of monosaccharide followed by Diels-Alder condensation with a glyoxylic ester and subsequent functionalisation of the resulting dihydropyran ring (S. David, J. Eutache, and A. Lubineau, J. chem. Soc. Perkin I, 1974, 2274; S. David, A. Lubineau, and J.M. Vatèle, Chem. Comm., 1975, 701, J.chem. Soc. Perkin I, 1976, 1831). In a related work Grynkiewicz has described ways of linking a racemic dihydropyran derivative with suitably protected monosaccharide derivatives (G. Grynkiewicz, Carbohydrate Res., 1980, 80, 53).

(b) Physical Methods

(i) ^{1}H-Nuclear Magnetic Resonance

The value of ^{1}H-n.m.r. spectroscopy in elucidating the structures of carbohydrates is well recognised (L.D. Hall, Adv. carbohydrate Chem., 1964, 19, 51, Adv. carbohydrate Chem. Biochem., 1974, 29, 11; G. Kotowycz and R.U. Lemieux, Chem. Rev., 1973, 73, 669; S.J. Angyal, Angew. Chem. international Edn., 1969, 8, 157; R.U.Lemieux and J.D.Stevens Canad. J. Chem., 1966, 44, 249; B. Casu, *et al.*, Tetrahedron, 1966, 22, 3061). ^{1}H-n.m.r. spectra of D-*gluco*-oligosaccharides and D-glucan have been studied (J.M. Van Der Ween, J. org. Chem., 1963, 25, 564; T. Usi, *et al.*, Carbohydrate Res., 1974, 33, 105; A. De Bruyn, M. Anteunis and G. Verhegge, Bull. Soc. chim. Belg., 1975, 84, 721).

In the ^{1}H-n.m.r. spectra of several glucobioses and glucotrioses the anomeric and the inter-glycosidic proton resonances have been assigned (Usi *et al., loc.cit.*) In order to observe the signals of anomeric protons it was generally found necessary to heat the n.m.r. solution to shift the interferring HOD signals. The signals due to the anomeric proton of the reducing end of the molecule can readily be distinguished from those of the other protons because it displays the effect of mutarotation (α- and β-anomers). With the exception of (1→ 2) linked glucobioses, adjacent peaks having chemical shifts and coupling constants close to those observed for α- and β-D-glucopyranose may be assigned to H-1α and H-1β . As expected, the coupling constants for the *trans* configuration were ∿7.5 Hz and for the *cis* ∿3 Hz. The anomeric compositions of glucobioses and related compounds have been determined. In general, the β-anomer predominated except in (1 → 2) linked glucobioses. In sophorose, 2-*O*-β-D-glucopyranosyl-D-glucopyranose, the α-anomer constituted about two-thirds of the mixture (Usi *et al., loc.cit.*) These results indicate that the hydroxyl group at C-1 prefers an axial orientation when the 2-hydroxyl group is substituted. The inter-glycosidic anomeric proton resonances ($J_{1',2'}$ or $J_{1'',2''}$ ∿3Hz) for 1,2-*cis*-glucosides appear in the region of 5.00 - 5.45 p.p.m. and those of the corresponding 1,2-*trans*-glucosides ($J_{1',2'}$ or $J_{1'',2''}$ ∿7.5Hz) in the region of 4.50 - 4.80 p.p.m. The anomeric proton resonances of the inter-glycosidic linkage overlapped in the anomers except in the case of (1→2) linked glucobioses and laminarabiose (3-*O*-β -D-glucopyranosyl-D-glucopyanose).

300 MHz ^{1}H.n.m.r. spectra of a large number of glucobioses (A.De Bruyn *et al., loc.cit.*), melezitose and raffinose (M.Anteunis, A.De Bruyn and G.Verhegge, Carbohydrate Res., 1975, 44, 101) have been reported. The chmical shifts were compared with those α- and β-D-glucopyranose. The reasonances due to individual protons and the position of the glycosidic linkage were established by homo-INDOR experiments and by using increment rules. The authors noted that all protons on or vicinal to the carbon atom involved in the glycosidic linkage were shifted downfield.

Per-*O*-(trimethylsilyl) (Me_3Si) ethers of oligosaccharides have proved useful in determining the configuration of the glycosidic linkage by ^{1}H.n.m.r. spectroscopy (C.G.Hellerqvist, O.Larm and B.Lindberg, Acta. chem. Scand., 1971, 25, 743; J.P.Kamerling, D.Rosenberg and J.F.G.Vliegenthart, Biochem. biophys. Res. Comm., 1970, 38, 794; J.P. Kamerling,

M.J.A. De Bie and J.F.G.Vliegenthart, Tetrahedron, 1972, 28, 3037). Chemical shifts and coupling constant values of the anomeric protons of thirty Me_3Si derivatives of oligosaccharides have been reported (Vliegenthart and coworkers, *loc.cit.*) On the basis of these data a method for the determination of the configuration of the glycosidic linkage in oligosaccharides has been suggested. The ^{1}H-n.m.r. spectrum of octa-*O*-(trimethylsilyl) maltose shows three anomeric signals, with coupling constants of $\sim$ 3 Hz (axial-equatorial arrangement) and one anomeric signal with a coupling constant of $\sim$ 7 Hz (axial-axial arrangement).Hence it is concluded that the configuration of the glycosidic linkage in the above compound is α. The ^{1}H-n.m.r. spectrum of octa-*O*-trimethylsilyl) lactose shows three anomeric signals, with coupling constants of $\sim$ 7 Hz, and one anomeric signal with a coupling constant of $\sim$ 3 Hz, indicating β-configuration for the glycosidic linkage.

The application of this method is limited to oligosaccharides that contain monosaccharide units having an axial proton on C-2 (for example D-glucose and D-galactose). In the ^{1}H-n.m.r. spectra of Me_3Si derivatives the non-anomeric proton resonances are usually not well resolved. Therefore for the determination of the conformation of monosaccharide residues, the Me_3Si derivatives are not suitable. For this purpose, peracetylated derivatives are preferred. The non-anomeric protons of peracetylated derivatives of oligosaccharides are spread over a large δ-range ($\sim$2 p.p.m.) and therefore make their identification possible (K.Koizumi and T. Utamura,Carbohydrate Res., 1978,63, 283; J.O.Deferrari,I.M.E. Thiel and R.A.Cadenas, *ibid.*, 1973,26, 244, Anales Asoc.Quim. Argentina,1973,61,107; W.W.Binkley,D. Horton and N.S.Bhacca, Carbohydrate Res., 1969, 10, 245; H. Friebolin, G.Keilich and E. Siefert, Org.mag.Res.,1970,2,457; G.Keilich,E.Siefert and H.Friebolin, *ibid.*, 1971, 3, 31; A.De Bruyn, M.Antenuis and G.Verhegge, Carbohydrate Res., 1975, 42, 157; G.R.Newkome *et al.*,*ibid.*1976,48,1; T.Otake, Bull.chem.Soc.Japan, 1974, 47,1938). However, the anomeric proton of the glycosyl residue is not well resolved and appears in the region of non-anomeric protons.

A useful way of identifying the position in a disaccharide at which a reaction has occurred is by comparing the ^{1}H-n.m.r. spectra of the peracetylated starting compound with that of the peracetylated product. The position of isolated hydroxyl groups in sucrose and maltose has been ascertained using this approach. In the ^{1}H-n.m.r. spectrum of 2,3,6,1',3',4',6'-hepta-*O*-acetylsucrose, separate signals for H-1, 2, 3, 3',4' are observed, as expected, at δ 5.64, 4.3, 5.42, 5.5 and 5.37, respectively (R. Khan, Adv. carbohydrate Chem.Biochem., 1976, 33, 235). The fact that the signal of H-4 is shifted upfield from where it usually occurs (δ ~5.29) in sucrose octa-acetate indicates the location of the free hydroxyl group in the sucrose hepta-acetate. Addition of trichloro-acetyl isocyanate to the hepta-acetate in chloroform-d solution results in the appearance of a singlet at δ 8.78 due to the immino proton, thereby confirming the presence of one hydroxyl group in the compound. Simiarly, in the ^{1}H-n.m.r. (200 MHz) spectrum of methyl 2,6,2',3',4',6'-hexa-*O*-acetyl-β-maltoside, the H-3 resonance appears as a wide triplet ($J_{2,3}$ = $J_{3,4}$ = 9.4 Hz) at ~1.6 p.p.m. at higher field than that of the corresponding resonance for methyl-β-maltoside hepta-acetate. This indicates that the hydroxyl group in the hexa-acetate is located at C-3. The resonances due to other protons on the reducing ring are also moved upfield with respect to the corresponding resonances for the peracetylated derivative, but to a smaller extent (W.E. Dick, Jr., B.G. Baker and J.E. Hodge, Carbohydrate Res., 1968, 6, 52; B. Koeppen, *ibid*., 1970, 13, 193, P.L. Durette, L. Hough and A.C. Richardson, J. chem. Soc. Perkin I,1974,88).

The upfield shift of resonances due to protons on adjacent carbon atoms bearing acetaloxy (R.Khan, Carbohydrate Res., 1974, 32, 375; R. Khan and K.S.Mufti, *ibid*., 1975, 43, 247; R. Khan, K.S. Mufti, and M.R. Jenner, *ibid*., 1978, 65, 109), methoxyl (M.G. Lindley, G.G.Birch and R. Khan, *ibid*., 1975, 43, 360), methylsulphonyloxy (R. Khan and K.S. Mufti, *loc. cit*., P.L. Durette, L. Hough and A.C. Richardson, J. chem. Soc. Perkin I, 1974, 97), azidodeoxy (Hough and coworkers, *loc. cit*.), and chlorodeoxy group (R. Khan, M.R.Jenner and K.S.Mufti, Carbohydrate Res., 1975, 39, 253) in comparison with the corresponding acetoxyl substituent has also been observed.

A Fourier-transform method has been developed to measure the spin-lattice relaxation time (T_1 values) of the anomeric protons of several oligo- and polysaccharides. Data for cellobiose, maltose, lactose, gentiobiose and melibiose reveal that proton T_1-values may provide a useful method for eval-uating conformations of oligosaccharides (L.D.Hall and

C.M. Preston, Carbohydrate Res., 1976, 49, 3).

(ii) ^{13}C-Nuclear Magnetic Resonance Spectroscopy

Carbon-13 n.m.r. spectroscopy is now an established technique for the determination of the composition, linkage, and in a few cases, the sequence in complex carbohydrates (G. Kotowycz and R.U. Lemieux, Chem. Rev., 1973, 73, 669; N.K. Wilson and J.B. Stothers, in "Topics in Stereochemistry", eds.E.L. Eliel and N. Allinger, Wiley-Interscience, New York, 1974; A.S. Perlin, in "MTP International Review of Science, Organic Chemistry", Series 2, ed. G.O. Aspinall, Butterworth, London, 1976, vol. 7, p.1; B. Coxon, in "Developments in Food Carbohydrate - 2", ed. C.K. Lee, Applied Science Publishers, 1980, p.351; H.J. Jennings and I.C.P. Smith, Methods Enzymol. Part C, 1978, 39).

The ^{13}C-n.m.r. spectra of oligosaccharides are tentatively assigned using the chemical shift correlation technique which involves comparison of the chemical shift values with those of the monomer components (D.E. Dorman and J.D. Roberts, J. Amer. chem. Soc., 1970, 92, 1355; 1971, 93, 4463; D. Doddrell and A. Allerhand, *ibid.*, 1971, 93, 2779; T. Usi, *et al.*, J. chem. Soc. Perkin I, 1973, 2425; P. Colson, H.J. Jennings and I.C.P. Smith, J. Amer. chem. Soc., 1974, 96, 8081; J. Haverkamp, M.J.A. De Bie and J.F.G.Vliegenthart, Carbohydrate Res., 1974, 37, 111; D.D.Cox, *et al.*, *ibid.*, 1978, 67, 23; P.Kováč and J. Hirsch, *ibid.*, 1980, 85, 117; A. Lipták, *et al.*, Tetrahedron, 1980, 36, 1261).

More definite methods of ^{13}C assignments are listed below.

Chemical substitution (P. Colson, *et al.*, Canad. J. Chem., 1975, 53, 1030; P. Colson, H.J. Jennings and I.C.P. Smith, *loc. cit.*; J. Haverkamp, M.J.A. De Bie, and J.F.G. Vliegenthart, *loc.cit.*).

Selective heteronuclear spin decoupling (D.Y. Gagnaire, F.R. Taravel, and M.R. Vignon, Carbohydrate Res., 1976, 51, 157; J. Feeney, D. Shaw, and P.J.S. Pauwels, Chem. Comm., 1970, 554; N.S. Bhacca, F.W. Wehrli, and N.H. Fischer,J.org. Chem., 1973, 38, 3618).

Selective spin labelling by ^{13}C (G. Excoffier,D. Y.Gagnaire and F.R. Taravel, Carbohydrate Res., 1977, 56, 229;

D.Y. Gagnaire, *et al.*, Nouveau, J. Chim., 1977, 1, 423; T.E. Walker, *et al.*, J. Amer. chem. Soc., 1976, 98, 5807; J.A. Schwarez and A.S.Perlin, Canad. J. Chem., 1972, 50, 3667).

Selective spin labelling by ^{2}H (A.S. Perlin and B. Casu, Tetrahedron Letter, 1969, 2921; H.J. Koch and A.S. Perlin, Carbohydrate Res., 1970, 15, 403; P.A.J. Gorin, Canad. J. Chem., 1974, 52, 458, Carbohydrate Res., 1975, 39, 3; H.J. Koch, and R.S. Stuart, *ibid.*, 1977, 59, C1; G.K.Hamer, *et al.*, Canad. J. Chem., 1978, 56, 3109: A. Heyraud,*et al.*, Biopolymers, 1979, 18, 167).

Deuterium-induced differential isotope technique (P.E. Pfeffer, K.M. Valentine, and F.W. Parrish, J. Amer. chem. Soc., 1979, 101, 1265).

^{13}C-N.m.r. spectra of glucobioses and glucotrioses reveal that C-1 chemical shifts for most of α- and β-anomeric forms of each reducing sugar in aqueous solution are in magnitude to those of C-1 in α- and β-D-glucopyranose and the C-1 resonance of the β-form is shifted downfield by about 3.9 p.p.m. from that of the α-form except in the case of 1,2-linked gluobioses (T. Usi, *et al.*,*loc. cit.*). The C-1 resonances usually appear at 93.2 p.p.m. for α-anomers and at 97.1 p.p.m. for β-anomers. The C-1 resonances of α-kojiobiose (90.8 p.p.m.) and β-sophorose (95.8 p.p.m.) appear at slightly higher field than those of the other reducing sugars. The α-linked, anomeric carbon nuclei,except in α, α -trehalose, resonate in the region of 97-101 p.p.m. and β-linked anomeric carbon nuclei, except in β , β -trehalose, are downfield in the region of 103-105 p.p.m.through out the series of reducing gluobioses, the signals of the glycosidic carbon atoms appear at lowest field among those of the remaining carbon nuclei and are observed as overlapped peaks of an anomer pair, except in the case of 1,2-linked glucobioses.

The positions of linkages in di-and tri-saccharides can be ascertained by observing which ^{13}C resonances of their component monosaccharides have undergone a large downfield shift on *O*-glycosylation. ^{13}C-N.m.r. spectra of (1 ⟶ 1),(1 ⟶ 2) (1 ⟶ 3), (1 ⟶ 4), and (1 ⟶ 6) gluobioses and glucotrioses indicate that this downfield shift (4 - 10 p.p.m.) is, with the exception of trehaloses, comparable with that

(8 - 11 p.p.m.) undergone by the carbon signals of monosaccharides on methylation of a hydroxyl group directly attached to the carbon (Usi *et al.*, *loc.cit.)*

All the ^{13}C resonances of maltose and cellobiose have been unambiguously assigned using selective proton irradiation, ^{13}C selective spin decoupling and isotopic chemical shifts of hydroxyl bearing carbon atoms induced by deuterium (A.Heyraud, *et al.*,*loc.cit.*) In the ^{13}C-n.m.r. spectrum of maltose the α/β ratio corresponding to mutarotation is equal to the value obtained for D-glucose. Hence, the signals due to C-1α, C-1β, C-1', C-6β, C-6α and C-6' are readily assigned. The two latter signals are overlapping. By analogy with cellobiose the signals at 78.9, 78.6 and 101 p.p.m. (corrected to external tetramethylsilane as a reference) are assigned to C-4α, C-4β and C-1' respectively. In comparison with cellobiose the resonances due to the C-1' and C-4 α/β in maltose appear at a higher field. The resonances for the C-2α, C-3α and C-5α from the C-2β, C-3β and C-5β are distinguished on the basis of α/β anomeric ratio. The signals at 71.4, 74.0 and 76.0 p.p.m. are assigned to C-5α,C-5' and C-5β, respectively by a β-isotopic effect. Selective spin-decoupling of H-2', H-2α and H-2β permits the assignment of C-2', C-2α and C-2β. Selective labelling of maltose with ^{13}C at C-1' confirms the assignment for C-2' (73.1 p.p.m.) because of a coupling with C-1'. The existence of a geminal coupling ^2J(C-1' ⟶ C-5') ∿1.5 Hz, which is characteristic of 1,2-*cis*-linkage, confirms that the resonance at 74.0 p.p.m. is due to C-5'.

^{13}C-N.m.r. spectra of oligosaccharides and polysaccharides containing D-fructose have been examined (A.Allerhand and D.Doddrell, J.Amer.chem.Soc., 1971, 93, 2777; D.Doddrell and A.Allerhand, *ibid.*, 1971, 93, 2779; W.W.Binkley, *et al.*, Carbohydrate Res., 1972, 23, 301; L. Hough, *et al.*, *ibid.*, 1976, 47, 151; R.Khan, Adv. Carbohydrate Chem. Biochem., 1976, 33, 235; S. Ho, H.J.Koch, and R.S.Stuart, Carbohydrate Res., 1978, 64, 251; F.R. Seymour, *et al.*, *ibid.*, 1979, 72, 57; H.C. Jarrell,*et al.*, *ibid.*, 1979, 76, 45; A.J. Jones, P. Hanisch and A.K. McPhail, Austral. J.Chem., 1979, 32, 2763; P.E.Pfeffer, K.M. Valentine, and F.W. Parrish, *loc.cit.*) The chemical shift values observed for sucrose almost equal the sum of the values for α-D-glucopyranose and β-D-fructofuranose.

The resonance assignments for sucrose have been confirmed by deuterium isotope shift (Pfeffer, *et al.*, *loc.cit.*). A deuterium induced ^{13}C differential isotope shift (d.i.s.) is defined as the chemical shift difference in parts per million between the ^{13}C shift as observed in H_2O and the upfield D_2O-induced ^{13}C shift. The differential shift positions have been measured simultaneously in the magnetic field with a dual coaxial n.m.r. cell. By means of a linear regression analysis of the contribution to the d.i.s. values it has been determined that these contributions are additive and that measurable shifts are principally induced on ^{13}C resonances by directly bonded hydroxyl groups (β-shift, -0.11 to -0.15 p.p.m., e.g. β-shift at C-1, -0.11 and at C-6, -0.15 p.p.m.) and hydroxyl group on vicinal carbons (γ-shift, -0.03 to -0.06 p.p.m., e.g. γ-shift from a *trans*-anomeric OD, -0.06 and *cis*-anomeric OD, -0.03 p.p.m.). Application of this technique to other disaccharides has led to ^{13}C shift re-assignments (or confirmatory evidence for previous assignments) for C-2, 3, 5 of α-lactose, C-3, 5 of methyl β-lactopyranoside, C-3', 5' of α,β-cellobiose, C-2, 3, 4 of α-cellobiose, C-2, 3, 4, 5 of β-cellobiose, C-2, 3, 5 of α,α-trehalose and of the α- and β-anomers of α,β-trehalose. In sucrose, the near coincidence of the d.i.s. values observed at carbon sites 6, 1' and 6' (0.14, 0.15 and 0.14, respectively) prevents unambiguous assignments by this technique.

The C-deuteriation technique has been used extensively to facilitate the analysis of ^{13}C-n.m.r. spectrum. Replacement of a ^{13}C bound proton by a deuterium usually causes a substantial reduction in the intensity of the α-^{13}C resonance from the spectrum (H.J.Koch and A.S.Perlin, Carbohydrate Res., 1970, 15, 403) and shift the signal of a β-carbon atom, thereby identifying these resonances, or it can demonstrate the presence or absence of coupling between the proton and various atoms of the molecule. A simple C-deuteriation method employing D_2O and Raney nickel catalyst has been developed (H.J.Koch and R.S.Stuart, *ibid.*, 1977, 59, C 1). This technique cannot be used for carbohydrates having groups sensitive to reduction and therefore reducing sugars are converted into glycosides prior to the ^1H-^2H exchange reaction. The possibility of inversion of configuration during the exchange reaction has been considered (H.J.Koch and R.S.Stuart, *loc.cit.*; F. Balza, *et al.*, Carbohydrate Res., 1977, 59, C 7). Inversion is a much slower process than deuteriation and therefore can be avoided by optimising the rate of the ^1H-^2H exchange reaction.

The ^{13}C-n.m.r. spectrum of the catalytically deuteriated methyl β-maltoside displays resonances only for C-1,4, 5, 1',5' and CH_3 (plus a residual signal due to slow exchange of H-2') that allowed the earlier assignments for C-3, 5 (D.E.Dorman and J.D.Roberts, J.Amer.chem.Soc., 1971, 93, 4463) and for C-2', 5' (Usi, *et al.*,*loc.cit.*) to be reversed. The ^1H-^2H exchange reaction with methyl β-cellobioside caused deuteriation at C-3,6,2',3',4',6'. On the basis of these results the earlier assignments for C-3,5,3',5' have been reversed (G.K.Hamer, *et al.*, Canad.J.Chem.,1978,56, 3109). The marked simplification of the ^{13}C-spectrum of deuteriated β-cellobioside has facilitated conformational studies of the interglycosidic linkage.

The most important coupling constants of disaccharides are the three-bond ^{13}C-proton and ^{13}C-^{13}C couplings measured across the glycosidic bonds. The torsional angles ϕ and ψ about the glycosidic bonds of disaccharides (Figure 1) are of considerable interest in studies of structural differences in solution and in the solid state. Synthesis of oligosaccharides labelled extensively with deuterium has led to a marked simplification of their proton coupled ^{13}C-n.m.r. spectra which allows measurement, with the aid of computer simulation of the coupling constants J(C-1',H-4) 4.3Hz and J(C-4,H-1')

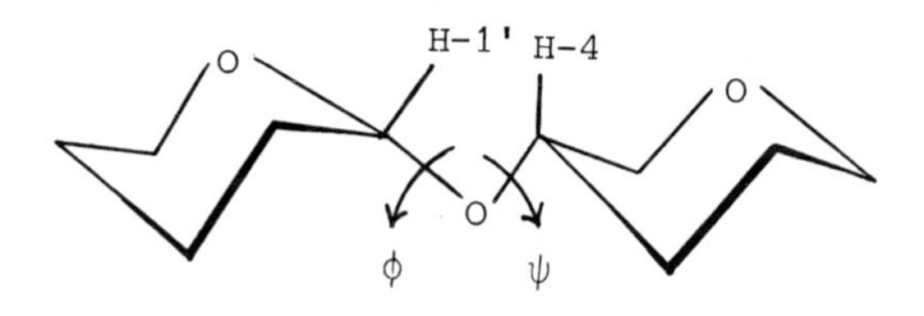

Maltose

Figure 1

4.2 Hz for methyl β-cellobioside-d_8 (Hamer,*et al.*,*loc.cit.*) The inter-residue couplings of methyl β-cellobioside-d_8 correspond to torsional angles ϕ and ψ of about 25-30°. Although these values reflect a thermodynamically-averaged conformation, they suggest that both ϕ and ψ are of the same order as their values (ϕ = 25° and ψ = 38°) in the solid state (J.T.Ham and D.G.Williams, Acta Crystallog.,1970,B26, 1373).

(iii) Mass Spectrometry

Mass spectrometry has become an important technique in the determination of the structures of carbohydrates (N.K. Kochetkov, and O.S.Chizhov,Adv.carbohydrate Chem.,1966,21, 39; O.S.Chizhov and N.K.Kochetkov,Methods carbohydrate Chem. 1972,6,540; H.Budzikiewicz,C. Djerassi and D.H.Williams, "Structure Elucidation of Natural Products by Mass Spectrometry", Holden-Day,San Francisco,1964,Vol.II, p.203, N.K. Kochetkov,O.S.Chizhov and A.F.Bochkov,in "MTP International Review Science, Organic Chemistry", Series I,ed. G.O.Aspinall, Butterworth,London,1973, Vol.7, p.147; J.Lönngren and S. Svensson,Adv.carbohydrate Chem.Biochem.,1974,29, 41; E.C. De Jong, *et al.*,in "Recent Development in Mass Spectrometry in Biochemistry and Medicine",ed. A.Frigerio, Plenum, New York,1978,p.483; A.L.Burlingame, *et al.*, Analyt.Chem., 1980,52, 214R).

Mass spectra of permethylated oligosaccharides have been reported (O.S.Chizhov,L.A.Polyakova and N.K.Kochetkov,Dokl. Akad.Nauk SSSR, 1964, 158, 685; J.Kärkkäinen, Carbohydrate Res.,1971,17, 1; K.G.Das and B.Thayumanavan, Org.mass.Spec., 1972, 6, 1063; J. Moore and S.E.Waight, Biomed. mass Spec., 1975, 2, 36; I.Mononen, J. Finne and J.Kärkkäinen, Carbohydrate Res., 1978, 60, 371; B.A. Fraser, F.P. Tsui and W. Egan, *ibid.*, 1979, 73, 59; V. Kovácik, *et al.*, Biomed. mass Spec., 1978, 5, 136).

The electron impact mass spectra of a large number of permethylated di-, tri- and tetra-saccharides have been described (Moore and Waight, *loc.cit.*) Distinct molecular ions $(M)^{+\cdot}$ are observed for all the compounds, except octa-*O*-methyl sophorose. For the sophorose derivative the species of highest mass (m/e 423) corresponds to the loss of a methoxyl radical from $(M)^{+\cdot}$. In all the compounds studied, the ion

$(M - 45)^{+}$, due to loss of $CH_3OCH_2^{\cdot}$, is more abundant than $(M)^{+\cdot}$, and this can be used to determine the molecular weights.

The fission of interglycosidic linkages in permethylated oligosaccharides follows principles similar to those for fission of the glycosidic linkage in permethylated glycosides (scheme 9). Pathway a results in a glycosyl cation $(G1)^{+\cdot}$, which is resonance stabilised and should therefore be more favoured than pathway b. All permethylaldohexosylaldohexose disaccharides give a peak at m/e 219, trisaccharides reveal a homologous diglycosyl cation at m/e 423. Characteristic differences in the fragmentation patterns and the intensity of fragment ions of permethylated oligosaccharides enable the identification of the position of the glycosidic linkages (Moore and Waight, *loc.cit.*) The presence of a (1 → 1) glycosidic linkage is characterised by an intense peak at m/e 101 and weak peaks at m/e 88, 219 and 279. The peaks at m/e 380 and 305 are characteristics of (1 → 2) and (1 → 4) linkages. Further fragmentation of the ion m/e 380 allows to differentiate between (1 → 2) and (1 → 4) linkages. A relatively abundant ion at m/e 353 is characteristic of (1 → 6) linkage.

The mass spectra of trimethylsilyl (Me_3Si) ethers of oligosaccharides have been studied (N.K.Kochetkov,O.S.Chizhov, and N.V.Molodtsov,Tetrahedron, 1968, 24, 5587; J. P. Kamerling, *et al.*, Tetrahedron Letters, 1971, 2367; Tetrahedron, 1971, 27, 4275, 1972, 28, 4375; W.W.Binkley, *et al.*, Carbohydrate Res., 1971, 17, 127; R. Bentley, *ibid.*,1977, 59, 274; G.Puzo and J.C. Prome, Biomed. mass Spec., 1978, 5, 146). These derivatives are easy to prepare. Characteristic ions that are diagnostic of linkage position, though not of stereochemistry, have been identified. The main disadvantages of the use of trimethylsilyl derivatives are that they give very weak molecular ions upon electron impact and that the large increase in mass over the parent sugar results in difficulties in establishing a mass scale. Use of peracetates of disaccharides for mass spectral study has generally been preferred,because of the ease of their preparation and purification, (J.Karliner,Tetrahedron Letters,1968,3545; G.S.Johnson, W.S.Ruliffson, and R.G.Cooks, Chem. Comm., 1970,587; W.W.Binkley, *et al.*,*loc.cit.*; J. Guerrera and C.E.Weil, Carbohydrate Res.,1973,27,471; P.L. Durette, L. Hough and A.C.Richardson, *ibid.*, 1973, 31, 114; R.G.Edwards, *et al.*, *ibid.*, 1974, 35, 111; B.A. Fraser, F.P.Tsui and W. Egan, *loc.cit.*; C. Bosso, *et al.*, Org.

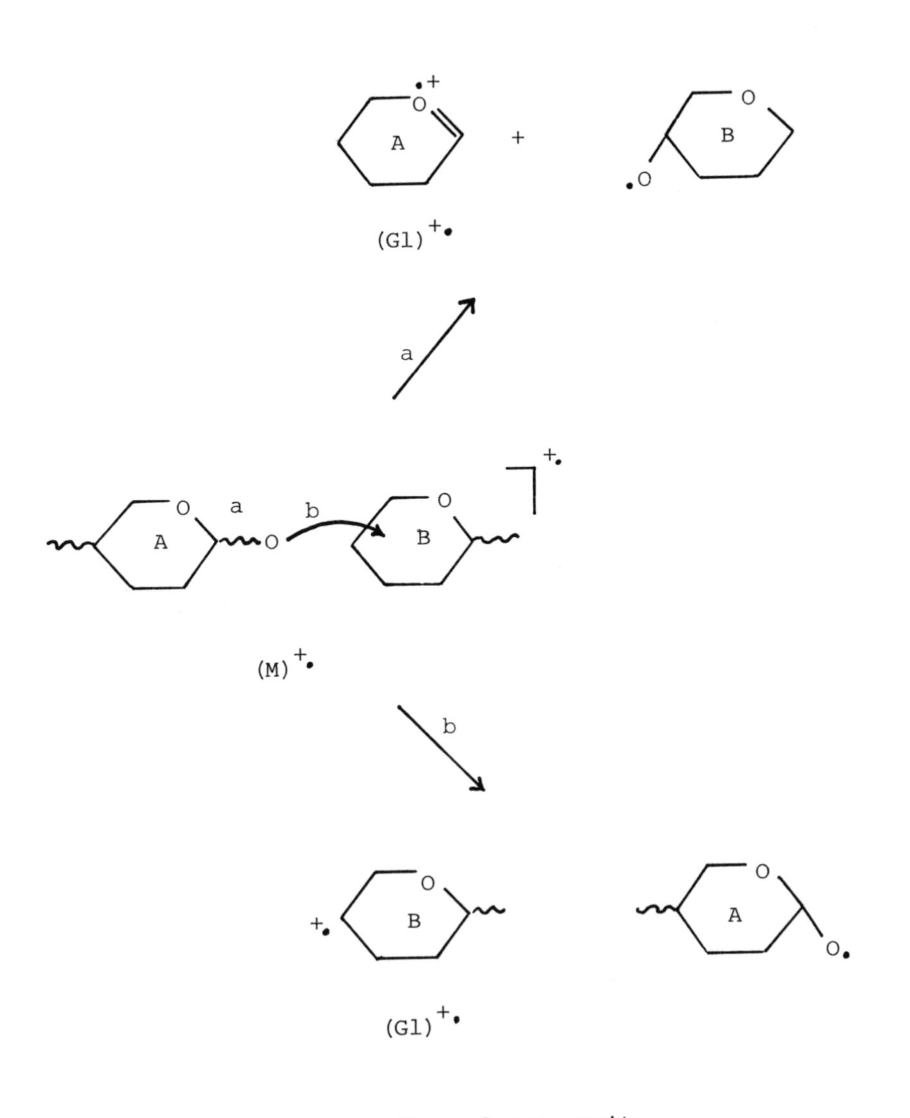

Gl = glycose unit

Scheme 9

mass. Spectrom., 1978, 13, 477; R. Khan, Adv. carbohydrate Chem. Biochem., 1976, 33, 235). High-resolution, electron-impact, mass spectra of peracetate derivatives of oligosaccharides containing D-fructofuranosyl residues have been reported (W. W. Binkley, *et al.*, *loc. cit.*). In sucrose octa-acetate, the initial fragmentation has been shown to proceed by localised ionisation at a single glycosyl residue to give a peak at m/e 331 due to the hexopyranosyl (14), and the keto-furanosyl (15) cations. The cleavage of the fructosidic bond is the most favoured initial fragmentation, as this leads to a tertiary carbonium ion.

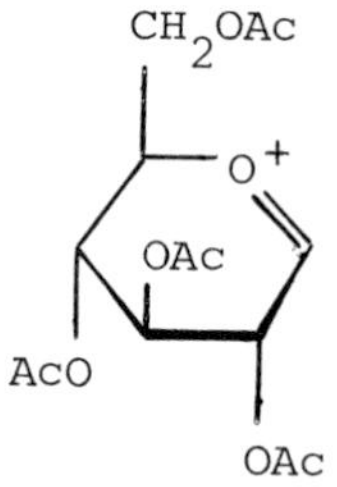

(14)

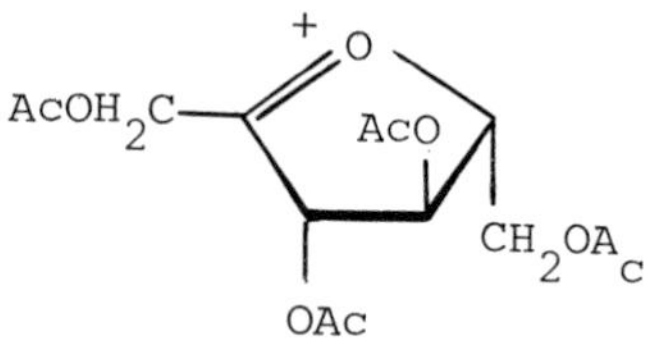

(15)

Mass spectral data for a large number of maltose derivatives has been reported (see above: Durette, *et al.*; Edwards, *et al.*; Guerrera and Weill, and G.O. Aspinall, *et al.* Canad. J. Chem., 1975, 53, 2182). In the mass spectrum of methyl 2,3,6,2'3'-penta-*O*-acetyl-4',6'-di-*O*-methyl- β -maltoside, a weak molecular ion peak at m/e 594 is observed, the fragmentation ions at m/e 563 (M - $CH_3O^\bullet$), 562 (M - CH_3OH), 535 (M - $CH_3COO^\bullet$), 534 (M - CH_3COOH), 503 (M - $CH_3O^\bullet$ - CH_3COOH), 502 (M - CH_3OH - CH_3COOH), and 475 (M - $CH_3COO^\bullet$ - CH_3COOH) provide evidence of the molecular weight (Aspinall and coworkers, *loc. cit.*). The major fragmentation pathway is initiated by cleavage of the two bonds attached to the interglycosidic oxygen atom (⋊O⋉) to give (16, m/e 275) and (17, m/e 303). The presence of ion (16) indicates that the non-reducing unit in the disaccharide contains two methoxyl groups, and the successive formation of ions at m/e 215 and 173 are consistent with the proposed location of the *O*-methyl and the *O*-acetyl substituents.

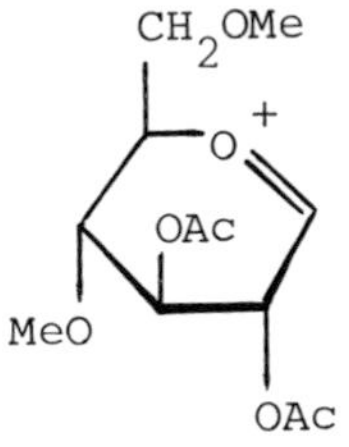

(16) 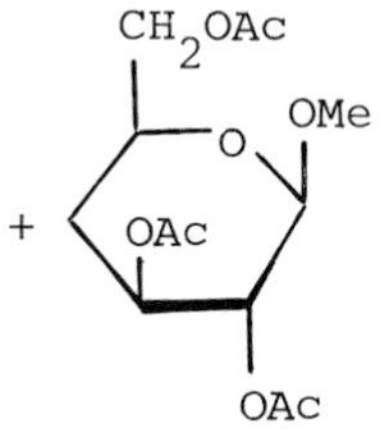 (17)

The fragmentation patterns under electron-impact of *N*-aryl-glycosylamines acetates (O.S. Chizhov, *et al.*, Org. mass Spec., 1971, 5 1157; Carbohydrate Res., 1973, 28, 21), *N*-phenylflavazole acetates (G.S. Johnson, W.S. Ruliffson, and R.G. Cooks, Chem. Comm., 1970, 587; Carbohydrate Res., 1971, 18, 233, 243), and *N*-phenylosotriazole acetates (O.S. Chizhov, *et al.*, Org. mass Spec., 1971, 5, 1145) of oligosaccharides have been described. The results indicate that the sequence of monosaccharide units, the position of glycosidic bonds, and other details of structure of oligosaccharides can be elucidated by this method.

A direct technique for obtaining electron-impact mass spectra of underivatised sucrose and lactose using a conventional electron ionisation mass spectrometer has been described (W.R. Anderson, Jr., W. Frick and G.D. Daves, Jr., J. Amer. chem. Soc., 1978, 100, 1974). The technique involves rapid sample heating, to effect vapourisation of polar, thermally unstable compounds while minimising decomposition and simultaneous detection of all ions formed by use of a photographic plate in the focal plane of a mass spectrometer of Mettauch Herzog geometry. Spectra of relatively involatile compounds obtained by this technique invariably exhibit the protonated molecular $(MH)^+$ ion rather than the molecular $(M)^+$ ion. The mass spectrum of sucrose shows ions at m/e 343 $(MH)^+$, 325 $(MH^+ - H_2O)$, 307 $(MH^+ - 2H_2O)$, 289 $(MH^+ - 3H_2O)$, 271 $(MH^+ - 4H_2O)$, and 163 (Hexopyranosyl and ketofuranosyl ions).

Chemical ionisation (c.i.) mass spectra of oligosaccharides peracetates using ammonia and isobutane as the reagent gas have been reported (R.C. Dougherty, *et al.*, J. org. Chem., 1974, 39, 451). For all the di-, tri- and tetra-saccharide peracetates examined, the oligosaccharide - ammonium complex (o.a.c.) was the base peak in the spectrum. In the spectra of the peracetate derivatives of sucrose, raffinose and stachyose, o.a.c. ions at m/e 696 (18), 984 and 1272, respectively, above the reagent-gas (m/e 60) were identified. As virtually all of the ions obtained under these conditions are even-electron ions, the even-mass ions must contain nitrogen. Ways in which

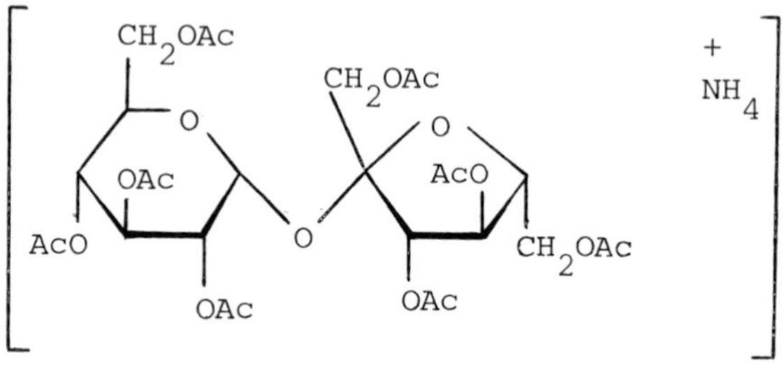

(18)

nitrogen-containing fragment could have arisen, such as thermolysis followed by attachment of $^{+}NH_4$ have been suggested. The glycosyl ions (m/e 331) resulting from the cleavage of the interglycosidic linkage of the disaccharide peracetates are also observed as the major fragments. Further fragmentation of the glycosyl ion proceeds in the manner precisely analogous to the electron-impact mass spectra of the disaccharide peracetates. Sequential elimination of acetic acid and ketene, often with corresponding metastable ions, accounts for the odd-mass ions between m/e 331 and 109. The relatively simple spectra obtained with c.i., together with the formation of molecular and high-mass ions, indicates the potential of this technique for structural studies of oligosaccharides.

The potential of negative chemical-ionisation (n.c.i.) techique for structural studies on very small samples of underivatised oligosaccharides has been demonstrated (A.K. Ganguly, *et al.*,Chem.Comm.,1979,148). The n.c.i. spectra of several oligosaccharides have been recorded, using methane samples into ion-source. Fairly intense molecular ions, M^{-}, are observed in each case. A small number of fragment ions are also observed, but their structures have not been deduced. Use of dichlorodifluoromethane as the reagent gas in n.c.i. mass spectrometry has proved valuable for structural studies of oligosaccharides, especially those having a number of free hydroxyl groups. Oligosaccharides produce intense (M + Cl) ions and few fragment ions corresponding to the loss of one or two discrete sugar units from either end of the oligosaccharide chain.

Field desorption (f.d.) mass spectra have been obtained for a number of di, tri- and tetra-saccharides and for some permethylated and peracetylated derivatives (J.Moore and E.S.Waight,Org.mass Spec.,1974,9, 903). The free sugars usually show intense $(M + 1)^{+}$ ions thus enabling the determination of molecular weights. Permethylated oligosaccharides are less suitable for f.d. studies, beacause of their increased volatility. Permethylated ethers show intesnse molecular ions, in contrast peracetylated oligosaccharides show only weak molecular ions, but strong $(M - 60)^{+\cdot}$ ions. Field desorption is particularly suitable for the determination of molecular weights of underivatised sugars. Fragmentation which occurs at higher emitter currents also give some structural information, but generally more detailed information can be obtained from the electron-impact mass spectrometry of suitable derivatives. The two ionisation techniques are therefore complementary.

A cationisation technique involving addition of alkali metal salts to samples has been used to obtain f.d. mass spectra of several oligosaccharides. (J.-C.Prome and G.Puzo, *ibid.*, 1977,12, 28). In the presence of sodium iodide, the mass spectra of di-, tri- and tetra-saccharides exhibit a cationised ion $(M - Na)^{+}$, which represents more than 85% of the total ion (Σ_{100}) for di- and tri-saccharides and about 60% for the tetra-saccharide stachyose. Alkali tetraphenyl borates have also been used as cation donors in f.d. mass spectrometry of oligosaccharides (H.J.Veith, Tetrahedron, 1977,33, 2825). The simple mass spectra which exhibit only $(M + cation)^{+}$ ion permit the determination of molecular weights.

Laser-induced desorption mass spectra of oligosaccharides have been determined (M.A.Posthumus, *et al.*, Anal.Chem., 1978,50,985). In this technique a thin layer of sample coated on a metal surface is exposed to a sub-microsecond laser pulse. This produces cationised species of the intact molecules and some other ions which appear to represent building blocks and other essential fragments. The results indicate that this technique has promising potential for the mass spectroscopic analysis of non-volatile and thermally labile compounds.

(iv) X-Ray Crystallography

The value of X-ray crystallography in determining configurations, conformations, bond lengths, valence angles, hydrogen bonding and molecular packing in carbohydrates is well recognised. (G.A.Jeffrey and R.D.Rosenstein, Adv.carbohydrate Chem.,1964,19, 7; R.H.Marchessault and A.Sarko,*ibid.*, 1967,22,421; G.Strahs, Adv.carbohydrate Chem.Biochem., 1970, 25, 53; G.A.Jeffrey and M.Sundaralingam, *ibid.*, 1974,30, 445; 1975,31, 347; 1976,32, 353; 1977, 34, 345; 1980, 37, 373; R.H.Marchessault and P.R.Sundararajan, *ibid.*,1976,33, 387; P.R.Sundararajan and R.H.Marchessault,*ibid.*,1978,35, 377; 1979,36,315).

In the crystal structure of α-lactose (4-*O*-β-D-galactopyranosyl-α-D-glucopyranose) monohydrate, both of the pyranose conformations are $^4C_1$, and the linkage torsional angles are -94°, +96° (Gal*p* ⟶ Glc*p*). An intramolecular hydrogen-bond between O - 3 of the D-glucose residues and the ring-oxygen atom (O - 5') of the D-galactopyranosyl group has been observed (D.C.Fries, S.T.Rao and M.Sundaralingam, Acta Crystallog. Sect.B, 1971, 27, 994; C.A.Beevers and H.N. Hansen, *ibid.*, p.1323. A similar intramolecular hydrogen-bonding (3-O—H-----O-5) in β-cellobiose (4-*O*-β-D-glucopyranosyl-β-D-glucopyranose) has also been established (S.S.C.Chu and G.A.Jeffrey, Acta Crystallog. Sect.B., 1968, 24, 830). There is a small twist in the backbone of the molecule that brings the hydrogen bond atoms close together.

The crystal structures of α-maltose (F.Takusagawa and R.A.Jacobson, Acta Crystallog.Sect.B.,1978, 34,213), β-maltose monohydrate (G.J.Quigley, A.Sarko and R.H.Marchessault, J.Amer.chem. Soc., 1970, 92, 5834) and methyl β-maltoside monohydrate (19), (Chu and Jeffrey, Acta Crystallog. 1967, 23, 1038) have been determined. In α-maltose, both the D-glucose moieties are slightly distorted from $^4C_1$ to a skew conformation. Such a distortion is not observed for β-maltose or methyl β-maltoside, in which both moieties assume $^4C_1$ conformation. The torsional angles O-4' —— C-1' —— O-1' —— C-4 and O-1 —— C-4 —— O-1' —— C-1' in α-maltose are + 172.1° and - 177.6°, respectively. This small twist in the backbone of the molecule is such as to favour intramolecule hydrogen-bonding between OH-2' and O-3. This interesting feature has been observed in all known crystal structures that contain a maltose residue, and appears to be one of the principal forces controlling conformations based on α-(1 ⟶ 4) linked D-glucoses. In methyl β-maltoside monohydrate (19), the water molecule is hydrogen bonded to both of the D-glucopyranose residues and may be considered to be part of the molecule. A small twist in the backbone of the molecule moves the donor atoms O-6 and O-6' apart, to accommodate the molecule of water, and brings O-2 and O-3' closer to form the intramolecular hydrogen-bonding.

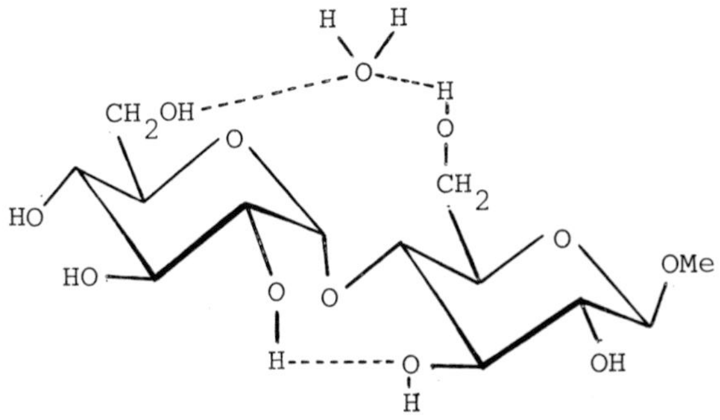

Methyl β -maltoside

(19)

An X-ray diffraction study of sucrose sodium bromide dihydrate confirms the chemically assigned, relative configurations of the asymmetric carbon atoms of the molecule (C.A. Beevers and W. Cochran, Proc. roy. Soc. London, Ser. A., 1947, 190, 257). Neutron diffraction has been used to elucidate the precise molecular structure of crystalline sucrose (G.M.Brown, and H.A.Levy, Science,1963,141, 921). Seven of the hydroxyl groups in the molecule are hydrogen—bonded, including two intramolecular bonds (O-2 --- H —— O-1' and O-5 --- H —— O-6') as shown in (20). Although the 4-hydroxyl group is not

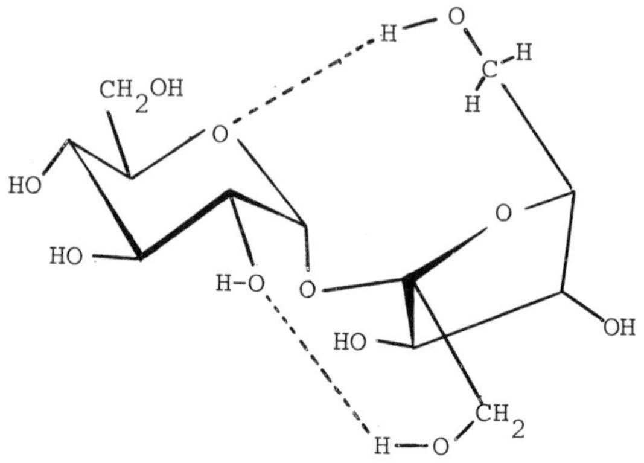

(20)

involved in hydrogen bonding, its hydrogen atom has been found to be loosely fixed in position through two fairly close contacts with oxygen atoms in other sucrose molecules. This is the only sugar thus far studied that has two direct intramolecular hydrogen-bonds.

An interesting feature noted in the crystal structure of α,α-trehalose dihydrate is the absence of direct intramolecular hydrogen bonding. However, indirect intramolecular hydrogen bonding through water (O-2 --- H_2O --- O-4' and O-2 --- H_2O --- O-6') has been shown (G.M.Brown, *et al.*, Acta Crystallog.Sect.B., 1972,28, 3145; T.Taga,M.Senma, and K. Osaki, *ibid.*,1972,28, 3258). The two D-glucopyranose rings have $^4C_1$ conformation. In the solid state the molecule has only an approximate two-fold symmetry with some significant structural differences between the two halves. The presence of the two indirect intramolecular hydrogen bonds probably explains the deviation from symmetry found in the torsion angles about the glycosidic linkages (Taga,*et al.*, *loc. cit.*)

2. *Disaccharides and related compounds*

a. *Pentosylpentoses, $C_{10}H_{18}O_9$ and glycosides*

5-*O*-α-L-Arabinofuranosyl-α-L-arabinofuranose. *p*-Nitrophenyl glycoside, $[\alpha]_D$ -163° (*c* 1.2, methanol) (D.Arndt and A. Graffi,Carbohydrate Res., 1976, 48, 128). *p*-Nitrophenyl 2,3-di-*O*-acetyl-5-*O*-(2,3,5-tri-*O*-benzoyl-α-L-arabinofuranosyl)-α-L-arabinofuranoside, m.p. 60-65°, $[\alpha]_D$ -62° (*c* 1.2, chloroform).

4-*O*-β-D-Xylopyranosyl-β-D-xylopyranose. Hexa-acetate, m.p. 155°, $[\alpha]_D$ -72.2° (*c* 1.28, chloroform) (J.-P.Utille and P.J.A. Vottero, *ibid.*,1977, 53, 259).

4-*O*-α-D-Xylopyranosyl-β-D-xylopyranose. Hexa-acetate, m.p. 135°, $[\alpha]_D$ +44.4° (*c* 0.7, chloroform) (Utille and Voterro, *loc.cit.)*

b. *Pentosylhexoses, Hexosylpentoses, $C_{11}H_{20}O_{10}$ and glycosides*

6-*O*-α-D-Xylopyranosyl-D-mannopyranose,$[\alpha]_D$ +95.7° (*c* 1.14, water) is prepared by the reaction of β-D-xylopyranosyl chloride 2,3,4-tris(chlorosulphate) with 1,2,3,4-tetra-*O*-acetyl-β-D-mannopyranose in chloroform in the presence of drierite and a catalytic amount of silver perchlorate, followed by de-chlorosulphation and de-esterification (H.J.Jennings, Chem. Comm., 1967, 722).

2-*O*-β-D-Xylopyranosyl-D-glucose, m.p. 202-203°, $[\alpha]_D$ +32° (5 min) ⟶ +17° (48 h) (*c* 1, water), is synthesised by condensation of 2,3,4-tri-*O*-acetyl-α-D-xylopyranosyl bromide and benzyl 3,5,6-tri-*O*-benzyl-α-D-glucofuranoside in acetonitrile in the presence of mercury (II) cyanide and mercury (II) bromide, followed by debenzylation (B.Erbing and B. Lindberg, Acta chem. Scand., 1969, 23, 2213).

4-*O*-β-D-Galactopyranosyl-D-xylose, $[\alpha]_D$ +18° (*c* 0.5, water), is synthesised by condensation of 2,3,4,6-tetra-*O*-benzoyl-β-D-galactopyranosyl bromide with benzyl 2,3-anhydro-β-D-ribopyranoside in nitromethane in the presence of silver triflate and 2,4,6-collidine, followed by removal of the protecting groups (B.Erbing, B. Lindberg and T.Norberg,*ibid.*, 1978, B 32, 308). Benzyl β-glycoside, m.p. 40-42°, $[\alpha]_D$ -38° (*c* 0.5, water).

3-*O*-α-D-Glucopyranosyl-α-D-arabinopyranose, m.p. 120-121°, $[\alpha]_D$ +53.7° ⟶ +46.6° (*c* 4, water), and 3-*O*-α-D-glucopyranosyl-β-D-arabinopyranose, m.p. 155-157°, $[\alpha]_D$ +14.2° ⟶ +49.2° (*c* 4, water), are prepared by the Ruff degradation of calcium maltobionate (H.S.Isbell, H.L.Frush, and J.D. Moyer, J.Res.nat.Bur.Stand.Sect.A., 1968, 72, 769). α-Hepta-acetate, m.p. 127-128°, $[\alpha]_D$ +62.4° (*c* 2.5, chloroform). β-Hepta-acetate, m.p. 194.5-195.5°, $[\alpha]_D$ +13.4° (*c* 2.5, chloroform).

c. Deoxyhexosyldeoxyhexoses, $C_{12}H_{22}O_9$, *and glycosides*

2-*O*-α-L-Fucopyranosyl-L-fucopyranose, m.p. 189-192°, $[\alpha]_D$ -158° (5 min) ⟶ -163° (20 h) (*c* 0.6, water), is synthesised by the reaction of 2,3,4-tri-*O*-acetyl-α-L-fucopyranosyl bromide with a suitable nucleophile (H.M.Flowers, *et al.*, Carbohydrate Res., (a) 1967, 4, 189; (b) 1979, 74, 177). The apparently stereospecific formation of an α-linked disaccharide is rather unexpected. Benzyl β-glycoside, m.p. 242-244°, $[\alpha]_D$ -81.8° (*c* 0.74, ethanol). Methyl α-glycoside , m.p. 190-192°, $[\alpha]_D$ -227° (*c* 0.57, methanol). Methyl-β-glycoside, m.p.210-212°, $[\alpha]_D$ -91.4° (*c* 0.7, methanol) (H.M.Flowers, *loc.cit.* (b)).

2-*O*-β-L-Fucopyranosyl-L-fucose, m.p. 152-154°, $[\alpha]_D$ -58.6° ⟶ 50.2° (after 20 h, *c* 0.9, water) (H.M.Flowers, *loc.cit.*, (b)). Methyl α-glycoside, m.p. 236-238°, $[\alpha]_D$, -88.9° (*c* 0.9, methanol). Methyl β-glycoside, m.p. 188-190°, $[\alpha]_D$ -1.2° (*c* 1, methanol).

3-*O*-α-L-Fucopyranosyl-L-fucose, m.p. 200-202°, $[\alpha]_D$ -190° (*c* 0.9, water), is prepared by treatment of methyl 2, 4-di-*O*-benzyl-α-L-fucopyranoside with 2-*O*-benzyl-3,4-di-*O*-*p*-nitrobenzoyl-α-L-fucopyranosyl bromide in the presence of mercury (II) cyanide (M.Dejter-Juszynski and H.M.Flowers, Carbohydrate Res., 1974, 37, 75). Methyl α-glycoside, m.p. 103-106°, $[\alpha]_D$ -272°, (*c* 0.9, absolute methanol). A similar treatment of the sugar nucleophile with a bromide having a particpating group at C-2 (i.e. 2,3,4-tri-*O*-acetyl-α-L-fucopyranosyl bromide) followed by de-acetylation and catalytic hydrogenolysis gives methyl 3-*O*-β-L-fucopyranosyl-α-L-fucopyranoside, $[\alpha]_D$ -123° (Dejter-Juszynski and Flowers, *loc.cit.*).

4-*O*-α-L-Fucopyranosyl-L-fucose, $[\alpha]_D$ -171° (*c* 0.55,H_2O) is prepared by using partially benzylated sugars as protected intermediates. The reaction of methyl 2,3-di-*O*-benzyl-α-L-fucopyranoside with 2-*O*-benzyl-3,4-di-*O*-*p*-nitrobenzoyl-α-L-fucopyranosyl bromide in the presence of mercury (II) cyanide in nitromethane and benzene, followed by removal of the protecting groups gives the above disaccharide (M.Dejter-Juszynski and H.M.Flowers, *ibid.*, 1975, 41, 308). Methyl α-glycoside, m.p. 102-104°, $[\alpha]_D$ -240° (*c* 0.95, water). Methyl α-glycoside penta-acetate, m.p. 90-93°, $[\alpha]_D$ -212° (*c* 0.83, chloroform).

4-*O*-β-L-Fucopyranosyl-α-L-fucopyranose. Methyl glycoside, m.p. 175-177°, $[\alpha]_D$ -132.5° (*c* 1.05, water) (Dejter-Juszynski and Flowers, *loc.cit.*).

2-*O*-α-L-Rhamnopyranosyl-L-rhamnopyranose, m.p. 80-85°, $[\alpha]_D$ -21° (*c* 1, water) is prepared by treatment of methyl 4-*O*-acetyl-α-L-rhamnopyranoside with 2,3,4-tri-*O*-acetyl-α-L-rhamnopyranosyl bromined in the presence of mercury (II) cyanide in acetonitrile (D.Schwarzenbach and R.W.Jeanloz, Carbohydrate Res., 1980, 81, 323). α-Hexa-acetate, m.p. 57-60°, $[\alpha]_D$ -47° (*c* 1, chloroform).

3-*O*-α-L-Rhamnopyranosyl-L-rhamnopyranose, m.p. 141-142°, $[\alpha]_D$ -41° (*c* 0.9, water) (Schwarzenbach and Jeanloz, *loc. cit.*). Methyl α-glycoside, m.p. 181-183.5° $[\alpha]_D$ -59° (*c* 1, chloroform). α-Hexa-acetate, m.p. 78-80.5°, $[\alpha]_D$ -37° (*c* 1, chloroform).

4-*O*-α-L-Rhamnopyranosyl-L-rhamnose, $[\alpha]_D$ -68° (*c* 2.2, water) is prepared by condensation of methyl 2,3-*O*-isopropylidene-α-L-rhamnopyranose with 2,3,4-tri-*O*-acetyl-α-L-rhamnopyranosyl bromide in the presence of mercury (II) cyanide in acetonitrile (G.M.Bebault, G.G.S Dutton and C.K.Warfield, Carbohydrate Res., 1974, 34, 174). Methyl α-glycoside, $[\alpha]_D$ -109° (*c* 2.5, water). α-Hexa-acetate m.p. 162-163°, $[\alpha]_D$ -63.6° (*c* 2.1, chloroform).

3-*O*-(3,6-Dideoxy-α-D-*xylo*-hexopyranosyl)-α-L-rhamnopyranose. *p*-Nitrophenyl α-glycoside, $[\alpha]_D$ -30° (*c* 1, water), is prepared by condensation of 3,6-dideoxy-2,4-di-*O*-*p*-nitrobenzoyl-α-D-*xylo*-hexopyranosyl bromide with *p*-nitrophenyl 2,4-di-*O*-benzoyl-α-L-rhamnopyranoside in the presence of mercury (II) cyanide in dichloromethane followed by de-esterification (P.J.Garegg and H.Hultberg, Carbohydrate Res., 1979, 72, 276). Penta-acetate, m.p. 180-181°, [α]$_D$ -60° (*c* 1.2, chloroform).

d. Deoxyhexosylhexoses, $C_{12}H_{22}O_{10}$

2-*O*-α-L-Fucopyranosyl-D-galactose, [α]$_D$ -56.7° (*c* 1.2, water), is synthesised by the condensation of benzyl 6-*O*-benzoyl-3,4-*O*-isopropylidene-β-D-galactopyranoside with 2,3,4-tri-*O*-acetyl-α-L-fucopyranosyl bromide in the presence of mercury (II) cyanide (A.Levy, H.M.Flowers and N.Sharon, Carbohydrate Res., 1967, 4, 305). The disaccharide has been shown to be a component of blood-group substances where it serves as an important antigenic determinant. It has also been synthesised by the reaction of 1,3,4,6-tetra-*O*-acetyl-α-D-galactopyranose with tri-*O*-benzyl-α-L-fucopyranosyl bromide in dichloromethane in the presence of tetraethylammonium bromide and molecular sieve (R.U.Lemieux and H. Driguez, J.Amer.chem.Soc., 1975, 97, 4069). Benzyl β-glycoside, m.p. 205-207°, [α]$_D$ -97.8° (*c* 0.92, water) (Levy, *et al., loc.cit.*) Methyl β-glycoside, [α]$_D$ -70.1° (*c* 0.88, water) (H.M.Flowers, Carbohydrate Res., 1979, 74, 177). 2-*O*-β-L-Fucopyranosyl-D-galactose, [α]$_D$ +48.3° (*c* 1, water) (Levy, *et al., loc.cit.*; K.L.Matta, Carbohydrate Res., 1973, 31, 410). Methyl-α-glycoside, [α]$_D$ +35.8° (*c* 0.74, chloroform), (H.M.Flowers, *loc.cit.*) 3-*O*-α-L-Fucopyranosyl-D-glucose, [α]$_D$ -86° (*c* 1, water) is prepared by condensation of tri-*O*-benzoyl-α-L-fucopyranosyl bromide with 1,2:5,6-di-*O*-isopropylidene-α-D-glucofuranose in dichloromethane in the presence of tetraethylammonium bromide (R.U.Lemieux, *et al.*, J. Amer.chem. Soc., 1975, 97, 4056). 6-*O*-α-L-Fucopyranosyl-D-galactose, [α]$_D$ -61.1° (*c* 1, water), is prepared by the reaction of 2-*O*-benzoyl-3,4-di-*O*-*p*-nitrobenzoyl-α-L-fucopyranosyl bromide with 1,2,3-tri-*O*-benzoyl-α-D-galactopyranose in nitromethane and benzene in the presence of mercuric cyanide (K.L.Matta, E.A.Z. Johnson and J.J. Barlow, Carbohydrate Res., 1974, 32, 418). It has also been prepared by condensation of tri-*O*-benzyl-α-L-fucopyranosyl bromide with 1,2:3,4-di-*O*-isopropylidene-α-D-galactopyranose in the presence of tetraethylammonium bromide and Hünigs base in dichloromethane followed by removal of the protecting groups (Lemieux and coworkers, *loc.cit.*).

1-*O*-α-L-Rhamnopyranosyl-D-fructose, $[\alpha]_D$ -4.85° (*c* 0.58, chloroform), is prepared by reacting 2,3,4-tri-*O*-benzoyl-α-L-rhamnopyranosyl bromide with 2,3:4,5-di-*O*-isopropylidene-β-D-fructopyranose in acetonitrile in the presence of mercury (II) cyanide and mercury (II) bromide (S.Kamiya, S.Esaki and M.Hama, Agric.biol.Chem., 1967, 31, 261). 3-*O*-α-L-Rhamnopyranosyl-D-glucose, $[\alpha]_D$ +4° (water), is prepared by condensation of 2,3,4-tri-*O*-acetyl-α-L-rhamnopyranosyl bromide with 1,2:5,6-di-*O*-isopropylidene-α-D-glucofuranose in benzene in the presence of mercury (II) cyanide (F.Imperato, J.org.Chem. 1976, 41, 3478). 3-*O*-α-L-Rhamnopyranosyl-D-galactose, $[\alpha]_D$ +8.1° (equil., *c* 1.3, water), $[\alpha]_D$ -10.5° (equil., *c* 1.85, methanol) (V.I.Betaneli, *et al.*, Carbohydrate Res., 1980, 84, 211). 4-*O*-α-L-Rhamnopyranosyl-D-glucose, $[\alpha]_D$ -6° (*c* 1, water) (F.Imperato, *ibid.*, 1977, 58, 217). Phenylosazone, m.p. 165-167°. 6-*O*-α-L-Rhamnopyranosyl-D-mannose, $[\alpha]_D$ -8.2° (*c* 1, water) (Kamiya, *et al.*, *loc.cit.*) Hepta-acetate, m.p. 160° (ethanol), $[\alpha]_D$ -5.2° (*c* 3.8, chloroform). 2-*O*-(6-Deoxy-α-L-glucopyranosyl)-D-glucose, $[\alpha]_D$ -56.94° (*c* 1.44, methanol), is synthesised by the condensation of 2,3,4-tri-*O*-acetyl-6-deoxy-α-L-glucopyranosyl bromide with 1,3,4,6-tetra-*O*-acetyl-α-D-glucopyranose in the presence of mercury (II) cyanide and mercury (II) bromide in absolute acetonitrile (S.Kamiya, S.Esaki and F.Konishi, Agr.biol.Chem., 1976, 40, 273). β-Hepta-acetate, m.p. 168-171°, $[\alpha]_D$ -87.5° (*c* 0.8, dichloromethane). 2-*O*-(6-Deoxy-β-L-glucopyranosyl)-D-glucose, $[\alpha]_D$ +79.6° (*c* 1.28, water) (Kamiya, *et al.*, *loc.cit.*). α-Hepta-acetate, m.p. 195-196°, $[\alpha]_D$ +63.7° (*c* 0.8, chloroform). 2-*O*-(6-Deoxy-α-L-glucopyranosyl)-D-galactose, $[\alpha]_D$ -59.3° (*c* 1.4, methanol) (Kamiya, *et al.*, *loc.cit.*). α-Hepta-acetate, m.p. 195°, $[\alpha]_D$ -18° (*c* 2, chloroform). 2-*O*-(6-Deoxy-β-L-glucopyranosyl)-D-galactose, $[\alpha]_D$ +71.35° (*c* 0.95, methanol-water) (Kamiya, *et al.*, *loc.cit.*). α-Hepta-acetate, m.p. 209°, $[\alpha]_D$ +62.5° (*c* 1.6, chloroform).

4-*O*-(6-Deoxy-α-D-galactopyranosyl)-D-glucose, $[\alpha]_D$, +128.3° (*c* 0.72, water), is prepared from 1,6-anhydro-4', 6'-di-*O*-mesyl-β-maltose tetra-acetate by way of selective displacement of the 6'-mesylate by iodide, reduction, displacement of the 4'-mesylate by benzoate and removing the protecting groups (M. Mori, M. Haga and S. Tejima, Chem. pharm. Bull., 1975, 23, 1480). α-Hepta-acetate, m.p. 186-189°, $[\alpha]_D$ +137.5°, (*c* 1.1, chloroform).

e. Dideoxyhexosylhexosides

3-*O*-(3,6-Dideoxy- α-D-*arabino*-hexopyranosyl)-D-mannose. Methyl β-glycoside, $[\alpha]_D$ +23° (*c* 0.5, water) (P.J.Garegg and N.-H. Wallin, Acta chem. Scand., 1972, 26, 3892). *p*-Isothiocyanatophenyl α-glycoside, $[\alpha]_D$ +180° (G.Ekborg, P.J.Garegg, and B. Gotthammar, *ibid.*, 1975, B 29, 765).

3-*O*-(3,6-Dideoxy- α -D-*ribo*-hexopyranosyl)- α -D-mannopyranose. Methyl glycoside penta-acetate, $[\alpha]_D$ +85°, is synthesised by the condensation of 2,3,4,6-tetra-*O*-benzoyl-D-*arabino*-hex-1-enopyranose with methyl 2-*O*-benzyl-4,6-*O*-benzylidene- α-D-mannopyranoside in the presence of boron trifluoride catalyst (G. Alfredsson and P.J. Garegg, *ibid.*, 1973, 27, 556). *p*-Isothiocyanatophenyl glycoside, $[\alpha]_D$ +174° (*c* 0.3, water) (P.J. Garegg, and B. Gotthammar, Carbohydrate Res., 1977, 58, 345).

3-*O*-(3,6-Dideoxy- α -D-*xylo*-hexopyranosyl)- α -D-mannose. Methyl glycoside, $[\alpha]_D$ +103° (water), is synthesised by condensation of methyl 2-*O*-benzyl-4,6-*O*-benzylidene- α -D-mannopyranoside with 2-*O*-benzyl-3,6-dideoxy-4-*O*-*p*-nitrobenzoyl- α -D-*xylo*-hexopyranosyl bromide using Helferich procedure (K. Eklind, P.J. Garegg, and B. Gotthammar, Acta chem. Scand., 1976, B 30, 300). *p*-Isothiocyanatophenyl glycoside , $[\alpha]_D$ +158° (K. Eklind, P.J. Garegg, and B. Gotthammar, *ibid.*, 1976, B 30, 305).

f. Hexosyldeoxyhexoses, $C_{12}H_{22}O_{10}$

3-*O*- β -D-Galactopyranosyl-L-rhamnose, $[\alpha]_D$ +8.5° (*c* 2, ethanol), is prepared by the condensation of 2,3,4,6-tetra-*O*-acetyl- α -D-galactopyranosyl bromide with benzyl α -L-rhamnopyranoside. 4-*O*- β -D-Galactopyranosyl-L-rhamnose, $[\alpha]_D$ -4.2° (*c* 2, methanol), was also formed (R.R. King, and C.T. Bishop, Carbohydrate Res., 1974, 32, 239).

4-*O*- α -D-Galactopyranosyl-L-rhamnose, $[\alpha]_D$ +114° (*c* 0.93, water), is prepared by the halide ion catalysed reaction of benzyl *exo*-2,3-*O*-benzylidene- α -L-rhamnopyranoside with tetra-*O*-benzyl- α -D-galactopyranosyl bromide followed by removal of the protecting groups (P. Fügedi, *et al.*, *ibid.*, 1980, 80, 233).

4-*O*-β-D-Galactopyranosyl-L-rhamnose, $[\alpha]_D$ -8° (*c* 2.2, water), is synthesised by the condensation of 2,3,4,6-tetra-*O*-acetyl-α-D-galactopyranosyl bromide with methyl 2,3-*O*-isopropylidene-α-L-rhamnopyranoside in the presence of mercury (II) cyanide in acetonitrile (G.M. Bebault, G.G.S. Dutton, and N.A. Funnell, Canad. J. Chem., 1974, 52, 3844). α-Hepta-acetate, $[\alpha]_D$ -55° (*c* 2, chloroform). Methyl-α-glycoside, $[\alpha]_D$ -49° (*c* 1.6, methanol). Alditol, m.p. 179-180°, $[\alpha]_D$ +2.8° (*c* 2.2, water). Alditol octa-acetate, m.p. 114-115°, $[\alpha]_D$ -62.8° (*c* 2.1, chloroform).

2-*O*-β-D-Glucopyranosyl-L-rhamnose, $[\alpha]_D$ +9.5° (*c* 2, water), is obtained from the reaction of 2,3,4,6-tetra-*O*-acetyl-α-D-glucopyranosyl bromide with benzyl α-L-rhamnopyranoside in the presence of silver carbonate (R.R. King and C.T. Bishop, Canad. J. Chem., 1974, 52, 3913). The reaction mixture also yields 3-*O*-β- and 4-*O*-β-D-glucopyranosyl-L-rhamnoses. Benzyl α-glucoside, $[\alpha]_D$ -48.8° (*c* 1, ethanol).

3-*O*-β-D-Glucopyranosyl-L-rhamnose, $[\alpha]_D$ -11.4° (*c* 2. water) (King and Bishop, *loc. cit.*). Benzyl α-glycoside, $[\alpha]_D$ -68.4° (*c* 1, ethanol).

4-*O*-β-D-Glucopyranosyl-L-rhamnose, $[\alpha]_D$ -24.5° (*c* 1, water) (King and Bishop, *loc. cit.*). Benzyl α-glycoside, m.p. 96-98°, $[\alpha]_D$ -75.6° (*c* 1, ethanol). α-Pyranose hepta-acetate, m.p. 98-100° (N.K. Kochetkov, and coworkers, Carbohydrate Res., 1974, 33, C 5).

4-*O*-α-D-Glucopyranosyl-L-rhamnose, $[\alpha]_D$ +10° (*c* 1, water), is prepared by condensation of 2-*O*-benzyl-3,4,6-*p*-nitrobenzoyl-β-D-glucopyranosyl bromide with methyl 2,3-*O*-isopropylidene-α-L-rhamnopyranoside in the presence of mercury (II) cyanide in nitromethane (J.M. Berry and G.G.S. Dutton, Canad. J. Chem., 1974, 52, 681). Methyl α-glycoside, $[\alpha]_D$ +43° (*c* 1.2, methanol). Methyl α-glycoside hexa-acetate, m.p. 137-139°, $[\alpha]_D$ +62.3° (*c* 2.6, chloroform). Hepta-acetate, $[\alpha]_D$ +55° (*c* 2, chloroform).

4-*O*-α-D-Glucopyranosyl-6-deoxy-L-idose, $[\alpha]_D$ +91.2° (*c* 0.62, water), is prepared by catalytic reduction of tri-*O*-acetyl-4-*O*-(tetra-*O*-acetyl-α-D-glucopyranosyl)-β-D-*xylo*-hex-5-enopyranose (H. Goto, M. Mori and S. Tejima, Chem. pharm. Bull. Japan, 1978, 26, 1926).

4-*O*-β-D-Glucopyranosyl-6-deoxy-L-idose, $[\alpha]_D$ -28.7° (*c* 0.65,water) is obtained by catalytic reduction of tri-*O*-acetyl-4-*O*-(tetra-*O*-acetyl- β-D-glucopyranosyl)-β-D-*xylo*-hex-5-enopyranose (*idem. ibid.*,1978,29,1926). α-Hepta-acetate, m.p.142-144°, $[\alpha]_D$ -60.3° (*c* 0.82, chloroform).

4-*O*-α-D-Mannopyranosyl-L-rhamnose, m.p.143-145°, $[\alpha]_D$ +53° ⟶ +60.3° (2 h) (*c* 1,water) (G.M.Bebault and G.G.S. Dutton, Canad.J.Chem.,1974,52,678). Methyl α-glycoside, $[\alpha]_D$ +13° (*c* 2.2, water). Methyl α-glycoside hexa-acetate, $[\alpha]_D$+4° (*c* 1.3,chloroform). α-Hepta-acetate, m.p. 149.5-150.5°, $[\alpha]_D$ -5.5° (*c* 4.3, chloroform). Alditol, $[\alpha]_D$ +57° (*c* 3.5, water). *p*-Nitrophenyl α-glycoside, m.p. 126-128°, $[\alpha]_D$ -90° (*c* 0.5, water) (G.Ekborg,J.Lönngren, and S. Svensson, Acta chem.Scand.,1975, B 29, (1031).

4-*O*-β-D-Mannopyranosyl-L-rhamnose,$[\alpha]_D$ -46° (*c* 2.5, (water) (G.M.Bebault and G.G.S.Dutton, Carbohydrate Res., 1974,37,309). Methyl α-glycoside, m.p. 108.5-110°,$[\alpha]_D$ -72.4° (*c* 2.2, water). Methyl α-glycoside hexa-acetate, $[\alpha]_D$ -57° (*c* 2.1, chloroform). α-Hepta-acetate, m.p.164-165°, $[\alpha]_D$ -67.8° (*c* 1.3,chloroform).

g. Hexosylhexoses, $C_{12}H_{22}O_{11}$, and hexosylhexosides

(i) Fructosylfructose

1-*O*-β-D-Fructofuranosyl-D-fructose, m.p. 189-192.5°, $[\alpha]_D$ -66° (water), is isolated from inulin by partial hydrolysis and paper chromatography of the resultant hydrolysate (A.G. Dickerson and J. Moore,Carbohydrate Res., 1975,39, 162).

(ii) Galactosylgalactoses

3-*O*-α-D-Galactopyranosyl-D-galactose, $[\alpha]_D$ +184° (*c* 1.25, water), is prepared by the reaction of tetra-*O*-benzyl-α-D-galactopyranosyl bromide with 1,2:5,6-di-*O*-isopropylidene-α-D-galactofuranose in the presence of tetraethylammonium bromide and Hünig's base in dichloromethane. (R.U.Lemieux, *et al.*, J.Amer.chem.Soc., 1975,97, 4056; 4069).

3-*O*-β-D-Galactopyranosyl-D-galactose, m.p.165-168°, $[\alpha]_D$ +71° ⟶ +62° (*c* 1,water), is prepared by the reaction of 2,3,4,6-tetra-*O*-acetyl-α-D-galactopyranosyl bromide with 2,3-di-*O*-acetyl-1,6-anhydro-β-D-galactopyranose in the presence of mercury (II) cyanide followed by acetolysis and removal of the protecting group. 3-*O*- α -D-galactopyranosyl-D-galactose is also formed (M.E. Chacón-Fuertes and M. Martin-Lomas, Carbohydrate Res., 1975, <u>43</u>, 51).

4-*O*- α -D-Galactopyranosyl-D-galactose, m.p. 212-213°, $[\alpha]_D$ +167° ⟶ +170° (*c* 1, water), and 4-*O*- β -D-galactopyranosyl-D-galactose, m.p. 204-206°, $[\alpha]_D$ +76° ⟶ + 69° (*c* 1.1, water) (Chacón-Fuertes and Martin-Lomas, *loc. cit.*).

5-*O*- β -D-Galactopyranosyl-D-galactofuranose, $[\alpha]_D$ -64° (*c* 2, water), is synthesised from benzyl 2,3-di-*O*-acetyl-6-*O*-benzoyl- α,β -D-galactofuranoside by its reaction with 3,5,6-tri-*O*-acetyl- α -D-galactofuranose 1,2-(methyl orthoacetate) in nitromethane in the presence of mercury (II) cyanide followed by removal of the protecting groups (W.A.R. van Heeswijk, H.G.J. Visser, and J.F.G. Vliegenthart, Carbohydrate Res., 1977, <u>59</u>, 81). Benzyl α , β -glycoside, $[\alpha]_D$ -113° (*c* 8, methanol). Alditol, m.p. 148-150°, $[\alpha]_D$ -63° (*c* 2, water).

6-*O*- α -D-Galactopyranosyl-D-galactose, $[\alpha]_D$ +149° (*c* 0.36, water), is prepared by the reaction of dimeric tri-*O*-acetyl-2-deoxy-2-nitroso- α -D-galactopyranosyl chloride with 1,2:3,4-di-*O*-isopropylidene- α -D-galactopyranose in *N,N*-dimethylformamide, followed by cleavage of the blocking groups (R.U. Lemieux, *et al.*, Canad. J. Chem., 1973, <u>51</u>, 42).

6-*O*- β -D-Galactopyranosyl-D-galactose, $[\alpha]_D$ +29° (*c* 0.72, water), is prepared by melt-polymerisation of 1,2,3,4-tetra-*O*-acetyl- β -D-galactopyranose in the presence of 5% zinc chloride (E.E. Lee, *et al.*, Carbohydrate Res., 1976, <u>49</u>, 475). $[\alpha]_D$ +34° (J.F. Stoddart and J.K.N. Jones, *ibid.*, 1968, <u>8</u>, 29). 6-*O*- β -D-Galactofuranosyl-D-galactose, m.p. 170-174°, $[\alpha]_D$ -25° (*c* 0.35, water), is synthesised by condensation of 3,5,6-tri-*O*-acetyl-1,2-*O*-(1-methyl ethylidene)- α -D-galactofuranose with benzyl 2,3,4-tri-*O*-benzyl- α -D-galactopyranoside (J.-C. Jacquinet and P. Sinaÿ, *ibid.*, 1974, <u>34</u>, 343).

(iii) Glucosylglucoses

Sophorose, 2-*O*- β -D-Glucopyranosyl-D-glucose, m.p. 195-197°, $[\alpha]_D$ +33.6° ⟶ +19° (equil., *c* 3.2, water), is prepared by the condensation of tetra-*O*-acetyl- α -D-glucopyranosyl bromide with 1,3,4,6-tetra-*O*-acetyl- α -D-glucopyranose in acetonitrile in the presence of mercury (II) cyanide and mercury (II) bromide (B.H. Koeppen, *ibid.*, 1968, 7, 410). α -Acetobromo-sophorose, m.p. 189-191°, $[\alpha]_D$ +97.7° (*c* 1.6, chloroform). β-Octa-acetate, m.p. 191-192°, $[\alpha]_D$ -3.5° (*c* 2.5, chloroform). Synthesis of sophorose by alternative routes has been described (S. Hirano, *et al.*, *ibid.* 1976, 68, 147). Benzyl α-glycoside, m.p. 214-215°, $[\alpha]_D$ +90.6° (*c* 1.6, water) (K. Takeo, *ibid.*, 1980, 86, 151). Methyl β-glycoside, m.p. 195-196°, $[\alpha]_D$ -38.4° (*c* 1.5, water) (*idem. ibid.*, 1979, 77, 131).

Nigerose, 3-*O*- α-D-Glucopyranosyl-D-glucose, $[\alpha]_D$ +136° (*c* 0.45, water) (R.U. Lemieux, *et al.*, *loc.cit.*). β -Octa-acetate, m.p. 151°, $[\alpha]_D$ +74° (*c* 1, chloroform) (G. Excoffier, D.Y. Gagnaire and M.R. Vignon, Carbohydrate Res., 1976, 51, 280).

Laminarabiose, 3-*O*- β -D-Glucopyranosyl-D-glucose. β-Octa-acetate, m.p. 161°, $[\alpha]_D$ -22° (G. Excoffier, D.Y.Gagnaire and M.R. Vignon, *ibid.*, 1976, 51, 280). Benzyl α-glycoside, +77.2° (*c* 1.8, water) (K. Takeo, *ibid.*, 1980, 86, 151).

Maltose, 4-*O*- α -D-glucopyranosyl-D-glucose, chemistry has been reviewed (L. Hough and A.C.Richardson, Pure appl.Chem., 1977, 49, 1069; E. Tarelli, in Development in Food Carbohydrate 2, ed. C.K. Lee, Applied Science Publishers, 1980, p.187; R. Khan, Adv. carbohydrate Chem.Biochem., 1981, 39, 214).

Cellobiose, 4-*O*- β -D-glucopyranosyl-D-glucose, chemistry has been reviewed (Hough and Richardson, *loc.cit.*; R.G.Edwards, in Developments in Food Carbohydrate - 2, ed. C.K. Lee, Applied Science Publishers, 1980, p.229). Methyl α-cellobioside hepta-acetate, m.p. 185-186°, $[\alpha]_D$ +55.4° (*c* 1.1, chloroform), is prepared from β-cellobiosyl *N,N*-dimethyldithiocarbamate by treatment with methanol in the presence of mercury (II) and cadmium carbonate (C.T.Gi, H. Ishihara, and S. Tejima, Chem. pharm. Bull. Japan, 1978, 26, 1570). *o*-Nitrophenyl β-cellobioside, m.p. 220°, $[\alpha]_D$ -87.5° (*c* 1, water) (J. Lehmann and E. Schröter, Carbohydrate Res., 1979, 71, 65). Methyl 6,6'-di-*O*-tosyl- β-cellobioside hexa-acetate, m.p. 156-159°, $[\alpha]_D$ -5.1° (*c* 1, chloroform), (J. Thiem, *ibid.*, 1979, 68, 287). 6,6'-Di-*O*-trityl, m.p. 149-150°, $[\alpha]_D$ +36° (*c* 2.5, methanol). 6-*O*-Trityl, m.p. 153-154°, $[\alpha]_D$ +33° (*c* 2, methanol). 6'-*O*-Trityl, m.p. 148-149° $[\alpha]_D$ +21° (*c* 1, methanol) (K. Koizumi and T. Utamura, Yakugaku Zasshi, 1978, 98, 327).

5-*O*-α-D-Glucopyranosyl-D-glucofuranose, $[\alpha]_D$ +71° (*c* 2, water), and 5-*O*-β-D-glucopyranosyl-D-glucofuranose, $[\alpha]_D$ -28°, (*c* 2,water), are obtained from the reaction of 1,2-*O*-cyclohexylidene-α-D-glucofurano-6,3-lactone with 2,3,4,6-tetra-*O*-acetyl-α-D-glucopyranosyl bromide in the presence of mercury (II) cyanide and mercury (II) bromide in acetonitrile (W.A.R.von Heeswijk, F.R.Wassenburg and J.F.G. Vliegenthart, Carbohydrate Res., 1978,62, 281).

Isomaltose, 6-*O*-α-D-glucopryanosyl-D-glucose, $[\alpha]_D$ +108° (*c* 0.14, water), has been synthesised on a light-sensitive solid support (U.Zehavi and A.Patchornik,J.Amer. chem.Soc., 1973, 95, 5673). β-Octa-acetate, m.p. 144-145°, $[\alpha]_D$ +100° (chloroform) (G.Excoffier, D.Y.Gagnaire and M.R.Vignon, Carbohydrate Res., 1976,46, 215).

Gentiobiose, 6-*O*-β-D-glucopyranosyl-D-glucose $[\alpha]_D$, +9.8° (*c* 2, water), is prepared by condensation of 2,3,4,6-tetra-*O*-acetyl-α-D-glucopyranosyl bromide with benzyl β-D-glucopyranoside in nitromethane in the presence of silver carbonate (R.R.King and C.T. Bishop, Canad.J.Chem., 1975,53, 1970). It has also been prepared by the β-D-glucosyltransferase activity of extracts of *Tetrahymena pyriformis* using D-glucose as acceptor and cellobiose as donor (D.J. Manners, J.R.Stark and D.C.Taylor, Carbohydrate Res., 1971, 16, 123). 6-*O*-β-D-Glucopyranosyl-3-*O*-methyl-D-glucose, m.p. 80°, $[\alpha]_D$ +1° (*c* 1.5 water) (G.O.Aspinall. *et al.*, Canad.J.Chem., 1975, 53, 2171). β-Octa-acetate, m.p. 194-196°, $[\alpha]_D$ -6° (*c* 1, chloroform) (G.Excoffier, D.Y.Gagnaire, and M.R.Vignon, Carbohydrate Res., 1976, 46, 215). *p*-Hydroxyphenyl β-glycoside, m.p. 265.5-167.5°, $[\alpha]_D$ -58.3° (*c* 0.75, water) (K.Takiura, *et al.*, Chem.pharm.Bull.Japan, 1974, 22, 2451).

(iv) Mannosylmannoses

2-*O*-α-D-Mannopyranosyl-D-mannose. β-Octa-acetate, m.p.102-103°, $[\alpha]_D$ +5.7° (*c* 0.78, chloroform), is prepared by condensation of tetra-*O*-acetyl-α-D-mannopyranosyl bromide with 1,3,4,6-tetra-*O*-acetyl-β-D-mannopyranose in acetonitrile in the presence of mercury (II) cyanide and mercury (II) bromide (C.M.Reichert, Carbohydrate Res., 1979,77, 141). *p*-Nitrophenyl α-glycoside $[\alpha]_D$ +85.1° (*c* 1.2, water).

3-*O*-α-D-Mannopyranosyl-D-mannose. Methyl α-glycoside, $[\alpha]_D$ +91° (*c* 0.8, water) is synthesised by the condensation of methyl 4,6-*O*-benzylidene-α-D-mannopyranoside with tetra-*O*-acetyl-α-D-mannopyranosyl bromide using the Helferich modification of the Koenigs-Knorr reaction followed by removal of the protecting groups (E.E.Lee and J.O.Wood, *ibid.*, 1979, 75, 322).

4-*O*-α-D-Mannopyranosyl-D-mannose, $[\alpha]_D$ +66.2° (*c* 0.5, water), is prepared by the condensation of 1,6-anhydro-2,3-*O*-isopropylidene-β-D-mannopyranose with tetra-*O*-acetyl-α-D-mannopyranosyl bromide in nitromethane in the presence of mercury (II) cyanide (P.L.Durette and T.Y.Shen, *ibid.*, 1978, 65, 314). α-Octa-acetate, $[\alpha]_D$ +57.9° (*c* 0.63, chloroform). Benzyl 1-thio-α-glycoside, m.p. 187-192° (dec.), $[\alpha]_D$ +249° (*c* 0.9, methanol).

6-*O*-α-D-Mannopyranosyl-D-mannose. α-Octa-acetate, m.p. 162°, $[\alpha]_D$ +68° (*c* 0.97, chloroform) (C.M.Reichert, *ibid.*, 1979, 77, 141). *p*-Nitrophenyl α-glycoside, $[\alpha]_D$ +114° (*c* 0.29, water).

(v) Glucosylfructoses

1-*O*-α-D-Glucopyranosyl-D-fructose, $[\alpha]_D$ +49° (*c* 1, water), is prepared by the enzymic reaction of phenyl α-D-glucopyranoside and D-fructose in the presence of brewer's yeast, α-glucosidase (S.Chiba and K.Shimomura, Agr.biol.Chem., 1971, 35, 1292), phenylosazone, m.p. 200-203°.

5-*O*-α-D-Glucopyranosyl-D-fructopyranose, m.p. 156-157.5° $[\alpha]_D$ -7° (*c* 2, water), is formed on base-catalysed isomerisation of 5-*O*-α-D-glucopyranosyl-D-glucofuranose (W.A.R.van Heeswijk, F.R.Wassenburg and J.F.G.Vliegenthart, Carbohydrate Res., 1978, 62, 281).

(vi) Galactosylglucoses

4-*O*-α-D-Galactopyranosyl-D-glucose, m.p. 227-229° (dec.), $[\alpha]_D$ +159.6° (*c* 1.02, water), is prepared from 1,6-anhydro-4',6'-di-*O*-mesyl-β-maltose by S_N2 displacement reaction using sodium benzoate as nucleophile followed by removal of the protecting groups (M.Mori, M. Haga and S.Tejima, Chem. pharm.Bull. Japan, 1975, 23, 1480). α-Octa-acetate, $[\alpha]_D$ +117° (*c* 1.14, chloroform).

Lactose, 4-*O*-β-D-galactopyranosyl-D-glucose, chemistry has been reviewed (L. Hough and A.C.Richardson, Pure appl. Chem., 1977, 49, 1069; L.A.W.Thelwall, in Development in Food Carbohyrate - 2, ed. C.K.Lee, Applied Science Publishers, 1980, p.275).

(vii) Glucosylgalactoses

2-*O*-α-D-Glucopyranosyl-D-galactose. *p*-Trifluoroacetamidophenyl tri-*O*-acetyl-2-*O*-(α-D-glucopyranosyl)-β-D-galactopyranoside, $[\alpha]_D$ +34° (acetone), is prepared by the condensation of *p*-nitrophenyl 3,4,6-tri-*O*-acetyl-β-D-galactopyranoside with tetra-*O*-benzyl-α-D-glucopyranosyl bromide in dichloromethane in the presence of tetraethylammonium bromide and molecular sieve 4 Å (P.J.Garegg, I.J.Goldstein and T. Iverson, Acta chem.Scand., 1976, B 30, 876).

3-*O*-β-D-Glucopyranosyl-D-galactose, m.p. 203-205°, $[\alpha]_D$ +40° (*c* 0.8, water), is synthesised by the condensation of benzyl 2,6-di-*O*-acetyl-β-D-galactopyranoside with tetra-*O*-acetyl-α-D-glucopyranosyl bromide in a mixture of nitromethane and benzene in the presence of mercury (II) cyanide followed by removal of the protecting groups (H.M.Flowers, Carbohydrate Res., 1967, 4, 312). Benzyl β-glycoside, m.p. 200-202°, $[\alpha]_D$, -20.7°, (*c* 0.9, water).

6-*O*-α-D-Glucopyranosyl-D-galactose, $[\alpha]_D$ +125° ⟶ +138° (water), is prepared by the reaction of tetra-*O*-benzyl-α-D-glucopyranosyl bromide with 1,2:3,4-di-*O*-isopropylidene-α-D-galactopyranose in the presence of tetraethylammmonium bromide in dichloromethane (R.U.Lemieux, *et al.*, J.Amer. chem.Soc., 1975, 97, 4056; Canad.J.Chem.,1975, 51, 42).

(viii) Others

3-*O*-β-D-Glucopryanosyl-D-mannose, m.p. 195-197°, $[\alpha]_D$ -10.5° (5 min) ⟶ -26.2° (*c* 2, water), is prepared by condensation of 3,4,6-tri-*O*-acetyl-1,2-*O*-methylorthoacetyl-α-D-glucopyranose with benzyl 2-*O*-benzyl-4,6-*O*-benzylidene-α-D-mannopyranoside in the presence of mercury (II) bromide in nitromethane (G.Alfredsson, H.B.Borén, and P.J.Garegg, Acta chem.Scand.,1972, 26, 3431).

6-*O*-α-D-Glucopyranosyl-D-mannose, $[\alpha]_D$ +96° (*c* 0.5, water), is prepared by the condensation of 1,2,3,4-tetra-*O*-

acetyl-D-mannopyranose with 3,4,6-tri-*O*-acetyl-β-D-glucopyranosyl chloride in the presence of mercury (II) succinate in benzene followed by de-acetylation (B.Helferich and W.M. Müller, Ber.,1973,106, 2508). Alditol, m.p. 122-125°, $[\alpha]_D$ +87.3° (*c* 0.2, water).

6-*O*-α-D-Mannopyranosyl-α-D-galactose, m.p. 166-168°, $[\alpha]_D$ +115° ⟶ +96° (water) is prepared by the reaction of tri-*O*-acetylglucal with 1,2:3,4-di-*O*-isopropylidene-α-D-galactopyranose in the presence of boron trifluoride in benzene followed by hydroxylation and de-acetylation (R.J. Ferrier and N. Prasad, Chem. Comm., 1968, 476).

6-*O*-β-D-Mannopyranosyl-D-galactose, m.p. 178-180°, $[\alpha]_D$ +23° (5 min) ⟶ +0.7° (24 h) (*c* 0.75, water) (G.Ekborg, B.Lindberg and J.Lönngren, Acta chem.Scand., 1972, 26,3287).

4-*O*-β-D-Galactopyranosyl-D-allose, m.p. 210-212°, $[\alpha]_D$ +49° (3 min) ⟶ +50.2° (30 min, constant value) (*c* 1, water) (R.S.Bhatt, L. Hough and A.C.Richardson, Carbohydrate Res., 1976, 51, 272).

4-*O*-β-D-Idopyranosyl-D-glucose, $[\alpha]_D$ +31° (*c* 1.1, water), is synthesised from 1,6-anhydro-4',6'-*O*-benzylidene-β-lactose (T. Chiba and S. Tejima,Chem. pharm.Bull., 1976, 24, 1684; 1977, 25, 1049).

3-*O*-α-D-Talopyranosyl-D-galactopyranose, $[\alpha]_D$ +128° (water) (R.U.Lemieux and R.V.Stick, Austral.J.Chem., 1978, 31, 901).

6-*O*-α-D-Talopyranosyl-D-galactose, $[\alpha]_D$ +101° (*c* 0.64, water) is obtained from 6-*O*-(tri-*O*-acetyl-2-oximino-α-D-*lyxo*-hexopyranosyl)-1,2:3,4-di-*O*-isopropylidene-α-D-galactopyranose by de-oximation, reduction and removal of the protecting groups (R.U.Lemieux, K.James and T.L.Nagabhushan, Canad.J.Chem., 1973,51, 42).

3-*O*-α-D-Talopyranosyl-D-glucose, $[\alpha]_D$ +115° (*c* 0.65, water) (Lemieux and coworkers, *loc.cit.*).

h. Non-reducing disaccharides

α-D-Xylopyranosyl β-D-xylopyranoside, m.p. 208-210° (P. Kovác, J.Hirsch and V. Kovácik, Carbohydrate Res.,1979, 75, 109). Hexa-acetate, m.p. 175-176°.

Sucrose, β-D-fructofuranosyl α-D-glucopyranoside, chemistry has been extensively reviewed (L. Hough in 'Sugar' ed. J.Yudkin, J.Edelman and L. Hough, Butterworth, London, 1972, p.49; R. Khan, Adv.carbohydrate Chem.Biochem.,1976, 33, 236; R. Khan in Sucrochemistry, ACS Symposium Series, ed. J.L.Hickson, American Chemical Society, Washington, D.C., 1977, 41, 40; L. Hough, *ibid.*, p.9; M.R.Jenner in Developments in Food Carbohydrates - 2, ed., C.K.Lee, Applied Science Publishers, 1980, p.91).

α-D-Fructofuranosyl α-D-glucopyranoside. Octa-acetate, m.p. 110-112°, $[\alpha]_D$ +83.5^{o}, is prepared by the condensation of trimethylsilyl 2,3,4,6-tetra-*O*-acetyl-α-D-glucopyranoside with 1,3,4,6-tetra-*O*-benzoyl -α-D-fructofuranosyl bromide in the presence of silver perchlorate in toluene (A.Klemer, K.Gaupp and E. Buhe, Tetrahedron Letters, 1969, 4585; G.A.Newkome, *et al.*, Carbohydrate Res., 1976, 48, 1).

α-D-Fructofuranosyl β-D-glucopyranoside, $[\alpha]_D$ +52.3^{o} (*c* 0.9, methanol) is prepared by the condensation of 2,3,4,6-tetra-*O*-acetyl-D-glucopyranose with 1,3,4,6-tetra-*O*-acetyl-D-fructofuranose in the presence of phosphorous pentoxide in anhydrous benzene followed by de-esterification (Newkome, *et al.*, *loc.cit.*). Octa-acetate, m.p. 129-130.5^{o}, $[\alpha]_D$ +20° (*c* 1, chloroform).

α-Allopyranosyl β-D-fructofuransoide. Octa-acetate, $[\alpha]_D$ +39° (*c* 1.45, chloroform) is synthesised from β-D-fructofuranosyl α-D-*ribo* hexopyranoside-3-ulose by reduction with sodium borohydride (L.Hough and E.O'Brien, Carbohydrate Res., 1980, 84, 95). Octabenzoate, m.p. 85-87°, $[\alpha]_D$ +70.1^{o} (*c* 0.33, chloroform).

α-D-Glucopyranosyl α-L-sorbofuranoside, m.p. 184-186°, $[\alpha]_D$ +32.5-33.9^{o} (*c* 5, water) is prepared by use of sucrose glucosyltransferase and sucrose or α-D-glucopyranosyl fluoride as the donor (J.C. Mazza, A.Akgerman and J.R.Edwards, Carbohydrate Res., 1975, 40, 402).

α,α-Trehalose, α-D-glucopyranosyl α-D-glucopyranoside has been reviewed (L.Hough and A.C.Richardson, Pure appl. Chem.,1977, 49, 1069; C.K.Lee in Developments in Carbohydrates - 2, ed. C.K.Lee, Applied Science Publishers, London, 1980, p.1).

(i) Nitrogen-containing disaccharides

(i) Disaccharides containing an amino sugar and a neutral sugar residue

2-Acetamido-2-deoxy-6-*O*-(α-D-xylopyranosyl)-α-D-glucopyranose, m.p. 100-104° (softening), 159° (dec.), $[\alpha]_D$ +105° (8 min) ⟶ +81.5° (24 h) (*c* 1.52, water). Benzyl β-glycoside, m.p. 228-228.5° $[\alpha]_D$ +15° (*c* 1.31, ethanol-water 1:4). 2-Acetamido-2-deoxy-6-*O*-(β-D-xylopyranosyl)-α-D-glucopyranose, m.p. 167-168°, $[\alpha]_D$ +20° (2 min) ⟶ -2° (24 h) (*c* 1.07, water), (J.-R.Pougny and P. Sinaÿ, Carbohydrate Res., 1974, 38, 161). Benzyl β-glycoside, m.p. 226-226.5°, $[\alpha]_D$ -67° (*c* 0.63, methanol-water, 1:4).

2-Acetamido-2-deoxy-3-*O*-(α-L-fucopyranosyl)-α-D-glucose, m.p. 218-220° (dec.), $[\alpha]_D$ -60° ⟶ -74° (*c* 0.83, water), is prepared by Koenigs-Knorr reaction of 2-*O*-benzyl-3,4-di-*O*-*p*-nitrobenzoyl-α-L-fucopyranosyl bromide with benzyl 2-acetamido-4,6-*O*-benzylidene-2-deoxy-α-D-glucopyranoside (M.Dejter-Juszynski and H.M.Flowers, *ibid.*, 1973, 30, 287).

2-Acetamido-2-deoxy-3-*O*-(β-L-fucopyranosyl)-α-D-glucose, m.p. 148-150°, $[\alpha]_D$ +8° ⟶ +1° (*c* 1.1, water) (Dejter-Juszynski and Flowers, *loc.cit.*)

2-Acetamido-2-deoxy-4-*O*-(α-L-fucopyranosyl)-α-D-glucose, m.p. 194-196°, $[\alpha]_D$ -79° ⟶ -99° (5 h) (*c* 0.8, methanol-water, 1:1), is prepared by the condensation of 2,3,4-tri-*O*-benzyl-α-L-fucopyranosyl bromide with benzyl 2-acetamido-3,6-di-*O*-benzyl-α-D-glucopyranoside in dichloromethane - *N*,*N*-dimethylformamide in the presence of tetraethylammonium bromide, diisopropylethylamine and molecular sieve (J.-C. Jacquinet and P. Sinaÿ, *ibid.*, 1975, 42, 251).

2-Acetamido-2-deoxy-6-*O*-(β-L-fucopyranosyl)-α-D-glucopyranose, m.p. 144-146°, $[\alpha]_D$ +31° ⟶ +29° (*c* 0.8, 70% ethanol), is prepared by the condensation of 2,3,4-tri-*O*-acetyl-α-L-fucopyranosyl bromide either with benzyl 2-acetamido-3-*O*-acetyl-2-deoxy-β-D-glucopyranoside or with benzyl 2-acetamido-3,4-di-*O*-acetyl-2-deoxy-β-D-glucopyranoside (E.S.Rachaman and R.W.Jeanloz, *ibid.*, 1969, 10, 435).

2-Acetamido-2-deoxy-6-*O*-(α-L-fucopyranosyl)-D-glucose, $[\alpha]_D$ -66° (*c* 1, water) (M.Dejter-Juszynski and H.M.Flowers, *ibid.*, 1971, 18, 219).

2-Acetamido-2-deoxy-4-*O*-(β-D-galactopyranosyl)-D-galactose, m.p. 160°, $[\alpha]_D$ +65.6° (*c* 1, methanol) (S.S.Rana, J.J.Barlow and K.L. Matta, *ibid.*, 1980, 84, 353).

2-Acetamido-2-deoxy-4-*O*-(β-D-galactopyranosyl)-α-D-glucopyranose, m.p. 170-171°, $[\alpha]_D$ +50° ⟶ +28.5° (12 h) (*c* 0.6, water-methanol, 9:1) (J.-C.Jacquinet and P. Sinaÿ, *ibid.*, 1976, 46, 138). Hepta-acetate, m.p. 224-225°, $[\alpha]_D$ +58° (*c* 1, chloroform).

2-Acetamido-2-deoxy-6-*O*-(β-D-galactopyranosyl)-D-glucose, $[\alpha]_D$ +27° (*c* 0.5, water), is prepared by the condensation of β-galactose penta-acetate with benzyl 2-acetamido-3,4-di-*O*-acetyl-2-deoxy-6-*O*-trityl-α-D-galactopyranoside in nitromethane in the presence of allyl bromide and silver perchlorate (V.A.Nesmeyanov, S.E.Zurabyan and A.Ya.Khorlin, Tetrahedron Letters, 1973, 3213). Benzyl β-glycoside, m.p. 232-234°, $[\alpha]_D$ +119° (*c* 1, methanol).

2-Acetamido-2-deoxy-4-*O*-(β-D-glucopyranosyl)-α-D-glucopyranose, m.p. 187-188°, $[\alpha]_D$ +52° ⟶ +32° (24 h) (*c* 0.7, water-methanol, 19:1) (M.A.M.Nassr, J.-C.Jacquinet, and P. Sinaÿ, Carbohydrate Res., 1979, 77, 99).

2-Acetamido-2-deoxy-4-*O*-(α-D-glucopyranosyl)-α-D-glucopyranose, m.p. 175-177° (softening), 205-207° (dec.), $[\alpha]_D$ +111° ⟶ +93° (24 h) (*c* 0.8, water-methanol, 19:1), is synthesised by the condensation of benzyl 2-acetamido-3,6-di-*O*-benzyl-2-deoxy-α-D-glucopyranoside with 2,3,4,6-tetra-*O*-benzyl-1-*O*-(*N*-methyl) acetmidyl-β-D-glucopyranose followed by catalytic hydrogenolysis (Sinaÿ and coworkers, *loc.cit.*) Hepta-acetate, m.p. 225°, $[\alpha]_D$ +92° (*c* 1, chloroform).

2-Acetamido-2-deoxy-6-*O*-(α-D-glucopyranosyl)-α-D-galactopyranose, m.p. 185° (dec.), $[\alpha]_D$ +145° ⟶ +138° (1 h) (*c* 0.63, water-ethanol), is prepared by the condensation of benzyl 2-acetamido-3,4-di-*O*-acetyl-2-deoxy-α-D-galactopyranoside with 2,3,4-tri-*O*-benzyl-6-*O*-*p*-nitro (or methoxy) benzoyl-α-D-glucopyranosyl bromide in benzene at 50° in the presence of pyridine (M.Petitou and P. Sinaÿ, *ibid.*, 1975, 40, 13). Hepta-acetate, m.p. 114-115°, $[\alpha]_D$ +133° (*c* 0.54, chloroform). *p*-Acetamidophenyl α-glycoside hexa-acetate, m.p. 229-229.5°, $[\alpha]_D$ +192° (*c* 0.42, chloroform).

2-Acetamido-2-deoxy-3-*O*-(α-D-mannopyranosyl)-D-glucose, m.p. 130-132°, $[\alpha]_D$ +56.5° (1 h) (*c* 1, water) (K.L.Matta, R.H.Shah and O.P.Bahl, *ibid.*, 1979, 77, 255; M.A.E.Shaban, and R.W.Jeanloz, *ibid.*, 1976, 52, 103). Benzyl α-glycoside, $[\alpha]_D$ +155° (*c* 1, 50% methanol).

2-Acetamido-2-deoxy-4-*O*-(α-D-mannopyranosyl)-D-glucopyranose, m.p. 146-150° (softens at ~100°), $[\alpha]_D$ +71° (1 h) (*c* 1, 50% ethanol) (Matta, *et al.*, *loc.cit.*) Benzyl α-glycoside, $[\alpha]_D$ +155° (*c* 1, 50% methanol)

2-Acetamido-2-deoxy-4-*O*-(β-D-mannopyranosyl)-D-glucopyranose, m.p. 162-165.5°, $[\alpha]_D$ +1.6° (no mutarotation) (*c* 0.86, water), is prepared by the condensation of benzyl 2-acetamido-3,6-di-*O*-benzyl-2-deoxy-α-D-glucopyranoside with 2-*O*-acetyl-3,4,6-tri-*O*-benzyl-α-D-glucopyranosyl bromide in the presence of silver trifluoromethanesulphonate and *s*-collidine followed by *O*-de-acetylation, oxidation of the 2-OH, stereo-selective reduction of the 2-keto group and catalytic hydrogenolysis (C.D.Warren, *et al.*, Carbohydrate Res., 1980, 82, 71).

2-Acetamido-2-deoxy-6-*O*-(α-D-mannopyranosyl)-D-glucose, m.p. 136-138°, $[\alpha]_D$ +50.5° (1 h) (*c* 1, water) (K.L.Matta, R.H.Shah and O.P. Bahl, *ibid.*, 1979, 77, 255).

3-*O*-(2-Acetamido-2-deoxy-α-D-galactopyranosyl)-D-galactopyranose, $[\alpha]_D$ +200° (water, equil.), is prepared by the condensation of 1,2:5,6-di-*O*-isopropylidene-α-D-galactofuranose with dimeric tri-*O*-acetyl-2-deoxy-2-nitroso-α-D-galactopyranosyl chloride followed by an established reductive sequence (R.U.Lemieux and R.V.Stick, Austral.J.Chem., 1978, 31, 901).

6-*O*-(2-Acetamido-2-deoxy-β-D-galactopyranosyl)-D-galactose, m.p. 182-184°, $[\alpha]_D$ +37.9° (*c* 1, water) is prepared by the reaction of 2-methyl-(3,4,6-tri-*O*-acetyl-1,2-dideoxy-α-D-galactopyrano)-[2,1-*d*]-2-oxazoline with 1,2,3-tri-*O*-benzoyl-α-D-galactose in anhydrous toluene and nitromethane containing toluene-*p*-sulphonic acid (K.L.Matta, E.A.Johnson and J.J.Barlow, Carbohydrate Res., 1973, 26, 215).

3-*O*-(2-Acetamido-2-deoxy-β-D-glucopyranosyl)-α-D-galactopyranose, m.p. 132-134° (shrinking at 125°), $[\alpha]_D$

$+44^{\circ} \longrightarrow +35^{\circ}$ (*c* 0.92, water), is synthesised by the condensation of benzyl 6-*O*-allyl-2,4-di-*O*-benzyl-β-D-galactopyranoside with 2-methyl-(3,4,6-tri-*O*-acetyl-1,2-dideoxy-α-D-glucopyrano)-[2,1-*d*]-2-oxazoline followed by removal of the protecting groups (C.Augé and A.Veyrières, *ibid.*,1977,54,45).

4-*O*-(2-Acetamido-2-deoxy-β-D-glucopyranosyl)-D-galactopyranose, m.p. 162-165° (dec.), $[\alpha]_D$ +8° (*c* 1, water). 4-*O*-(2-Acetamido-2-deoxy-β-D-glucopyranosyl)-D-glucopyranose, m.p. 190-195°, $[\alpha]_D$ +30° (*c* 0.7, water) (D.Shapiro, *et al.*, J.org.Chem., 1967, 32, 3767, 1970, 35, 1464).

5-*O*-(2-Acetamido-2-deoxy-α-D-glucopyranosyl)-β-D-glucofuranose, m.p. 128°, $[\alpha]_D$ +101° (equil.) (*c* 1.8, water), is prepared by the reaction of dimeric 3,4,6-tri-*O*-acetyl-2-deoxy-2-nitroso-α-D-glucopyranosyl chloride with 1,2-*O*-isopropylidene-α-D-glucofuranose-6,3-lactone (W.A.R. van Heeswijk, P. de Haan and J.F.G.Vliegenthart, Carbohydrate Res., 1976, 48, 187).

6-*O*-(2-Acetamido-2-deoxy-β-D-glucopyranosyl)-D-glucose, $[\alpha]_D$ +3° (*c* 1.88, water) (T.S. Antonenko, S.E.Zurabyan, and A. Ya Khorlin, Izvest Akad.Nauk SSSR, 1970, 2766). 6-*O*-(2-Acetamido-2-deoxy-α-D-glucopyranosyl)-D-galactopyranose, $[\alpha]_D$ +10° (10 min) (*c* 1.55, water) (Antonenko, *et al.*, *loc. cit.*). 6-*O*-(2-Acetamido-2-deoxy-α-D-mannopyranosyl)-D-glucopyranose, $[\alpha]_D$ +36° (*c* 0.45, water) (T.S.Antonenko, S.E. Zurabyan and A. Ya. Khorlin, Izvest Akad. Nautz SSSR, 1970, 2766).

2-*O*-(2-Acetamido-2-deoxy-β-D-glucopyranosyl)-D-mannopyranose, $[\alpha]_D$ -13° (8 min) $\longrightarrow$ -17° (equil.) (*c* 0.5,water) (R.Kaifu, T.Osawa and R.W.Jeanloz, Carbohydrate Res.,1975, 40, 111).

3-*O*-(2-Acetamido-2-deoxy-α-D-talopyranosyl)-D-galactopyranose, $[\alpha]_D$ +89° (equil., water) (R.U.Lemieux and R.V. Stick, *loc.cit.*)

(ii) Disaccharides containing two amino-sugar residues

2-Acetamido-6-*O*-(2-acetamido-2-deoxy-β-D-galactopyranosyl)-2-deoxy-D-galactose, $[\alpha]_D$ +42.5° (*c* 1, water) is prepared by the condensation of 2-acetamido-3,4,6-tri-*O*-acetyl-2-deoxy-α-D-galactopyranosyl chloride with benzyl 2-acetamido-

2-deoxy-β-D-galactopyranoside in the presence of cadmium carbonate and drierite in benzene-nitromethane (R.R.King and C.T.Bishop, Canad.J.Chem., 1975, 53, 1970). Benzyl β-glycoside, m.p. 244-246°, $[\alpha]_D$ -26.5° (*c* 1, water).

2-Acetamido-3-*O*-(2-acetamido-2-deoxy-α-D-galactopyranosyl)-2-deoxy-D-galactopyranose, $[\alpha]_D$ +177.6° (*c* 0.5, water) (H. Paulsen, Č.Kolář and W.Stenzel, Ber., 1978, 111,2358). α-Hexa-acetate, m.p. 229°, $[\alpha]_D$ +92.3° (*c* 1, chloroform). 2-Acetamido-3-*O*-(2-acetamido-2-deoxy-α-D-galactopyranosyl)-2-deoxy-D-glucopyranose, $[\alpha]_D$ +126° (*c* 0.535, water). α-Hexa-acetate, $[\alpha]_D$ +78.6° (*c* 1.01, chloroform). 2-Acetamido-3-*O*-(2-acetamido-2-deoxy-α-D-glucopyranosyl)-2-deoxy-D-glucopyranose, m.p. 203°, $[\alpha]_D$ +103° (*c* 0.96, water) (H. Paulsen and coworkers, Ber., 1978, 111, 2348, 2358).

2-Acetamido-3-*O*-(2-acetamido-2-deoxy-β-D-glucopyranosyl)-2-deoxy-D-glucopyranose. Benzyl α-glucoside penta-acetate, m.p. 254°, $[\alpha]_D$ +56° (*c* 1, chloroform), is synthesised on a polymeric support (G. Excoffier, *et al.*, Tetrahedron, 1975, 31, 549).

2-Acetamido-4-*O*-(2-acetamido-2-deoxy-α-D-glucopyranosyl)-2-deoxy-D-galactopyranose, m.p. 150°, $[\alpha]_D$ +88° (*c* 0.5, water). α-Hexa-acetate, m.p. 105-106°, $[\alpha]_D$ +93° (*c* 0.81, chloroform). β-Hexa-acetate, m.p. 200°, $[\alpha]_D$ +34° (*c* 0.5, chloroform). 2-Amino-6-*O*-(2-amino-2-deoxy-α-D-glucopyranosyl)-2-deoxy-D-glucose-dihydrochloride, $[\alpha]_D$ +85° (*c* 0.45, water) (H.Paulsen and coworkers, Ber., 1978, 111, 2334, 2348). β-Octa-acetate, m.p. 210-212°, $[\alpha]_D$ +51° (*c* 0.21, chloroform).

2-Acetamido-4-*O*-(2-acetamido-2-deoxy-β-D-glucopyranosyl)-2-deoxy-α-D-glucopyranose, m.p. 260-263° (dec.), $[\alpha]_D$ +32.3° (5 min) ⟶ +16° (equil.) (*c* 0.74, water) (U.Zehavi and R.W.Jeanloz, Carbohydrate Res., 1968, 6, 129). Methyl-β-glycoside, 232-236° (dec.), $[\alpha]_D$ -31° (*c* 0.28, water). Benzyl β-glycoside, m.p. 270-271°, $[\alpha]_D$ -37° (*c* 0.16, water).

2-Acetamido-2-deoxy-3-*O*-(2,6-di-acetamido-2,6-dideoxy-α-D-glucopyranosyl)-D-glucopyranose. α-Penta-acetate, m.p. 135°, $[\alpha]_D$ +124° (*c* 0.45, chloroform), is stereospecifically synthesised by the condensation of 2,6-diazido-3,4-di-*O*-benbenzyl-2,6-dideoxy-β-D-glucopyranosyl chloride with 1,6-anhydro-2-azido-4-*O*-benzyl-2-deoxy-β-D-glucopyranose in the presence of silver perchlorate, *s*-collidine and drierite in dichloromethane (H.Paulsen,*et al.*,Carbohydrate Res.,1979,68, 239). β-Penta-acetate,m.p. ~130°,$[\alpha]_D$ +91° (*c* 0.27,chloroform).

2,4-Diacetamido-2,4-dideoxy-3-*O*-(2,4-diacetamido-2,4-dideoxy-α-D-glucopyranosyl)-D-glucopyranose, $[\alpha]_D$ +79° (*c* 0.39, water), is prepared by the reaction of 6-*O*-acetyl-2,4-diazido-3-*O*-benzyl-2,4-dideoxy-β-D-glucopyranosyl chloride with 1,6-anhydro-2,4-diazido-2,4-dideoxy-β-D-glucopyranose in dichloromethane in the presence of silver carbonate, silver perchlorate and drierite (Paulsen and coworkers, *loc.cit.*).

j. Aldobiuronic acids

The value of β-elimination reactions in the structural study of acidic polysaccharides has been recognised (J. Kiss, Tetrahedron Letters, 1970, 1983; P.C. Wyss, J. Kiss and W. Arnold, Helv. Chim. Acta., 1975, 55, 1847; G.O. Aspinall and coworkers, Canad.J. Chem., 1975, 53, 2718; G.O. Aspinall and K.-G. Rosell, Carbohydrate Res., 1977, 57, C 23; J. Kiss, Adv. Carbohydrate chem. Biochem., 1974, 29, 229). A number of methylated aldobiuronic acids have been synthesised as model compounds for the study of base catalysed β-elimination reactions.

Methyl [benzyl 2,3-di-*O*-benzyl-4-*O*-(3,4,6-tri-*O*-acetyl-α-D-glucopyranosyl)-α-D-glucopyranosid] uronate, $[\alpha]_D$ +118° (*c* 0.66, ethyl acetate), is prepared by the condensation of methyl (benzyl 2,3-di-*O*-benzyl)-α-D-glucopyranosid) uronate with tri-*O*-acetyl-1,2-anhydro-α-D-glucopyranose in anhydrous toluene (J.Kiss and P. Taschner, J.Carbohydrate Nucleosides Nucleotides, 1977, 4, 101).

Methyl [methyl 3,4-*O*-isopropylidene-2-*O*-(α-D-glucopyranosyl)-α-D-galactopyranosid] uronate, m.p. 137-139°, $[\alpha]_D$ +134.3° (*c* 1, methanol), is prepared by the condensation of methyl (methyl 3,4-*O*-isopropylidene-α-D-galactopyranosid) uronate with α-acetobromoglucose in nitromethane in the presence of mercury (II) cyanide (P.Šipoš and Š. Bauer, Carbohydrate Res., 1968, 6, 494).

Methyl [methyl 3,4-*O*-isopropylidene-2-*O*-(2,3,4,6-tetra-*O*-acetyl-β-D-glucopyranosyl)-α-D-galactopyranosid] uronate, m.p. 151-152°, $[\alpha]_D$ +62.7° (*c* 1, chloroform) (Šipoš and Bauer, *loc.cit.*).

Methyl [benzyl 2-*O*-(β-D-xylopyranosyl)-D-galactopyranosid] uronate, m.p. 187-189° (S.B.Tjan, D.J.Poll and T.Doornbos, Carbohydrate Res., 1979, 73, 67).

Methyl 6-*O*-(methyl 2,3,4-tri-*O*-methyl-α-D-galactopyranosyluronate)-2,3,4-tri-*O*-methyl-β-D-glucopyranoside, m.p. 97-98°, $[\alpha]_D$ +82° (*c* 1.2, chloroform), is prepared by the chromium trioxide oxidation of methyl 2,3,4,2',3',4'-hexa-*O*-methyl-β-melibioside followed by methylation using diazomethane in ether (G.O. Aspinall, *et al.*, Canad.J.Chem., 1975, 53, 2182).

Methyl 3,4,6-tri-*O*-methyl-2-*O*-(methyl 2,3,4-tri-*O*-methyl-β-D-glucopyranosyluronate)-α-D-glucopyranoside, m.p. 70.5-72°, $[\alpha]_D$ +45° (*c* 1, chloroform) (P.Kováč, J.Hirsch and V. Kováčik, Carbohydrate Res., 1977, 58, 327).

Methyl 2,4,6-tri-*O*-methyl-3-*O*-(methyl 2,3,4-tri-*O*-methyl-β-D-glucopyranosyluronate)-α-D-glucopyranoside, m.p. 110-111.5°, $[\alpha]_D$ + 66° (*c* 1, chloroform), is prepared by the condensation of methyl 2,4,6-tri-*O*-methyl-α-D-glucopyranoside with methyl 2,3,4-tri-*O*-acetyl-1-bromo-1-deoxy-α-D-glucopyranosyluronate in the presence of silver carbonate followed by de-acetylation and methylation (Kovac, *et al.*, *loc.cit.*).

4-*O*-(Methyl α-D-glucopyranosyluronate)-β-D-glucopyranose hepta-acetate, m.p. 199-200°, $[\alpha]_D$ +71° (*c* 0.72, chloroform) (N. Roy and C.P.J.Glaudemans, J. org.Chem., 1968, 33, 1559).

Methyl 2,3,6-tri-*O*-methyl-4-*O*-(methyl 2,3,4-tri-*O*-methyl-β-D-glucopyranosyluronate)-α-D-glucopyranoside, m.p. 77.5-79.5°, $[\alpha]_D$ +68° (*c* 1, chloroform) (Kováč, *et al.*, *loc.cit.*).

6-*O*-(Methyl α-D-glucopyranosyluronate)-β-D-glucopyranose hepta-acetate, m.p. 169-170°, $[\alpha]_D$ +96.1° (*c* 0.5, chloroform), is prepared by potassium permagnate oxidation of 1,2,3,4,2',3',4'-hepta-*O*-acetylisomaltose followed by methylation (Roy and Glaudemans, *loc.cit.*).

Methyl 2,3,4-tri-*O*-methyl-6-*O*-(methyl 2,3,4-tri-*O*-methyl-β-D-glucopyranosyluronate)-α-D-glucopyranoside, m.p. 108-109°, $[\alpha]_D$ +53° (*c* 1, chloroform) (Kovac, *et al.*, *loc.cit.*).

6-*O*-(Methyl β-D-glucopyranosyluronate)-β-D-glucopyranose hepta-acetate, m.p. 200-201°, $[\alpha]_D$ -1.5° (*c* 1, chloroform) (Roy and Glaudemans, *loc.cit.*).

3. *Trisaccharides*

a. *Neutral sugar containing trisaccharides*

O-β-D-Xylopyranosyl-(1 ⟶ 2)-*O*-β-D-xylopyranosyl-(1 ⟶ 4)-D-xylopyranose. Methyl β-glycoside, m.p. 202-203°, $[\alpha]_D$ -80° (*c* 1, water) (J.Hirsch and P.Kovac, Carbohydrate Res., 1979, 77, 241).

O-α-D-Galactopyranosyl-(1 ⟶ 6)-*O*-α-D-galactopyranosyl (1 ⟶ 6)-D-glucose. Undeca-acetate, m.p. 106-107.5°, $[\alpha]_D$ +132.7° (*c* 0.98, chloroform) is prepared from melibiose and from stachyose (R. Adachi and T. Suami, Bull.chem.Soc. Japan, 1977, 50, 1901).

O-α-D-Glucopyranosyl-(1 ⟶ 2)-*O*-α-D-glucopyranosyl-(1 ⟶ 6)-D-glucose, $[\alpha]_D$ +150.5° (*c* 0.8, water), is synthesised from hepta-*O*-acetyl-β-kojiobiosyl chloride by condensation with benzyl 2,3,4-tri-*O*-benzyl-β-D-glucopyranoside in the presence of silver perchlorate, silver carbonate and drierite in dichloromethane (V.Pozsgay, P.Nánási and A. Neszmélyi, Carbohydrate Res.,1979, 75, 310).

O-α-D-Glucopyranosyl-(1 ⟶ 4)-*O*-α-D-glucopyraosyl-(1 ⟶ 4)-D-glucose. Methyl α-glycoside, m.p. 146-147.5°, $[\alpha]_D$ +203° (*c* 1.6, water) and methyl β-glycoside, $[\alpha]_D$ +111.9° (*c* 2.3, water), are prepared from 2',2",3,3',3",4", 6,6',6"-nona-*O*-acetyl-2-*O*-trichloroacetyl-β-maltotriosyl chloride (K.Takeo, K. Mine and T.Kuge, *ibid.*, 1976, 48, 197). 1-*S*-Acetyl-1-thio β-glycoside peracetate, m.p. 139-140°, $[\alpha]_D$ +93.3° (*c* 1.3, chloroform), is prepared from 1,6-anhydro-β-maltotriose nono-acetate (K.Takeo and T.Kuge, *ibid.*, 1976, 48, 282).

O-β-D-Glucopyranosyl-(1 ⟶ 3)-*O*-[β-D-glucopyranosyl-(1 ⟶ 6)] -D-glucopyranose. Methyl β-glycoside, m.p. 144-146°, $[\alpha]_D$ -38.4° (*c* 1.7, water), is formed when methyl 4, 6-*O*-benzylidene-β-D-glucopyranoside is condensed with α-acetobromoglucose in 1,1,2,2-tetrachloroethane in the presence of silver carbonate (K.Takeo, *ibid.*, 1979, 77, 131).

O-β-D-Glucopyranosyl-(1 ⟶ 6)-*O*-β-D-glucopyranosyl-(1 ⟶ 6)-*O*-D-glucose. *p*-Hydroxyphenyl β-glycoside,m.p.205-207°, $[\alpha]_D$ -59.5° (*c* 1.71,water) is prepared by condensation of *p*-acetoxyphenyl 2,3,4-tri-*O*-acetyl-β-D-glucopyranoside with acetobromogentiobiose in chloroform in the presence of silver oxide, drierite and iodine (K.Takiura, *et al.*, Chem. pharm.Bull.Japan, 1974, 22, 2451). β-Undeca-acetate, m.p. 220-221°, $[\alpha]_D$ -7° (*c* 1, chloroform) (G.Excoffier, D.Y. Gagnaire and M.R. Vignon, Carbohydrate Res., 1976, 46, 201).

b. Deoxysugar containing trisaccharides

O-β-D-Galactopyranosyl-(1 ⟶ 2)-*O*-[α-L-rhamnopyranosyl-(1 ⟶ 6)]-D-galactopyranose. Deca-acetate, m.p. 102-103°, $[\alpha]_D$ -33.7°, is synthesised by condensation of α-aceto-bromo-L-rhamnose with benzyl 3,4-di-*O*-benzyl-2-*O*-(2,3,4,6-tetra-*O*-benzyl-β-D-galactopyranosyl)-β-D-galactopyranoside in benzene-nitromethane (1:1) in the presence of mercury (II) cyanide (A. Lipták and P. Nánási, Tetrahedron Letters, 1977, 921).

O-α-D-Galactopyranosyl-(1 ⟶ 4)-*O*-[β-D-glucopyranosyl-(1 ⟶ 2)]-L-rhamnose, $[\alpha]_D$ +74.5° (*c* 1.2, water), is prepared by the reaction of benzyl 3-*O*-benzyl-4-*O*-(2,3,4,6-tetra-*O*-benzyl-α-D-galactopyranosyl)-α-L-rhamnopyranoside with α-acetobromoglucose and removal of the protecting group (P.Fugedi, A. Lipták and P.Nánási, Carbohydrate Res., 1980, 80, 233).

O-β-D-Mannopyranosyl-(1 ⟶ 4)-*O*-α-L-rhamnopyranosyl-(1 ⟶ 3)-D-galactopyranose, $[\alpha]_D$ -24° (*c* 0.8, water), is prepared by condensation of 1,2:5,6-di-*O*-isopropylidene-α-D-galactofuranose with 2,3-di-*O*-acetyl-4-*O*-(2,3,4,6-tetra-*O*-acetyl-β-D-mannopyranosyl)-α-L-rhamnopyranosyl bromide followed by removal of the protecting groups. Deca-acetate, $[\alpha]_D$ -11° (*c* 1, chloroform) (N.K.Kochetkov and coworkers, *ibid.*, (a) 1980, 84, 211; (b) 1975, 45, 283; (c) 1974, 33, C5)

O-α-D-Mannopyranosyl-(1 ⟶ 4)-*O*-α-L-rhamnopyranosyl-(1 ⟶ 3)-D-galactose, $[\alpha]_D$ +27.2° (equil.) (*c* 2.3, water) (Kochetkov, *et al.*, *loc.cit.*, (a)).

O-α-L-Rhamnopyranosyl -(1 ⟶ 4)-*O*-α-L-rhamnopyranosyl-(1 ⟶ 6)-D-galactose. Nona-acetate, m.p. 82-85°, $[\alpha]_D$ -32° (*c* 0.7, chloroform) (A.Lipták and P. Nánási, *ibid.*, 1975, 44, 313).

O-α-L-Fucopyranosyl-(1 ⟶ 2)-*O*-[α-D-galactopyranosyl-(1 ⟶ 3)]-*O*-D-galactose, $[\alpha]_D$ +35.2° (*c* 1.1, water) (R.U. Lemieux and H.Driguez, J.Amer.chem.Soc.,1975,97, 4069).

O-β-L-Fucopyranosyl-(1 ⟶ 3)-*O*-β-D-galactopyranosyl-(1 ⟶ 4)-D-glucopyranose, $[\alpha]_D$ +54.2° (initial) ⟶ +51° (4 h) (*c* 0.3, water), is prepared by a Koenigs-Knorr condensation of 1,2,3,6,2',3'-hexa-*O*-acetyl-β-lactose with 2,3,4-tri-*O*-acetyl-α-L-fucopyranosyl bromide. The reaction involves an acetyl migration from O - 3' to allow the 1 ⟶ 3 linkage to be formed (H.H.Baer and S.A.Abbas, Carbohydrate Res., 1979, 77, 117).

O-α-L-Fucopyranosyl-(1 ⟶ 3)-*O*-β-D-galactopyranosyl-(1 ⟶ 4)-D-glucose, $[\alpha]_D$ -39° (3 min) ⟶ -42.6° (2 h, final) (*c* 1, water) is synthesised by condensation of 1,2,3,6,2',6'-hexa-*O*-acetyl -α,β-lactose with 2,3,4-tri-*O*-benzyl-L-fucopyranosyl bromide with bromide ion catalysis followed by removal of the protecting groups (H.H.Baer and S.A. Abbas, *ibid.*, 1980, 84, 53).

c. Nitrogen-containing trisaccharides

O-β-D-Galactopyranosyl-(1 ⟶ 3)-*O*-(2-acetamido-2-deoxy-β-D-glucopyranosyl)-(1 ⟶ 3)-D-galactopyranose, m.p. 185-190°, $[\alpha]_D$ +19.8° (*c* 0.76, water, no mutarotation), is prepared by the condensation of 2-methyl-[4,6-di-*O*-acetyl-1,2-dideoxy-3-*O*-(2,3,4,6-tetra-*O*-acetyl-β-D-galactopyranosyl)-α-D-glucopyrano] -[2,1-*d*]-2-oxazoline with benzyl 6-*O*-allyl-2,4-di-*O*-benzyl-α-D-galactopyrananoside in the presence of toluene-*p*-sulphonic acid followed by removal of the protecting groups (C.Augé and A. Veyrières, J.chem.Soc.Perkin I, 1977, 1343).

O-β-D-Galactopyranosyl-(1 ⟶ 3)-(2-acetamido-2-deoxy-β-D-glucopyranosyl)-(1 ⟶ 6)-β-D-galactopyranose. *p*-Nitrophenyl glycoside, $[\alpha]_D$ -53° (*c* 0.5, water), is prepared by the reaction of 2-methyl-[4,6-di-*O*-acetyl-1,2-dideoxy-3-*O*-(2,3,4,6-tetra-*O*-acetyl-β-D-galactopyranosyl)-α-D-glucopyrano]-[2,1-*d*]-2-oxazoline with *p*-nitrophenyl 2,3-di-*O*-acetyl-β-D-galactopyranoside (K.L. Matta and J.J. Barlow, Carbohydrate Res., 1977, 53, 209).

O-(2-Acetamido-2-deoxy-β-D-galactopyranosyl)-(1 ⟶ 4)-*O*-β-D-galactopyranosyl-(1 ⟶ 3)-D-glucopyranose m.p. 175-178°, $[\alpha]_D$ +24.7° (*c* 0.9,water) (D.Shapiro, *et al.*, J. org. Chem., 1971, 36, 832).

O-(2-Acetamido-2-deoxy-β-D-galactopyranosyl)-(1 ⟶ 4)-*O*-β-D-galactopyranosyl-(1 ⟶ 4)-D-glucopyranose, m.p. 185-188°, $[\alpha]_D$ +30.3° (*c* 0.8, water) (Shapiro, *et al.*,*loc.cit.*).

O-β-D-Galactopyranosyl-(1 ⟶ 4)-*O*-β-D-glucopyranosyl-(1 ⟶ 4)-2-acetamido-2-deoxy-D-glucose, m.p. 200-203°, $[\alpha]_D$ +24° ⟶ +15° (*c* 0.35, water), is synthesised by the diphenylcyclopropenyl method (S.E.Zurabyan, G.G.Kolomeer and A. Ya. Khorlin, Biorg. Khim., 1978, 4, 654).

O-β-D-Galactopyranosyl-(1 ⟶ 4)-*O*-(2-acetamido-2-deoxy-β-D-glucopyranosyl)-(1 ⟶ 2)-D-mannose, $[\alpha]_D$ -13° (equil.) (*c* 0.86, water) is prepared by condensation of 2-methyl-[3,6-di-*O*-acetyl-1,2-dideoxy-4-*O*-(2,3,4,6-tetra-*O*-acetyl-β-D-galactopyranosyl)-α-D-glucopyrano]-[2,1-*d*]-2-oxazoline with 3,4:5,6-di-*O*-isopropylidene-D-mannose dimethylacetal in the presence of a catalytic amount of toluene-*p*-sulphonic acid (R. Kaifu and T.Osawa, Carbohydrate Res., 1976, 52, 179).

O-α-L-Fucopyranosyl-(1 ⟶ 2)-*O*-β-D-galactopyranosyl-(1 ⟶ 4)-2-acetamido-2-deoxy-D-glucopyranose, $[\alpha]_D$ -46.5° (*c* 0.5, water), is prepared by glycosylation of benzyl 2-acetamido-3,6-di-*O*-benzyl-2-deoxy-α-D-glucopyranoside by 3,4,6-tri-*O*-benzyl-1,2-*O*-(*t*-butoxyethylidene)-α-D-galactopyranose followed by *O*-de-acetylation, condensation with 2,3,4-tri-*O*-benzyl-α-L-fucopyranosyl bromide and hydrogenolysis (J.-C.Jacquinet and P. Sinaÿ, Tetrahedron, 1976, 32, 1693).

O-α-L-Fucopyranosyl-(1 ⟶ 2)-*O*-β-D-galactopyranosyl-(1 ⟶ 3)-2-acetamido-2-deoxy-α-D-galactopyranose. Phenyl glycoside, $[\alpha]_D$ +49.4° (*c* 0.5, methanol), is prepared by condensation of phenyl 2-acetamido-3-*O*-[3-*O*-benzoyl-4,6-*O*-(*p*-methoxybenzylidene)-β-D-galactopyranosyl]-2-deoxy-4,6-*O*-(*p*-methoxybenzylidene)-α-D-galactopyranoside with 2,3,4-tri-*O*-benzyl-α-L-fucopyrabosyl bromide in the presence of tetraethylammonium bromide. (S.S.Rana, J.J.Barlow and K.L.Matta, Carbohydrate Res., 1980, 87, 99).

O-(2-Acetamido-2-deoxy-β-D-glucopyranosyl-(1 ⟶ 3)-*O*-[2-acetamido-2-deoxy-β-D-glucopyranosyl-(1 ⟶ 6)]-D-galactose, m.p. 142-144°, $[\alpha]_D$ +6.5° (*c* 1.01, water), is synthesised by the reaction of 2-methyl-(3,4,6-tri-*O*-acetyl-1,2-dideoxy-α-D-glucopyrano)-[2,1-*d*]-2-oxazoline (6 moles) with benzyl 2,4-di-*O*-benzyl-β-D-galactopyranoside (1 mole) (S.David and A. Veyrières, Carbohydrate Res., 1975, 40, 23). Benzyl β-glycoside, m.p. 208-210° (dec.), $[\alpha]_D$, -33.8° (*c* 1.065, methanol).

O-(2-Acetamido-2-deoxy-β-D-glucopyranosyl)-(1 ⟶ 3)-*O*-[2-acetamido-2-deoxy-β-D-glucopyranosyl-(1 ⟶ 4)]-D-galactopyranose, m.p. 170-174°, $[\alpha]_D$ +24° (*c* 0.875, water), is obtained from the reaction of 2-methyl-(3,4,6-tri-*O*-acetyl-1,2-dideoxy-α-D-glucopyrano)-[2,1-*d*]-2-oxazoline with benzyl 6-*O*-allyl-2-*O*-benzyl-α-D-galactopyranoside, followed by removal of the protecting groups (C.Auge and A.Veyrières, *ibid.*, 1977, 54, 45).

O-(2-Acetamido-2-deoxy-α-D-galactopyranosyl)-(1 ⟶ 3)-*O*-β-D-galactopyranosyl-(1 ⟶ 3)-2-acetamido-2-deoxy-D-glucopyranose, $[\alpha]_D$ +135.1° (H.Paulsen, W. Stenzel and C. Kólar, Tetrahedron Letters, 1977, 2785).

O-(2-Acetamido-2-deoxy-β- D-glucopyranosyl)-(1 ⟶ 4)-*O*-(2-acetamido-2-deoxy-β-D-glucopyranosyl)-(1 ⟶ 6)-D-glucopyranose, $[\alpha]_D$ +1.4° (*c* 2.25, water) (T.S.Antonenko, S.E.Zurabyan and A. Ya. Khorlin, Izvest Akad.Nauk SSSR, 1970, 2776).

O-β-D-Mannopyranosyl-(1 ⟶ 4)-*O*-(2-acetamido-2-deoxy-β-D-glucopyranosyl)-(1 ⟶ 4)-2-acetamido-2-deoxy-D-glucopyranose, $[\alpha]_D$ +0.2° (*c* 1.5, water) is prepared by condensation of 2-methyl-[3,6-di-*O*-acetyl-1,2-dideoxy-4-*O*-(2,3,4,6-tetra-*O*-acetyl-β-D-mannopyranosyl)-α-D-glucopyrano]-[2,1-*d*]-2-oxazoline with benzyl 2-acetamido-3,6-di-*O*-benzyl-2-deoxy-α-D-glucopyranoside (C.D.Warren, *et al.*, Carbohydrate Res., 1980, 82, 71).

O-α-D-Mannopyranosyl-(1 ⟶ 6)-*O*-(2-acetamido-2-deoxy-β-D-glucopyranosyl)-(1 ⟶ 4)-2-acetamido-*N*-(L-aspartyl-4-oyl)-2-deoxy-β-D-glucopyranose, m.p. 160° (dec.), $[\alpha]_D$ +39° (*c* 0.4, methanol) (M.A.E.Shaban and R.W.Jeanloz, *ibid.*, 1973, 26, 315).

O-(2-Amino-2-deoxy-α-D-glucopyranosyl)-(1 ⟶ 6)-*O*-(2-amino-2-deoxy-α-D-glucopyransyl)-(1 ⟶ 6)-2-amino-2-deoxy-D-glucopyranose-trihydrochloride, $[\alpha]_D$ +92°, (*c* 1.03, water) (H.Paulsen and W. Stenzel, Ber., 1978, 111, 2334).

4. Tetrasaccharides

O-β-D-Mannopyranosyl-(1 ⟶ 4)-*O*-α-L-rhamnopyranosyl-(1 ⟶ 3)-*O*-[α-D-glucopyranosyl-(1 ⟶ 6)]-D-galactose, $[\alpha]_D$ +27° (*c* 2, water), is prepared by glycosylation of benzyl 2,4-di-*O*-benzyl-6-*O*-(2,3,4-tri-*O*-benzyl-6-*O*-*p*-nitrobenzoyl-α-D-glucopyranosyl)-β-D-galactopyranoside or benzyl 2-*O*-acetyl-6-*O*-(2,3,4-tri-*O*-benzyl-6-*O*-*p*-nitrobenzoyl-α-D-glucopyrano syl)-β-D-galactopyranoside with 3-*O*-acetyl-4-*O*-(2,3,4,6-tetra-*O*-acetyl-β-D-mannopyranosyl)-β-L-rhamnopyranose 1,2-(methyl orthoacetate) followed by removal of protecting groups (N.K.Kochetkov, *et al.*, Carbohydrate Res., 1977, 54, 269).

O-β-D-Mannopyranosyl-(1 ⟶ 4)-*O*-β-L-rhamnopyranosyl-(1 ⟶ 3)-*O*-[α-D-glucopyranosyl-(1 ⟶ 6)]-D-galactose, $[\alpha]_D$ +56° (*c* 1, water) (Kochetkov, *et al.*, *loc.cit.*).

O-α-D-Mannopyranosyl-(1 ⟶ 4)-*O*-α-L-rhamnopyranosyl-(1 ⟶ 3)-*O*-[α-D-glucopyranosyl-(1 ⟶ 6)]-D-galactose, $[\alpha]_D$ +39° (*c* 2, water) (Kochetkov, *et al.*, *loc.cit.*).

O-(2-Acetamido-2-deoxy-α-D-galactopyranosyl)-(1 ⟶ 3)-*O*-[α-L-fucopyranosyl-(1 ⟶ 2)]-*O*-β-D-galactopyranosyl-(1 ⟶ 3)-2-acetamido-2-deoxy-D-glucopyranose, $[\alpha]_D$ +53.8° (H.Paulsen and Č. Kolář, Angew, Chem., 1978, 90, 823).

O-α-D-Galactopyranosyl-(1 ⟶ 3)-*O*-[α-L-fucopyranosyl-(1 ⟶ 2)]-*O*-β-D-galactopyranosyl-(1 ⟶ 3)-2-acetamido-2-deoxy-D-glucopyranose, $[\alpha]_D$ +33.3° (Paulsen and Kolář, *loc.cit.*).

O-α- L-Fucopyranosyl-(1 ⟶ 2)-*O*-β-D-galactopyranosyl-(1 ⟶ 4)-*O*-[α-L-fucopyranosyl-(1 ⟶ 3)] -2-acetamido-2-deoxy-α-D-glucopyranose, m.p. 214-216° (dec.) $[\alpha]_D$ -113° (3 min) ⟶ -124.5° (18 h) (*c* 0.5, water-methanol, 9:1), (J.-C.Jacquinet and P. Sinaÿ, J.org.Chem., 1977, 42, 720).

5. Polysaccharide

Isolation and purification

Extraction of polymer compounds from material of plant, microbial or animal origin generally yields a mixture of substances some of which may be closely related chemically and further purification is normally required.

No general scheme can be given for the isolation, fractionation and purification of complex carbohydrates because of their different individual features. However, it is of the uttermost importance that the procedures employed are carried out with care in order to avoid a simultaneous denaturation and partial degradation.

In addition to traditional separation methods such as ion exchange chromatography, gel filtration and zone electrophoresis, the use of lectins, in the form of an affinity column has become a powerful tool for the isolation of carbohydrate compounds. By employing such columns pure substances may be obtained from crude mixtures in a single step. The lectin concanavalin A has been used extensively for isolation and structural studies on polysaccharides and glycoconjugates containing terminal non-reducing α-D-glucopyranosyl, α-D-mannopyranosyl or 2-acetamido-2-deoxy-α-D-glucopyranosyl residues (I.J. Goldstein, Methods carbohydrate Chem., 1972, 6, 106). The isolation of a number of glycoproteins by affinity chromatography has been described by several authors in Colowick and Kaplan: Methods Enzymol. ed. V. Ginsburg, 1978, 50, for a review on carbohydrate-binding proteins of plants and animals see I.J. Goldstein and C.E. Hayes, Adv. carbohydrate Chem. Biochem., 1978, 35, 127).

Analysis

The qualitative or quantitative determination of the component sugars of complex carbohydrates can be accomplished by several different methods. A common way involves acid-catalysed depolymerisation followed by separation and estimation of each sugar by gas liquid chromatography (g.l.c.) or another chromatographic method. G.l.c. has been widely used for this purpose, the sugars being, in general, chromatographed as pertrimethylsilylated methyl glycosides (J.R. Clamp, *et al.*, Methods biochem. Analysis, 1971, 19, 229), or as peracetylated alditol derivatives, (J.H. Sloneker, Methods carbohydrate Chem., 1972, 6, 20).

A variety of methods for the analysis of carbohydrate-containing biopolymers by colorimetric, enzymic and other procedures have appeared and surveys of several of these methods have been compiled (V.Ginsburg, ed. Methods Enzymol. 1972, 28 and 1978, 50). For the estimation of submicrogram amounts of carbohydrate sensitive immuno-chemical (P. Owen and M.R. J. Salton, Analyt. Biochem., 1976, 73, 20) and radio-chromatographic (J. Stadler, *ibid.*, 1976, 74, 62) techniques have been developed.

Structural methods

Methylation analysis is one of the most important methods for structural investigation of complex carbohydrates, and the procedure most frequently used is that of Hakomori (S.Hakomori, J. Biochem. (Tokyo), 1964, 55, 205; P.A. Sandford, and H.E. Conrad, Biochemistry, 1966, 5, 1508), employing sodium methylsulphinylmethanide and methyl iodide in methyl sulphoxide, whereby all hydroxyl groups are etherified and all carboxyl groups esterified, usually in one step. Incomplete methylation which is occasionally observed, may be due to only partial solubilization of the polysaccharide. *O*-Acetyl groups present in the original polysaccharide are split off by the strong base, whereas β-elimination due to esterified uronic acid residues has proved insignificant.

The separation and quantitation of the partially methylated sugars obtained on hydrolysis of a fully methylated polysaccharide is preferably performed by g.l.c. Partially methylated sugars must be converted in to more stable derivatives, such as the alditol acetates. The majority of these can be well separated, e.g. on a OV-225 column. The gas-liquid chromatographic separation of methylated sugars and their derivatives has been reviewed (G.G.S.Dutton, Adv. carbohydrate Chem.Biochem., 1974, 30, 9). Methylation analysis combined with mass spectrometry is of particular value since unambiguous identification may be achieved of methylated sugars obtained from submilligram amounts of carbohydrate material. The mass fragmentation of methylated and partially methylated alditol acetates has been reviewed (J.Lönngren and S.Svensson, Adv. carbohydrate Chem.Biochem., 1974, 29, 41).

Hakomori methylation of carbohydrates containing 2-acetamido-2-deoxyhexose residues leads normally to *N*-methylation as well as *O*-methylation (K. Stellner, *et al.*, Arch. biochem. Biophys., 1973, 155, 464; G.O.H. Schwarzmann and R.W. Jeanloz, Carbohydrate Res., 1974, 34, 161). However, if carbohydrates containing 2-acetamido-2-deoxy-hexitol residues are subjected to the methylation procedure the product recovered after acid hydrolysis is not (or only partly) *N*-methylated (S. Hase and E.T. Rietschel, Eur. J. Biochem., 1976, 63, 93). A possible explanation is that the *N*-acetamido group on reaction with methyl iodide gives a methyl acetamidate group ($-\dot{C}-N=C(CH_3)OCH_3$), exclusively or together with the expected *N*-methylacetamido group. On subsequent acid hydrolysis of the former the original acetamido group is reformed.

When acid-labile linkages are present in a polysaccharide the methylated polymer can be subjected to mild acid hydrolysis and the product reduced with a deuteriated reagent and realkylated with trideuteriomethyl iodide. This product is then hydrolysed, reduced, acetylated and analysed by g.l.c.-m.s. Thus the trideuteriomethyl groups in the alditol derivatives formed, mark the positions to which the acid-labile residues were linked in the original material. This procedure has been used to establish the mode of linkage of galactofuranose residues in the *Klebsiella* O group 9 lipopolysaccharide (B. Lindberg, *et al.*, Carbohydrate Res., 1972, 23, 47).

For the elucidation of the structure of polysaccharides and the carbohydrate portion of glycoconjugates a number of useful degradation methods have been developed (B. Lindberg *et al.*, Adv. carbohydrate Chem. Biochem., 1975, 31, 185).

Periodate oxidation is one of the important analytical techniques for the investigation of carbohydrate structures. It is essential that the oxidation of the glycol groupings in the polysaccharide is complete and that over-oxidation is avoided. The latter can be excluded by performing the reaction in the dark, at low temperature (0 - 5°), and at a pH of 3.5 - 4.0.

During periodate oxidation of polysaccharides inter-residue hemiacetals may be formed (T. Painter and B. Larsen, Acta chem. Scand., (a) 1970, 24, 813; (b) 1970, 24, 2366), thus retarding the oxidation. For alginic acid (Painter and Larsen, *loc. cit.* (a)) and guaran (M.F. Ishak and Painter, *ibid.*, 1973, 27, 1268) the inter-residue hemiacetals are so stable that consumption of the theoretical amount of periodate is virtually impossible under normal conditions.

However, if the protected residues are reduced with borohydride and then re-oxidized the theoretical consumption of periodate is observed.

The Smith degradation (I.J. Goldstein, *et al.*, Methods carbohydrate Chem., 1965, 5, 361), involves reduction of the periodate oxidized polysaccharide with borohydride, followed by mild acid hydrolysis. The oxidized sugar residues contain acyclic acetal groups, which are hydrolysed much faster than the glycosidic linkage. The hydrolysis products can beanalysed by g.l.c. of the trimethylsilyl derivatives of the products (G.G.S. Dutton and K.B. Gibney, Carbohydrate Res., 1972, 25, 99) or by methylation analysis (B. Lindberg, *et al.*, Acta chem. Scand., 1972, 26, 2231).

The polyalcohol obtained in a Smith degradation may be methylated before the mild acid hyrolysis and ethylated afterwards. The hydroxyl groups liberated on hydrolysis and subsequently ethylated may be located by analysis of the reaction products by g.l.c.-m.s. Through this modification structural information can be obtained which otherwise would be lost in a conventional Smith degradation. This technique was employed in studies of the *Klebsiella* O group 7 polysaccharide (B. Lindberg, *et al.*, Acta Chem. Scand., 1973, 27, 3787).

On esterification the carboxyl group of a uronic acid residue becomes electron-withdrawing, and on treatment with anhydrous base β-elimination will occur if the uronic acid unit is substituted at *O*-4. This reaction is the basis for a degradation method applicable to polysaccharides containing uronic acid (B.Lindberg, *et al.*, Carbohydrate Res.,1973, 28, 351). It involves methylation of hydroxyl and carboxyl groups, base-catalysed β-elimination and mild acid hydrolysis. The degraded product is alkylated with trideuteriomethyl or ethyl groups, hydrolysed with acid and the resulting mixture of etherified sugars is converted to alditol acetates and analysed by g.l.c.-m.s. Comparison of this analysis with the methylation analysis of the original polysaccharide reveals the nature of the sugar residues on either side of the uronic acid residue.

It has been shown that treatment of methylated uronic acid containing polysaccharides with *M* sodium methylsulphinyl methanide in methyl sulphoxide results in complete loss of hexuronic acid, and that the acid hydrolysis step is unnecessary since the uronosyl linkage is cleaved by the base treatment (G.O. Aspinall and K.G. Rosell, Carbohydrate Res., 1977, 57, C 23). Thus in a single operation the methylated polysaccharide can be treated with base and then directly alkylated with trideuteriomethyl or ethyl iodide to label the site(s) to which uronic acid residues are attached.

A degradation procedure useful for polymers containing amino sugar involves nitrous acid deamination of 2-amino-2-deoxyhexose residues carrying equatorially oriented amino groups. Nitrous acid cleaves the glycosidic linkage of such residues under mild conditions (J.M. Williams, Adv. carbohydrate Chem. Biochem., 1975, 31, 9). The amino sugars of complex carbohydrates generally occur as 2-acetamido-2-deoxyhexose residues, and thus *N*-deacetylation is a pre-requisite for nitrous acid degradation. Removal of the *N*-acetyl group can be accomplished by treatment with sodium hydroxide in methyl sulphoxide in the presence of sodium thiophenolate (C. Erbing, *et al.*, Carbohydrate Res., 1976, 47, C 5) or with trifluoroacetic acid-trifluoroacetic anhydride (B. Nilsson and S. Svensson, *ibid*, 1978, 62, 377).

If the 2,5-anhydrohexoses obtained on deamination is reduced with sodium borodeuteride, the mass spectrum of the 2,5-anhydrohexitol derivative obtained on methylation analysis also will reveal the position through which the amino sugar was linked (L. Kenne, *et al.*, Carbohydrate Res., 1977, 56, 363).

Peracetylated aldohexopyranosides in which the aglycone group occupies an equatorial position in the most stable chair conformation (generally the β-anomer) are oxidised to 5-hexulosonates by chromium trioxide in acetic acid whereas the corresponding anomer with an axially oriented aglycone is oxidised only slowly (S.J. Angyal and K. James, Austral. J. Chem., 1970, 23, 1209). This difference in reactivity has been utilized to establish the anomeric nature of sugar residues in oligosaccharides (J. Hoffman, *et al.*, Acta chem. Scand., 1972, 26, 661). The oligosaccharide is reduced to the alditol which is acetylated and treated with chromium trioxide in acetic acid in the presence of an internal standard (*myo*-inositol hexa-acetate). Sugar analyses of the reaction product and the original material shows which sugar residues have survived the oxidation.

A new method has been developed for degrading the peptide moiety of glycoproteins, leaving the carbohydrate portion largely intact (B.Nilsson and S.Svensson, Carbohydrate Res., 1979, 72, 183). The method is based upon stabilization of the glycosidic linkages of the sugar residues by trifluoroacetyl groups and subsequent cleavage of the peptide bonds by trans-amidation. The two reactions are carried out in a mixture of trifluoroacetic acid and trifluroacetic anhydride. After

O- and *N*-detrifluoroacetylation the carbohydrate portion may be isolated and re-*N*-acetylated.

When asialofetuin was subjected to the procedure carbohydrate chains attached by *N*-glycosidic as well as by *O*-glycosidic linkages were isolated.

A comprehensive procedure for sequence analysis of complex carbohydrates has been developed (B.S. Valent, *et al.*, *ibid.*, 1980, 79, 165), involving the following main steps.

1. Conventional methylation analysis of the total purified polysaccharide to establish the linkage positions of the component sugars.

2. Partial acid hydrolysis of the methylated polysaccharide followed by reduction with sodium borodeuteride and subsequent ethylation of the resulting oligosaccharides.

3. Fractionation and isolation of the peralkylated oligosaccharides by liquid chromatography.

4. Determination of the anomeric configuration of the glycosidic linkages of the peralkylated oligosaccharides by ^{1}H-n.m.r. spectroscopy.

5. Complete acid hydrolysis of the oligosaccharides followed by reduction with sodium borodeuteride and peracetylation of the partially alkylated alditols.

6. Identification by g.l.c.-m.s. of the partially alkylated alditol acetates obtained from the respective isolated oligosaccharides, and deduction of the oligosaccharide structure.

As oligosaccharide structures will overlap within the unhydrolysed polysaccharide, the oligosaccharide sequences may be pieced together to a complete structure. The applicability of this sequence method has been demonstrated by subjecting the polysaccharides lichenan and xanthan gum, and a nonasaccharide derived from a cell-wall xyloglucan, to the procedure. This technique also offers the possibility to distinguish between 4-*O*-substituted aldopyranosyl residues and 5-*O*-substituted aldofuranosyl residues in complex carbohydrates (A.G. Darvill, *et al.*, Carbohydrate Res., 1980, 86, 309).

Determination of the absolute configuration of monosaccharides has until recently been difficult for small amounts of material. Glycosides of a D and an L sugar prepared with a chiral alcohol are diastereomeric and it has been shown that they can be separated by gas liquid chromatography of suitable derivatives. On glass capillary g.l.c. on SE-30 of the trimethylsilylated (-)-2-butyl glycosides of D- and L-monosaccharides, both neutral sugars, uronic acids and amino sugars, give peak patterns which can be used for assignment of the D-, or the L-configuration. (G. J. Gerwig, *et al.*, *ibid.*, 1978, 62, 349; *ibid.*, 1979, 77, 1).

Enzymatic methods

The chemical techniques of structural analysis of complex carbohydrates have been supplemented to an increasing extent by enzymic methods (J.J.Marshall, Adv.carbohydrate Chem., Biochem., 1974, 30, 257). For the analysis of polysaccharide structures the hydrolytic enzymes have proved most important. It should be emphasised, however, that purity and specificity are general requirements for the use of enzymes in such studies.

exo-enzymes are particularly useful in carbohydrate sequence analysis, splitting off stepwise susceptible sugar residues until a non-susceptible linkage or residue is encountered. However, the presence of only trace *endo*-enzyme activity is sufficient to cause ruptures in the interior of the chain, resulting in the exposure of new terminal sugar residues. Thus the purity of an enzyme is essential.

A procedure has been developed for the detection of α-amylase in a glucoamylase preparation (J.J. Marshall and W.J. Whelan, Anal. Biochem., 1971, 43, 316). It involves preparation of a modifed amylose substrate containing blockages to *exo*-enzyme action, achieved by a limited (~5%) oxidation of amylose with periodate.

$$\text{G-G-G-G-G-}\underset{\text{ox}}{\text{G}}\text{-G-G}\overset{\downarrow}{\text{-}}\text{G-G-G-G-G-}\underset{\text{ox}}{\text{G}}\text{-G- } \ldots$$

$$\text{-G-} = \;\longrightarrow 4)\text{-}\ \alpha\text{-D-glucopyranosyl-}(1 \longrightarrow$$

$$\text{G}_{\text{ox}} = \text{ periodate oxidised } \alpha\text{-D-glucopyranosyl residue}$$

Action of glucoamylase results in stepwise release of glucose units from the non-reducing end of the molecule to the oxidised unit G_{ox} at which point the degradation will stop. The presence of contaminating α-amylase results in *endo*-cleavage for example at the point indicated by the arrow, with liberation of a new non-reducing chain-end, allowing glucose residues to be liberated until the next G_{ox} is reached. This is a general method, applicable to any suitable polysaccharide and the appropriate *exo*-enzyme (*loc.cit.*).

The rate of liberation of a monosaccharide unit from different glycoconjugates by a specific glycosidase varies, depending on the non-carbohydrate moiety and the length and sequential arrangement of the oligosaccharide chain. However, provided the appropriate enzymes are available for investigation of the carbohydrate side chains of a glycoconjugate, it may be possible to elucidate both the anomeric configuration and the sequence on a small amount of substrate. Further, if the enzymic analysis is combined with methylation analysis, the mode of linkage of the various sugar residues may be established as well. General directions for the structural investigation of the carbohydrate chains of glycoproteins have been described (Y.T. Li and S.C.Li, Methods carbohydrate Chem., 1976, 7, 221). The preparation, purification and assay of a number of both *exo*- and *endo*-glycosidases of importance in degradation studies on complex carbohydrates have been described in detail (V. Ginsburg, ed. Methods Enzymol., 1972, 28, 699-999; 1978, 50, 488-584).

Enzymic methods not involving glycosidases have been developed for the determination of two of the common alditols formed by Smith degradation, namely glycerol (D.W. Noble and R.J. Sturgeon, Carbohydrate Res., 1970, 12, 448), and erythritol (R.J. Sturgeon, *ibid.*, 1971, 17, 115). These methods are based on phosphorylation by specific kinases followed by coupling with other enzyme reactions resulting in a change in the oxidation state of NAD, present as a co-factor. Thus the extent of the reaction may be measured, and the amount of alditol quantitated, by uv absorption measurements. The determination may be performed directly on a fraction of the reaction mixture after the Smith degradation.

6. Plant Polysaccharides

Fructans are reserve polysaccharides present in various plant, especially in the *Graminae* and *Compositae*. Fructans are normally non-reducing, linear polysaccharides with either

(2 → 1) (inulin type) or (2 → 6) (levan type) linkages and contain a small proportion ofnon-reducing D-glucopyranose end groups. Branched fructans present in wheat endosperm, consists of a backbone of levan type fructan with fructofuranose residues attached to *O*-1 of some of the hexose units of the chain.

A new type of fructans has been described. The rhizomes of *Polygonatum odoratum* contain four different fructans possessing a non-reducing, linear structure composed of (2 → 1)-linked β-D-fructofuranose residues with one D-glucopyranose residue linked in the middle of the molecule as in *neo*-kestose (M. Tomoda, *et al.*, Chem. pharm. Bull. Japan, 1973, 21, 1806).

Fru*f*(2 → 1)Fru*f*(2 → 6)Glc*p*(1 ← 2)Fru*f*(1 ← 2)Fru*f*

Tubers from *Asparagus cochinchinensis* also contain fructans having *neo*-kestose as part of the linear chain. Periodate oxidation and methylation studies have revealed a *neo*-kestosyl residue at one end of the molecule, the rest of which is composed of β(2 → 1)-linked β-D-fructofuranosyl residues. (M. Tomoda and N. Satoh, Chem.pharm. Bull. Japan, 1974, 22, 2306).

(a) Glucomannans

Glucomannans are present in seeds, bulbs and fleshy leaves of various land plants, in close association with cellulose in coniferous woods and also in some hardwoods.

The carbohydrate moiety of these polymers consists mainly of 1 → 4-linked α-D-glucopyranose and α-D-mannopyranose. Methylation studies and partial hydrolysis show that up to 4 mannose- and 2 glucose-residues are present in blocks while mannose and glucose are also alternating along the chain. A few glucomannans are thought to have branching on C_2, C_3 or C_6.

The glucomannans are mainly acetylated. Those reported as not containing acetate groups have been treated with alkali during the isolation procedure.

The mannosyl residues of glucomannan acetates from *Lilium* sp. are monoacetylated on *O*-2, *O*-3 and *O*-6 (M. Tomoda and N. Satoh, Chem.pharm. Bull. Japan, 1972, 27, 468, and refs.

cited therein). The glucomannan acetate from *L. Longiflorum* has acetate groups on the mannosyl residues at positions 2,3, 6; 2,6; 3,6 and 2; and on the glucosyl residues at 2,3,6 and 6. The acetyl groups on the glucomannan from *Aloe plicatilis* are present on positions 2; 3; 2,3; and 2,3,6 on both sugars (B. Smestad Paulsen, *et al.*, Carbohydrate Res., 1978, 60, 345).

Work on glucomannan from *Pinus silvestris* shows that this also is acetylated. *O*-Acetyl groups are present mainly at positions 2- or 3- of mannose and are irregularly distributed along the chain. The polysaccharide does also contain approx. 5% galactose (L. Kenne, *et al.*, Carbohydrate Res., 1975, 44, 69). Most of the glucomannans studied in recent years (1975-1980) are acetylated. Earlier work on this group of polysaccharides must be open to doubt until it has been re-investigated.

(b) Galactoglucomannans

Polysaccharides isolated from various trees were thought to be glucomannans, but it has now been shown that some of them are galactoglucomannans, and it is suggested that all softwood polysaccharides earlier referred to as glucomannans are actually galactoglucomannans. (V.D. Harwood, Svensk Papperstidning, 1973, 76, 377; B. Lindberg, *et al.*, *ibid.*, 1973, 76, 383; L.Kenne, *et al.*, Carbohydrate Res., 1975, 44, 69; K.S. Jiang and Timell, Cellulose chem. Technol., 1972, 6, 503). These polysaccharides are to a certain extent acetylated. Kenne, *et al.*, studied the distribution of acetate groups along the chain of a polymer isolated from *Pinus silvestris*, and suggested that they are irregularly distributed along the chain. The main chain consists of (1 $\longrightarrow$ 4)-linked mannosyl and glucosyl units. α-D-galactopyranosyl units are attached through *O*-6 of the hexose units in the main chain. Approximately half of the mannosyl residues are substituted having *O*-acetyl groups equally distributed between *O*-2 and *O*-3.

Polysaccharides from seeds of *Liliaceae* and *Iridaceae* appear to be galactoglucomannans (N. Jakimow-Barras, Phytochem., 1973, 12, 1331). The one from the endosperm of *Asparagus officinalis* consists of a linear chain of β-(1 $\longrightarrow$ 4)-linked D-glucose and D-mannose residues to which on average, every fifth or sixth hexose unit in the main chain has an α (?)-D-galactopyranosyl unit attached at *O*-6.

Another type of polysaccharide having a linear chain of (1 ⟶ 4)-linked β-D-glucopyranosyl and β-D-mannopyranosyl residues as the main chain has been isolated from de-fatted seeds of *Cassia tora* (S.C.Varshney, *et al.*,J.chem.Soc., Perkin I, 1976, 1621). D-galactopyranosyl residues are attached to D-mannopyranosyl residues with α-(1 ⟶ 6) linkages. D-Xylopyranosyl residues are attached to D-glucopyranosyl residues, also by α(1 ⟶ 6) linkages. A few side chains consists of β(1 ⟶ 6) glucopyranose units attached to *O*-6 of D-glucose.

(c) Galactomannans

The presence of galactomannans in the Plant Kingdom is confined to a few plant families. They are widely distributed in the seeds of the family *Leguminoseae*. They are also found amongst the *Annonaceae*, *Convolvulaceae*, *Ebenaceae*, *Loganiaceae* and *Palmae*, and are mainly present as reserve polysaccharides. Review articles on the structure, intermolecular interaction of galactomannans with other polysaccharides (I.C.M.Dea and A.Morrison, Adv.Carbohydrate chem. Biochem.,1975,31,241) and the biochemistry (P.M.Dey,*ibid.*, 1978,35,341) of the galactomannans are available.

The ratio of mannose to galactose in the galactomannans varies with its source. α-D-Galactopyranosyl units are distributed along a β(1 ⟶ 4)-linked D-mannopyranoside chain, and are located on *O*-6. Treatment of galactomannans with a α-D-galactosidase and β-D-mannosidase indicates a block structure for the various galactomannans investigated. Certain areas consists of up to six α-D-galactopyranoside units linked to adjacent β-mannopyranoside units, other areas are devoid of galactose and consists of up to six β-D-mannopyranoside units.

The distribution of D-galactopyranosyl units along the mannan chain in guaran and locust bean gum, was shown by means of periodate oxidation to be nearly random. (L.Hoffman, *et al.*, Acta chem.Scand.,1976, B30, 365).

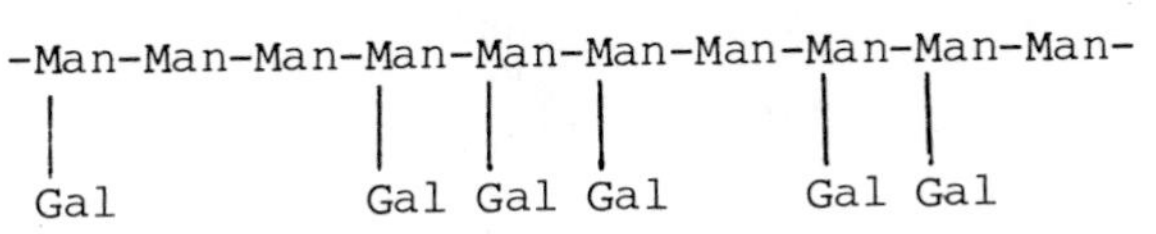

D-*Galacto*-D-*xylo*-glucans are present in seeds of *Impatiens balsamina* (2%), *Tropaeolum majus* (20%) and *Tamarindus indica* (40%), and are also referred to as amyloids. The amyloids from *T.majus* and *T.indica* consist of a backbone of β(1 ⟶ 4)-linked D-glucopyranosyl residues with branches on *O*-6 consisting of α-D-xylopyranose and β-D-galactosyl-(1 ⟶ 2) D-xylosyl groups (J.E. Courtois and P. LeDizet, An. Quim., 1974, 70, 1067). Studies on the structure of the amyloid from *I.balsamina*, after depolymerisation with a purified cellulase from *Penicillium notatum*, indicate that it has the following structure (J.E. Courtois, *et al*., Carbohydrate Res., 1976, 49, 439).

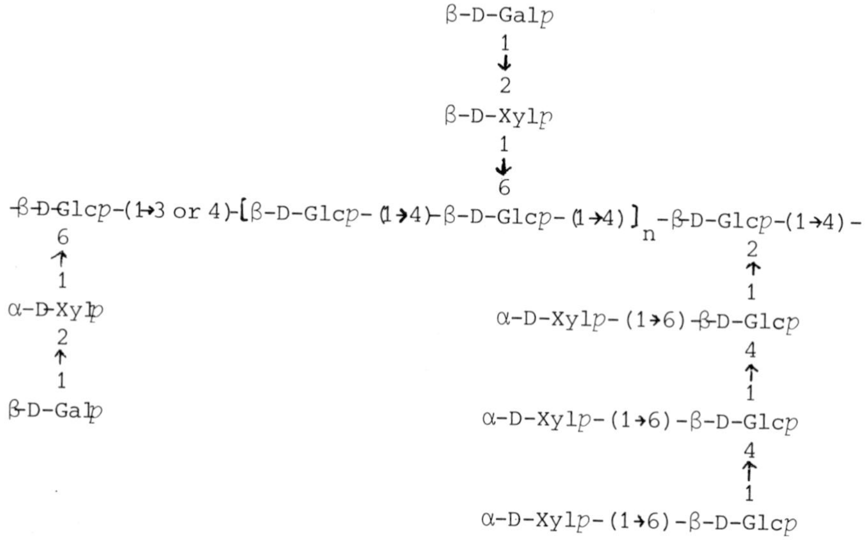

```
                                          β-D-Galp
                                             1
                                             ↓
                                             2
                                          β-D-Xylp
                                             1
                                             ↓
                                             6
-β-D-Glcp-(1→3 or 4)-[β-D-Glcp-(1→4)-β-D-Glcp-(1→4)]n-β-D-Glcp-(1→4)-
    6                                                    2
    ↑                                                    ↑
    1                                                    1
α-D-Xylp                               α-D-Xylp-(1→6)-β-D-Glcp
    2                                                    4
    ↑                                                    ↑
    1                                                    1
β-D-Galp                               α-D-Xylp-(1→6)-β-D-Glcp
                                                         4
                                                         ↑
                                                         1
                                       α-D-Xylp-(1→6)-β-D-Glcp
```

(d) Pectic substances

Pectic substances are complex glycanogalacturonans, often accompanied by neutral glycans, like arabans, arabinogalactans and galactans. Knowledge of their structure, composition and their connections with each other has increased to a large extent over the last decade. For reviews see M. McNeil, *et al*., Prog. chem. org. natural Prod., 1979, 37, 191, and Y.S. Ovodov, Pure appl. Chem., 1975, 42, 351.

(i) Arabans are associated with pectic substances, and structural investigations have shown that they are highly

branched, the residues are mainly present in the furanose form, the glycosidic linkage is in the α-configuration and arabinose is present as the L-isomer. Structural studies show that the arabinofuranosyl residues are mainly 5-, 3,5- and 2,5-linked. 2,3,5-Linked units are also reported (G.O. Aspinall, and I.W. Cottrell, Canad. J. Chem., 1971, 49, 1019 ; J. P. Joseleau, *et al.*, Carbohydrate Res., 1977, 58, 165; I.R. Siddiqui and P.J. Wood, *ibid.*, 1974, 36, 35). A number of pectic polysaccharides contain arabinosyl residues (G.O. Aspinall and J.A. Molloy, J. chem. Soc. (C), 1968, 2994; D.A. Rees and N.J. Wight, Biochem. J., 1969, 115, 431; I.R. Siddiqui and P.J. Wood, Carbohydrate Res., 1976, 50, 97; A. Hillestad, *et al.*, *ibid.*, 1977, 57, 135; B. Smestad Paulsen and P.S. Lund, Phytochem., 1979, 18, 569). These studies do not show whether the arabinosyl units are present in long chains or as mono- or oligosaccharides attached to the main chain of a polysaccharide. Evidence is given for the presence of arabinosyl residues as mono- and disaccharides in pectin from rape seed (G.O. Aspinall and K.S. Jiang, Carbohydrate Res., 1975, 38, 247). Evidence is also given for the presence of a homoaraban as a part of the sycamore cell wall (K.W. Talmadge, *et al.*, Plant physiol., 1973, 51, 158).

(ii) Arabinogalactans or arabinogalactan proteins are polymers present in most plant tissues. They have been found in such diverse sources as leaves, roots, seeds, fruits, pollen, in both filtrates and cells of plant tissues in cultures and in the stigmatic style exudate of various plants. (For review, see A.E. Clarke, *et al.*, Phytochem., 1979, 18 , 521. Also, G.O. Aspinall and K.G. Rosell, *ibid.*, 1978, 17, 919; P.A. Gleeson and A.E. Clarke, Biochem. J., 1979, 181, 607; S. Haavik, *et al.*, 1981, in preparation). These polymers are highly branched and those containing protein are termed glycoproteins or proteoglycans, depending on the ratio of carbohydrate to protein present in the molecule.

The arabinogalactans can be divided into two groups depending on nature of the linkage between the galactosyl residues.

(1) Arabino-4-galactans, as represented by the structure proposed for the soya bean arabinogalactan (G.O. Aspinall, Adv. carbohydrate Chem. Biochem., 1969, 24, 333).

→4)Gal*p*(1→4)Gal*p*(1→4)Gal*p*(1→4)Gal*p*(1→4)Gal*p*(1→

(2) The arabino-3,6-galactans, where the main chain normally consists of 1,3- and 1,6-linked β-D-galactopyranose residues to which are attached shorter chains of L-arabinofuranose residues. For review of the pure polysaccharides (gums and mucilages), see G.O. Aspinall, (*loc. cit.*).

In the arabinogalactan proteins the linkage between carbohydrate and protein consists of galactose to serin and arabinose to hydroxyproline. A part of the cell wall structure for suspension cultured sycamore cells probably consist of a hydroxyproline rich arabinogalactan protein. (K. Keestra, *et al.*, Plant Physiol., 1973, 51, 188).

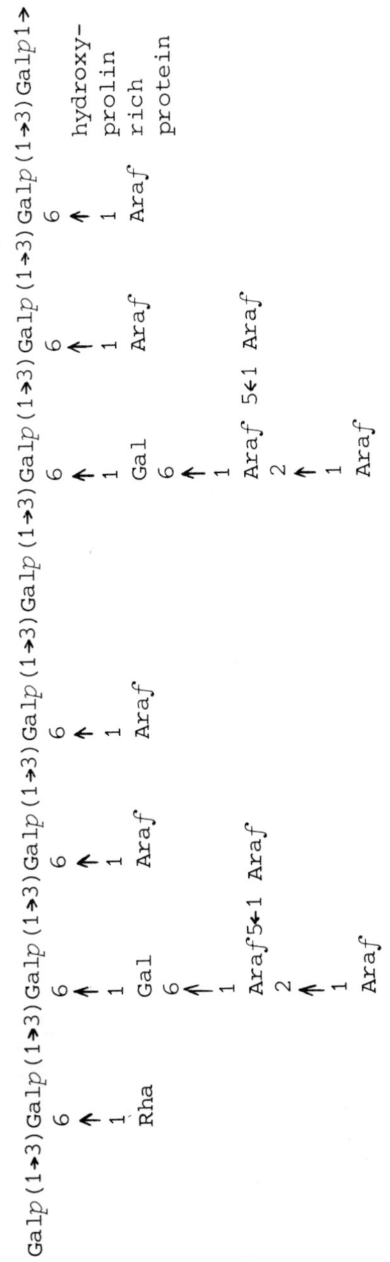

```
Galp(1→3)Galp(1→3)Galp(1→3)Galp(1→3)Galp(1→3)Galp(1→3)Galp(1→3)Galp(1→3)Galp(1→3)Galp1→
         6        6        6        6                 6        6        6
         ↑        ↑        ↑        ↑                 ↑        ↑        ↑     hydroxy-
         1        1        1        1                 1        1        1     prolin
         Rha      Gal      Araf     Araf              Gal      Araf     Araf  rich
                  6                                   6                       protein
                  ↑                                   ↑
                  1                                   1
                  Araf5←1 Araf                        Araf 5←1 Araf
                  2                                   2
                  ↑                                   ↑
                  1                                   1
                  Araf                                Araf
```

"Extensin" is the name given to another type of hydroxyproline rich arabinogalactan protein. Extensin is involved in the control of elongation growth. The following structure has been proposed (D.T.A.Lamport, *et al.*, Biochem. J., 1973, 133, 125; Y. Akiyama and K. Katz, Agric. Biol. Chem., 1977, 41, 79):

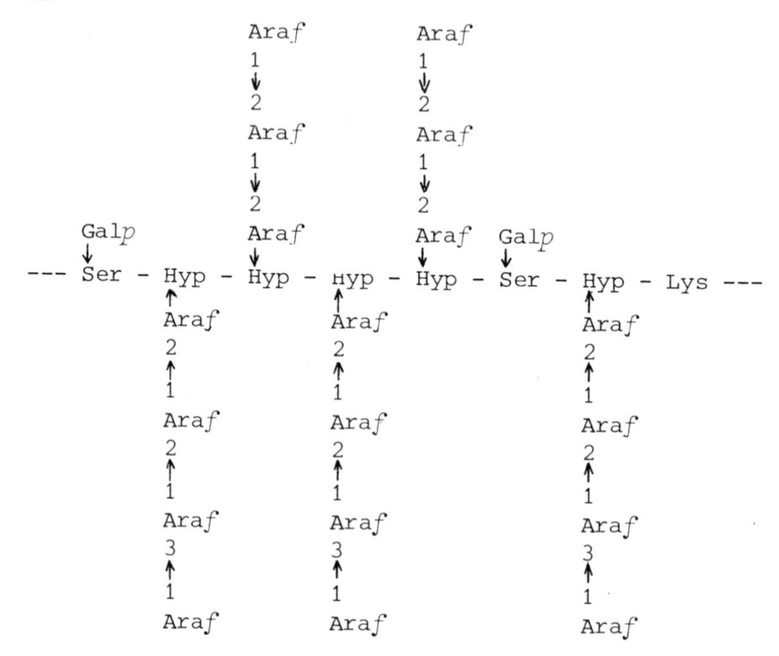

(iii) Galactans are primarily β-(1 ⟶ 4)-linked polymers (R. Toman, *et al.*, Carbohydrate Res., 1972, 25, 371) and have been isolated from citrus pectin (J.M. Labavitch, J. biol. Chem., 1976, 251, 5904) and white willow (Toman, *et al.*, *loc. cit.*). It is probable that pectic polysaccharides contain β-(1 ⟶ 4)-linked galactans, but most of the galactosyl residues present in pectic polymers have been shown to occur as β-(1 ⟶ 4)-linked disaccharides. (Aspinall and Jiang, Carbohydrate Res., 1979, 38, 247). Several pectic polysaccharides contain (1 ⟶3)- and (1 ⟶ 6)-linked galactosyl units. These are most likely part of an arabinogalactan which is discussed above.

(iv) Galacturonan. Homogalacturonans as such are rare. They occur as a part of the acidic pectic polysaccharides as regions consisting of galacturonic acid residues. These are found in sycamore (K.W. Talmadge, *et al.*, Plant Physiol., 1973, 51, 158), in the capsule of opium poppy (J. K. Wold, *et al.*, Acta chem. Scand., 1970, 24, 1262) and have been isolated from the primary cell wall of *Rosa* cell cultures *in vitro* (G. Chambat and J.R. Joseleau, Carbohydrate Res., 1980, 85, C10). The linkage is mainly α (1 ⟶ 4)-, the carboxyl groups are extensively methyl esterified, but it is not known how the methyl ester groups are distributed along the chain. It is probable that some regions devoid of methyl ester groups and other regions fully esterified are present in certain pectic substances (Mcneil, *et al.*, Prog. chem. org.natural Prod., 1979, 37, 191).

(v) Rhamnogalacturonan I (RGI) and II (RGII) are polymers obtained after treatment of suspension cultured sycamore cells with endopolygalacturonase (Mcneil, *et al.*, *loc.cit.*). These are not pure rhamnogalacturonans. RGI is thought to be a group of pectic polymers with the following structure as the backbone of the molecule.

$$\left[(1\rightarrow 4)\text{D-Gal}p\text{A}(1\rightarrow 4)\alpha\text{-D-Gal}p\text{A}(1\rightarrow 2)\ \alpha\text{-L-Rha}p(1\rightarrow 4)\text{D-Gal}p\text{A}\ (1\rightarrow 2)\ \alpha\text{-L-Rha}p(1\rightarrow 2)\text{L-Rha}p(1\rightarrow\right]_{100-200}$$

The *O*-4 of rhamnose and the *O*-3 of galacturonic acid are substituted with side chains probably consisting of arabinose and galactose (K.W. Talmadge, *et al.*, Plant Physiol., 1973, 51, 158). RGII is a new type of pectic polymer containing the rare sugars 2-*O*-methylfucose, 2-*O*-methylxylose and apiose (A.G. Darvill, *et al.*, Plant Physiol., 1978, 62, 418), in addition to the sugars normally found in pectic polymers. Rhamnose and galacturonic acid are the dominating monosaccharides.

(e) Hemicelluloses are a part of the primary cell wall in plants and have been extensively studied over the years. For reviews, see K.C.B. Wilkie, Adv. Carbohydrate Chem. Biochem., 1979, 36, 215 and Mcneil, *et al.*, Prog.chem.org.natural Prod., 1979, 37, 191. The hemicelluloses can be divided into several groups.

(i) Fucogalactoxyloglucans. This group of polymers was, in 1969, isolated from the culture medium and in 1973 shown to be part of the primary cell wall of suspension cultivated sycamore cells (G.O. Aspinall, *et al.*, Canad. J. Biochem., 1969, 47, 1063; W.D. Bauer, *et al.*, Plant Physiol., 1973, 51, 174). The sugars present are D-fucose, D-galactose, D-xylose and D-glucose. The structure has been elucidated by a combination of methylation studies and investigations of the oligosaccharides obtained after enzyme digestion. These polymers resemble the amyloid galactoxyloglucans (see p.79), the main difference being the presence of terminal fucose residues in the hemicellulose polymer. The tentative structures have been assigned to oligosaccharides composed of 7, 9 and 22 sugar units (McMcneil, *et al.*, Prog. chem. org. natural Prod., 1979, 37, 191). The proposed structure of the largest is given on p.87.

(ii) Xylans. These hemicelluloses have a backbone composed of a β(1 ⟶ 4)-linked xylan and are generally considered to be the major constituents of the secondary cell wall of dicotyledons. The polysaccharides of this group vary from plant to plant, the difference being the nature of the side chains glycosidically linked to the xylan backbone.

One type, having a side chain consisting of one 4-*O*-methylglucuronosyl unit linked through *O*-2 of the xylosyl residue, is present in tobacco, rapeseed and white willow (S. Eda, *et al.*, Agric. biol. Chem., 1977, 41, 429; I.R. Siddiqui and P.J. Wood, Carbohydrate Res., 1977, 54, 231; R. Toman, Cellulose Chem. and Technology, 1973, 7, 351).

Another group of polymers has (1 ⟶ 4)-linked glucosyl units interspersed with the (1 ⟶ 4)-linked xylosyl units in the backbone. These can have side chain through *O*-2 terminating in 4-*O*-methyl-D-glucuronosyl or D-xylosyl residues. (P.C. Bardalaye and G.W.Hay, Carbohydrate Res., 1974, 37, 339). Galacturonosyl and rhamnosyl units are also reported as parts of this group of the hemicelluloses (K. Shimizu and O. Samuelson, Svensk Papperstidn., 1973, 76, 150).

An interesting group of polymers amongst the hemicelluloses having the xylan backbone is the arbinoxylans. These are

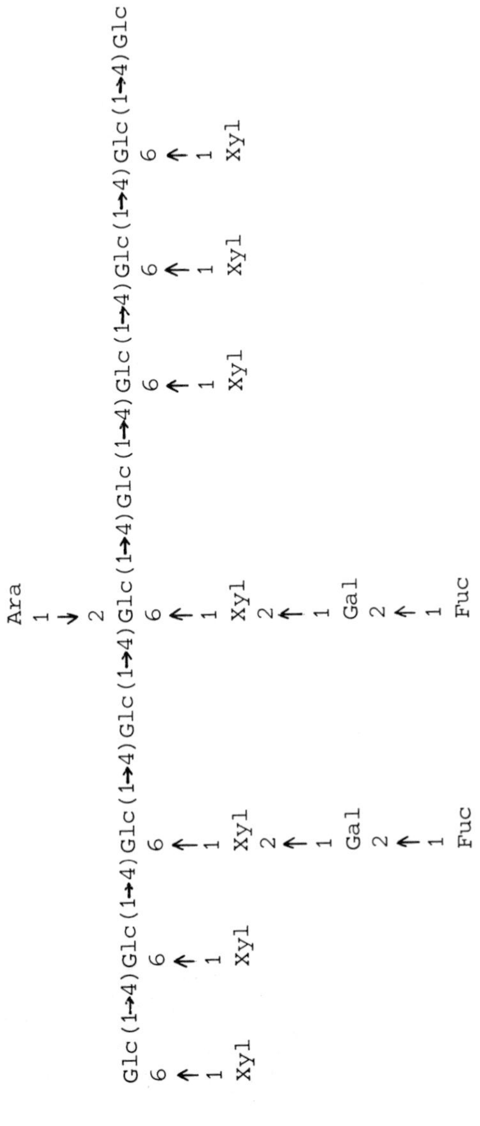

```
                                Ara
                                1
                                ↓
                                2
Glc(1→4)Glc(1→4)Glc(1→4)Glc(1→4)Glc(1→4)Glc(1→4)Glc(1→4)Glc(1→4)Glc(1→4)Glc
6       6       6               6               6       6       6
↑       ↑       ↑               ↑               ↑       ↑       ↑
1       1       1               1               1       1       1
Xyl     Xyl     Xyl             Xyl             Xyl     Xyl     Xyl
                2               2
                ↑               ↑
                1               1
                Gal             Gal
                2               2
                ↑               ↑
                1               1
                Fuc             Fuc
```

interlinked involving ferulic acid [3-(4-hydroxy-3-methoxyphenyl) prop-2-enoic acid]. Diferulic acid is thought to be a crosslink between the hemicelluloses in wheatgeram (H.V. Markwalder and H. Neukom, Phytochem., 1976, 15, 836).

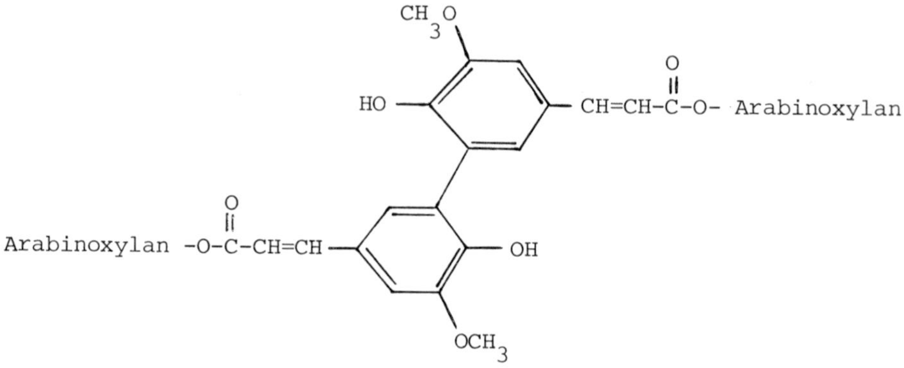

CH_3O
HO
CH=CH-C(=O)-O- Arabinoxylan
Arabinoxylan -O-C(=O)-CH=CH
OH
OCH_3

Ferulic acid may be a precursor for lignin. Lignin-arabinoxylan complexes are isolated from the hemicellulose of spruce. (D. Fengel and M. Pryklenk, Svensk Papperstidning, 1975, 78, 617). Lignin-carbohydrate fractions are also found in spruce-holocellulose.

Another type of lignin-carbohydrate complex is found in ripe cotton capsules. The carbohydrate moiety is composed of D-xylosyl, L-arabinosyl, D-glucosyl and D-galactosyl residues (L. S. Smirnova and K.A. Abdnazimov, Kleim.prirod. Svedinenii, 1974, 397).

7. *Seaweed Mucilages*

Seaweeds contain a variety of polysaccharides, some being of commercial interest due to their gel forming properties. Agar, carrageenan and alginic acid are those of most importance and these have been extensively studied. A review on the differences in the fine structure, how the conformation is influenced by minor differences, and the consequence of this on the physical properties of the polymers has appeared (E. Percival, J. Sci. Food. Agric., 1972, 23, 937). The secondary and tertiary structure of alginic acid, carrageenans and agarose have been discussed by Rees and Welsh (Angew. Chem. internat. Edn., 1977, 16, 214).

"The Chemistry and Enzymology of Marine Algal Polysaccharides" by E. Percival and R.H. McDowell (Academic Press, 1967) covers the earlier chemistry of seaweed mucilages and the chemistry and biochemistry of algal cell-wall polysaccharides is discussed by A. Haug (MTP Internat. Rev. Science, Biochem. Ser. One, 1974, 11, 51).

Physiological and biochemical method used for studies on polysaccharides from seaweeds are discussed in the "Handbook of Phycological Methods", ed. J.A. Hellebust and J.S. Craigie (Cambridge University Press, 1978).

(a) Agar and porphyran were earlier classified as two different polysaccharides. The main difference being that porphyran contained 6-*O*-methyl-D-galactose, which was not thought to be present in agar. Studies during the last decade (1970-1980) have revealed a greater similarity between these polysaccharides and it might be more correct to classify the red algal galactan-sulphates, containing both D- and L-galactose as agar-like polysaccharides.

The agars consist of (1 $\rightarrow$ 3)-linked α-D-galactosyl and (1 $\rightarrow$ 4)-linked β-L-galactosyl unit. The first unit can have sulphate half-ester groups at position 4 or 6, methyl at *O*-6 or the unit can be substituted with pyruvate giving 4,6-*O*-(1-carboxyethylidene)-D-galactosyl residues. The L-galactosyl unit can have sulphate half-ester group at position 6, some of the sulphated groups can be methylated at *O*-2 or the unit can be 3,6-anhydro-L-galactose. It is now considered that agar is a mixture of polysaccharides each having basically the same backbone. The various polymers are substituted in different ways with the groups mentioned above, masking the repeating structure.

Fractionation of the agar from *Gelidium* sp. by means of ion exchange chromatography shows (M. Duckworth and W. Yaphe, Carbohydrate Res., 1971, 16, 189), that the agar is composed of several types of polymers which can be classified as three extremes.

1. Neutral agarose.
   → 3) β-D-Gal*p*(1 → 4)3,6 anhydro-α-L-Gal*p*(1 →
2. Pyruvated agarose with small amounts of sulphate.
   1 of 20 β-D-Gal*p* is substituted with 4,6-*O*-(1-carboxyethylidene) groups.
3. Non-gelling galactansulphate, with none or few 3,6-anhydro-L-Gal*p* and 4,6-*O*-(1-carboxyethylidene)-D-Gal*p* residues.

The agar from *Gracilaria vernicosa* consists of various components with structures intermediate between a non-sulphated highly methylated and a non-methylated, highly sulphated galactan (K. Izumi, J. Biochem., 1972, 73, 135). Agartype polysaccharides have been isolated from *Polysiphonia lanosa* and *Ceramium rubrum*, both consisting of alternating units of a (1 ⟶ 3)-linked β-D-Gal*p* derivative and a (1 ⟶ 4)-linked α-L-Gal*p* derivative. The (1 ⟶ 3)-linked unit includes 6-*O*-methyl, 6-sulphate and 6-*O*-methyl 4-sulphate derivatives, the (1 ⟶ 4)-linked 6-sulphate, 2-*O*-methyl 4-sulphate derivatives, the (1 ⟶ 4)-linked 6-sulphate, 2-*O*-methyl 6-sulphate derivatives or 3,6-anhydro-L-galactose. The agar from *C.rubrum* also contains 2-*O*-methyl-3,6-anhydro-L-galactose, and differ from other agars in that most of the L-galactose units are non-sulphated (J.F. Batey and J.R. Turvey, Carbohydrate Res., 1975, 43, 133, and J.R.Turvey and E.L. Williams, *ibid*., 1976, 49, 419).

(b) Carrageenan consists of a group of sulphated D-galactans. The members of the carrageenan family differ in the amount and position of the sulphate half-ester groups and of the amount of 3,6-anhydro-D-galactose. Basically the carrageenans are built on the same repeating disaccharide unit.

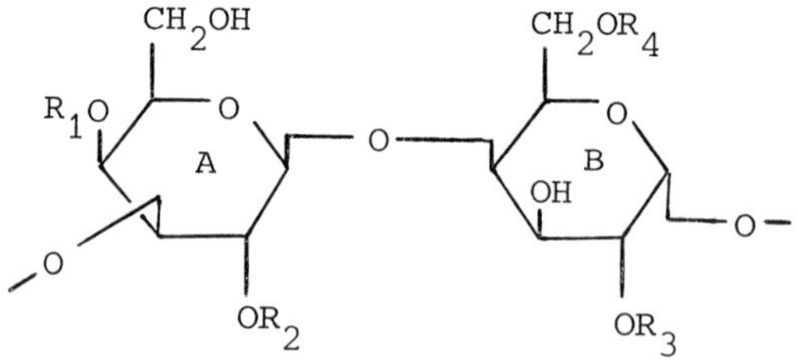

CH_2OH CH_2OR_4
R_1O O A O B OH O
O OR_2 OR_3

$R_{1,2,3,4}$ either H or SO_3H

Various carrageenans have been isolated on the basis of their solubility in potassium chloride solution before and after treatment with alkali. Thus the concept of today is six idealised carrageenans, namely $\ell, \kappa, \lambda, \mu, \nu$, and ξ carrageenan.

ℓ-Carrageenan consists of a repeating unit having R_1 and R_3 as $-SO_3H$ while R_2 is H. Part B of the repeating unit is 3,6-anhydro D-galactose (D.A.Rees, *et al.*, J.Polymer Sci. Part C. Polymer Symp., 1969, p.261).

κ-Carrageenan differs from ℓ-carrageenan only in that the R_3 is H and not SO_3H (D.A.Rees, *et al.*,*ibid.*)

λ-Carrageenan was originally the potassium chloride soluble fraction of the carrageenan isolated from certain alga, but this fraction was later shown to consist of various carrageenans. λ-Carrageenan was redefined in 1965 by T.C.S. Dolan and D.A.Rees (J.chem.Soc., 1965, 3534) and is the polymer with R_2, R_3 and R_4 = SO_3H and R_1 = H, i.e. no 3,6-anhydrogalactose present.

μ-Carrageenan is a part of the potassium chloride soluble fraction which upon alkali treatment is converted into an insoluble part identical to κ-carrageenan. The idealised structure of μ-carrageenan has $R_1 = R_4 = -SO_3H$ and $R_2 - R_3 = H$. Some of the D-galactose 6-sulphate residues are replaced with 3,6-anhydrogalactose, the extent of this is variable depending on the source (N.S.Anderson, *et al.*, Carbohydrate Res., 1968,7, 468).

ν-Carrageenan has been preoposed as a precursor for ℓ-carrageenan (D.J.Stancioff and N.F.Stanley, 6th Int.Seaweed Symp., 1969, p.595).

ξ-Carrageenan is still another type composed of galactose 2-sulphate which probably is linked alternately $\beta(1 \rightarrow 4)$ and $\alpha(1 \rightarrow 3)$ (A.Penman and D.A.Rees, J.chem. Soc. Perkin I, 1973, 218).

As mentioned these carrageenans are idealised forms, the natural ones as extracted from algal samples are not only mixtures of precursors and "finished" molecules, but do also contain molecules of an intermediate composition. The μ and ν-carrageenans are proposed as precursors for κ- and ℓ-carrageenans respectively (A. Haug, MTP Internat. Rev.Sci.Biochem.Ser.1, Vol.11, p.51).

Immunological methods have been developed for determining the various types of carrageenans. E.L.McCandless and her group have for instance shown that only λ-carrageenan is produced by tetra-sporangial plants of *Chondrus crispus*, while κ-carrageenan is produced by gametangial plants (S.P.C. Hosford and E.L. McCandless, Canad. J. Botany, 1975, 53, 28, 35). These immunological methods have been useful in obtaining very pure samples of the various carrageenans. (V. DiNinno and E.L. McCandless, Carbohydrate Res., 1978, 66, 85; 1978, 67, 235).

The carrageenans are of commercial interest due to their gel forming ability. The type of network is dependent on the polysaccharide conformation. Irregularities along the chain will cause "kinks" and thus give the gel different physical properties (E. Percival, J. Sci. Food Agric., 1972, 23, 933; D.A. Rees and E.J. Welsh, Angew. Chem. internat. Edn., 1977, 16, 214).

(c) Alginic acid, the major matrix polysaccharide of the brown algae, is composed of D-mannuronic acid and its C-5 epimer L-guluronic acid. The polymer comprises an unbranched chain of (1 ⟶ 4)-linked β-D-mannuronic acid and α-L-guluronic acid arranged in a blockwise fashion (B. Larsen, *et al.*, Acta chem. Scand., 1970, 24, 726). By use of ^{1}H- and ^{13}C-n.m.r. spectroscopy it has been shown that for alginate in solution the mannuronic acid residues are in the $^{4}C_{1}$ conformation and the guluronic acid residues in the $^{1}C_{4}$ conformation, independently of their nearest neighbouring units (A. Penman and G.R. Sanderson, Carbohydrate Res., 1972, 25, 273; H. Grasdalen, *et al.*, *ibid.*, 1977, 56, C11).

In studies on alginate synthesis it was shown that an enzyme from the bacterium *Azotobacter vinelandii* was capable in the presence of calcium, of converting algal polymannuronic acid into a copolymer of mannuronic and guluronic acid, indistinguishable from ordinary algal alginate (A.Haug and B. Larsen, Carbohydrate Res., 1971, 17, 297). Thus the C-5 epimerization from D-mannuronic to L-guluronic acid occurs at the polymer level, not at the nucleotide sugar level as believed previously. The epimerization is not reversible. Analysis of tissue from different parts of various brown algae shows that the older tissue is relatively richer in guluronic acid than younger tissue (A.Haug, *et al.*, *ibid.*, 1974, 32, 217).

(d) Fucan. A number of fucose-containing polysaccharides have been isolated from members of the *Phaeophyceae*. It is evident that these polymers belong to a family of structurally related polymers. This has been supported both by structural, immuno-chemical and metabolic studies (A.J.Mian and E. Percival, Carbohydrate Res., 1973, 26, 133, 147; V.Vreeland, and D.J.Chapman, Proc.Int.Seaweed Symp.9, 1979, 9, 337; R.G.S.Bidwell, *et al*., Canad.J.Bot., 1972, 50, 191). In addition to fucose, xylose, glucuronic acid, galactose and mannose are present. Some of the fucans are probably glycoproteins. Ion exchange chromatography has revealed the presence of several fractions having varying degrees of incorporation of uronic acid and half-ester sulphate groups in the polymer fraction isolated from *Himanthalia lorea, Bifurcaria bifurcata* and *Padina pavona*. At one extreme a fraction had a high glucuronic acid content and a low of sulphate content, while another fraction contained a small amount of glucuronic acid and a high amount of sulphate (A.J.Mian and E. Percival, *loc.cit*.) Fucan-4-sulphate is present mainly as (1 ⟶ 2)- and (1 ⟶ 3)-linked units and as branch points while glucuronic acid and xylose are mainly present as terminal and (1 ⟶ 4)-linked units. A fucan isolated from *Ascophyllum nodosum* has a relatively high galactose content. In this polymer all the galactose was present as branch points (D.G.Medcalf, *et al*., Carbohydrate Res., 1978, 66, 167). An ascophyllan like polymer (B.Larsen, *et al*., Acta chem.Scand., 1966, 20, 219), also present was shown to be a glycoprotein probably linked through the serine and/or the threonine of the protein chain (D.G.Medcalf and B.Larsen, Carbohydrate Res., 1977, 55, 539).

Sargassan is a similar type of polymer isolated from *Sargassum linifolium*. The polymer backbone is probably composed of (1 ⟶ 4)-linked β-D-glucuronic acid and β-D-mannose, while the branches comprise (1 ⟶ 4)-linked β-D-galactose, β-D-galactose 6-sulphate, β-D-galactose 3,6 disulphate, (1 ⟶ 2)-linked α-L-fucose 4-sulphate and (1 ⟶ 3)-linked β-D-xylose (A.F.Abdel-Fattah, *et al*., Carbohydrate Res., 1974, 33, 209). Fucan sulphates containing no other sugar are rare, but it has been shown that the *Fucus* embryos produce sulphated fucan (R.S.Quatrano and P.T.Stevens, Plant Physiol., 1976, 58, 224).

(e) Others. Water-soluble sulphated polysaccharides are also present in the green seaweeds. Studies on the polymer isolated from *Cladophora rupestris* indicate the presence of a backbone of (1 ⟶ 4)-linked arabinose interspersed with one (1 ⟶ 3)- or (1 ⟶ 3)(1 ⟶ 6)-linked galactose unit for approximately eight arabinose units (E.J. Bourne, *et al.*, J. chem. Soc., (c), 1970, 1561).

Acetabularia sp. contain a sulphated polymer composed of D-galactose, L-rhamnose, D-xylose, 4-*O*-methylgalactose and D-glucuronic acid, indicating a third group of green algal polysaccharides (E. Percival and B. Smestad, Carbohydrate Res., 1972, 25, 299).

The sulphated polymers isolated from *Urospora* sp. and *Codiolum pusillum* (another life form of *U. wormskioldii*) are similar to that isolated from *Ulva lactuca*. These polymers are basically similar, but differ in the finer structural details. A basic structure has been proposed where the three sugars are present in the main chain, and the side chains are mainly composed of xylosyl residues. (G.E.Carlberg and E. Percival, *ibid.*, 1977, 57, 223 and refs. cited therein).

Certain *diatoms* produce a (1 ⟶ 3) β-D-glucan as a reserve polysaccharide (S. Myklestad, in Handbook of Phycological Methods, see above, p.133). Detailed studies on some of these have shown that the glucan is branched, with branches at both *O*-6 and *O*-2 of different glucosyl residues (B.S.Paulsen and S. Myklestad, Carbohydrate Res., 1978, 62, 386).

The extracellular polysaccharide produced by certain *Chaetoceros* sp. is sulphated and composed of the sugars fucose, galactose and rhamnose. In one of the species, *C. curvisetus*, fucose is present both in the furanose and the pyranose form. The polymer is highly branched with fucofuranose responsible for the main part of the non-reducing units (B. Smestad, *et al.*, Acta chem. Scand. (B), 1975, 29, 337).

The unicellular alga *Porphyridium cruentum* produces an extracellular polysaccharide which is composed of D-xylose, D- and L-galactose, D-glucuronic acid, D-glucose, 2-*O*-methyl D-glucuronic acid and sulphate. The hexuronic acids are linked (1 ⟶ 3) to hexoses, while the main part of the units is (1 ⟶ 4) linked (D.G. Medcalf, *et al.*, Carbohydrate Res., 1975, 44, 87; J. Heaney-Kieras and D.J. Chapman, *ibid.*, 1976, 52, 169).

8. Bacterial carbohydrate polymers

A number of complex carbohydrate-containing polymers are associated with the bacterial cell wall or "envelope" that surrounds the cytoplasmic membrane (V. Braun and K. Hantke, Ann. Rev. Biochem., 1974, 43, 89). The rigid cell wall may vary in thickness from ca. 10 to 50 nm and it may account for as much as 20 to 30% of the dry weight of the bacterial cell. Based on chemical composition and structure and on morphology bacterial cell walls can be divided essentially into two classes, corresponding to the Gram stain classification. The Gram-positive cell wall consists of peptidoglycan (major component), teichoic acid and polysaccharide. The Gram-negative cell wall contains only a thin layer of peptidoglycan and no teichoic acid, but it is surrounded by an outer membrane composed of lipoprotein, phospholipid and lipopolysaccharide.

A characteristic feature of bacterial polysaccharides is their ordered structure composed of repeating oligosaccharide units (O. Lüderitz, *et al.*, Bacteriol. Rev., 1966, 30, 192). This is different from plant polysaccharides where complicated and irregular structures are frequently found, which cannot be formulated in terms of repeating units.

(a) Peptidoglycan - also known as murein, mucopeptide or glycopeptide - is responsible for the rigidity of the bacterial cell wall. It has a unique molecular architecture as it consists of linear polysaccharide chains crosslinked by oligopeptide units to form a single bag-shaped giant molecule of considerable mechanical strength.

Detailed studies on peptidoglycans isolated from cell walls of *Micrococcus luteus*, *Staphylococcus aureus* and *Echerichia coli* have revealed, although differences in chemical composition occur, that the respective peptidoglycans have the same basic organisation and structural features (K.H. Schleifer and O. Kandler, Bacteriol. Rev., 1972, 36, 407). The basic repeating unit of the *Staphylococcus aureus* peptidoglycan is illustrated overleaf.

CH_2OH CH_2OH

O O O OH O

CH_3-CH HNAc HNAc

CO

NH

L-Ala

D-Gln

L-Lys-Gly-Gly-Gly-Gly-Gly-NH-

D-Ala

Co

The polysaccharide chain consists of alternate units of 2-acetamido-2-deoxy-D-glucopyranose and its 3-*O*-D-lactyl ether derivative, *N*-acetyl muramic acid, linked together through 1,4- β-glycosidic bonds. To the *N*-acetyl muramic acid carboxyl groups are attached, through an amide linkage, tetrapeptide units of L-alanine, D-glutamine, L-lysine and D-alanine. The tetrapeptide is unusual in that it contains D-amino acids and that the D-glutamine residue forms a peptide linkage *via* its γ-carboxyl group. Finally the tetrapeptide side chains are crosslinked by pentaglycine bridges extending from the D-alanine residue of one chain to the ε-amino group of the L-lysine of another side chain.

Whereas the glycan strands of the peptidoglycan appear to be remarkably uniform in the great majority of bacterial species, the composition and structure of the peptide chains and particularly the nature of the crosslink between the peptide side chains may vary in several respects.

The biosynthesis of the peptidoglycans of bacterial cell walls and the action of penicillin on this process have been reviewed (J.M. Ghuysen, J. gen. Microbiol., 1977, 101, 13).

A different type of polymer, a teichuronic acid, has been obtained from the cell wall of *Micrococcus luteus* by precipitation with cetylpyridinium chloride after lysozyme digestion (Nasir-Ud-Din and R.W.Jeanloz, Carbohydrate Res., 1967, 47, 245). This acidic polymer consists of a linear chain with a repeating disaccharide unit.

$\rightarrow$ 4)-β-D-Man*p*NAcA-(1 $\rightarrow$ 6)-α-D-Clc*p*-(1 $\rightarrow$

The α-D-glucopyranosyl residue at the reducing end of the molecule is connected through a phosphodiester linkage to *O*-6 of an *N*-acetylmuramic acid residue of a peptidoglycan chain.

(b) Teichoic acids have been reviewed by J. Baddiley (Essays Biochem., 1972, 8, 35) and M. Duckworth ("Surface Carbohydrates of the Prokaryotic Cell", ed. I.W.Sutherland, Academic Press, 1977, p.177).

Walls of Gram-positive bacteria contain, in addition to peptidoglycan, one or more heteropolysaccharides. Most prominent among these are the teichoic acids which are polymers consisting of ribitol or glycerol phosphate, the residues being connected through phosphodiester linkages. The free hydroxyl groups may be glycosylated or esterified with D-alanine.

The term teichoic acid has been extended to cover all bacterial cell wall, capsular and membrane polymers that contain glycerol phosphate or ribitol phosphate residues. They can be divided into two major classes according to their cellular location. Thus wall teichoic acids comprise a major part (30-50%) of the purified walls of most Gram-positive bacteria and are covalently linked to the peptidoglycan. They exhibit considerable structural diversity and occur as either ribitol or glycerol phosphate chains, bearing a variety of substituents. Sugars or sugar-1-phosphate residues may form part of the main polymer chain.

Membrane teichoic acids - also known as lipoteichoic acids - are similar polymers found in nearly all Gram-positive bacteria. They appear on the outer surface of the cytoplasmic membrane and possess antigenic properties. The membrane teichoic acids are chemically more specialised and possess exclusively glycerol phosphate chains covalently linked to glycolipid molecules located within the cell membrane (P.A.Lambert, *et al*., Biochem. Biophys. Acta. 1977, 472, 1). Various functions have been ascribed to these glycerol phosphate polymers such as the binding of divalent cations required for optimal activity of membrane-bound enzymes, and acting as the lipoteichoic acid carrier in the biosynthesis of wall teichoic acid (J. Mauck and L. Glaser, Proc. natl. Acad. Sci.U.S., 1972, 69, 2386). Studies on the mode of linkage between wall teichoic acid and peptidoglycan in a *Staphylococcus aureus* mutant have indicated that the ribitol teichoic acid is attached to a N-acetylmuramic acid residue of the glycan through an oligomer of glycerol phosphate and a residue of 2-acetamido-2-deoxy-D-glucopyranose-1-phosphate (J. Coley, *et al*., F.E.B.S. Letters, 1977, 80, 405).

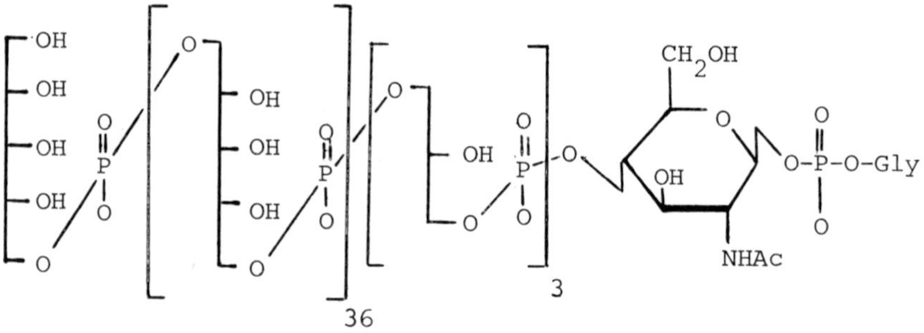

Detailed procedures for the isolation and characterisation of teichoic acids have been described (A.R. Archibald, Methods Carbohydrate Chem., 1972, 6, 162).

(c) Extracellular polysaccharides have been reviewed by I.W. Sutherland in "Surface Carbohydrates of the Prokaryotic Cell", ed. I.W. Sutherland, Academic Press, 1977,p.27).

Microbial extracellular polysaccharides are found in two forms. They may either form capsules attached to the cell surface or they may be secreted as soluble slime unattached to the microorganism. The composition of many expolysaccharides has revealed that the number of component sugars is fairly limited,the most common being D-glucose,D-mannose,D-galactose, L-fucose, L-rhamnose and D-glucuronic acid. In addition *O*-acetyl groups and pyruvate ketal groups are frequently found, and the presence of lactyl ether groups has also been reported (B. Lindberg, *et al.*, Carbohydrate Res., 1976, 49, 411; L.Kenne, *et al.*, *ibid.*, 1976, 51, 287).

Some bacteria such as *Streptococcus pneumoniae*, produce capsular material which is of considerable interest because of its virulent properties and antigenic nature. Each of the many different types of bacteria within this species can elaborate its own specific capsular polysaccharide (O.Larm, and B. Lindberg, Adv. carbohydrate Chem. Biochem., 1976, 33, 295). All of these are antigenic, and the Pneumococcus species was one of the first to be subclassified by immunological methods. Some eighty different pneumococcal polysaccharides have been identified and although the complete or almost complete chemical structure is known for only eight of these, partial structures are known for about twenty five.

For type 2 pneumococcal capsular polysaccharide the following structure has been proposed (L. Kenne, *et al.*, Carbohydrate Res., 1975, 40, 69).

```
→ 3)-α-L-Rhap-(1→3)-α-L-Rhap-(1→3)-β-L-Rhap-(1→4)-α-D-
                        2
                        ↑                            Glcp-(→
                        1

                    α-D-Glcp
                        6
                        ↑
                        1
                    α-D-GlcpA
```

Extracellular heteropolysaccharides are composed of oligosaccharide repeating units, varying in size usually from two to six monosaccharides. An exception to this is bacterial alginate that, as marine algal alginate, is composed of blocks of 1 ⟶ 4-linked β-D-mannopyranosyluronic acid and α-L-gulopyranosyluronic acid, respectively, and mixed blocks of both acids. In addition bacterial alginate is reported to be partly acetylated (P.A.J. Gorin and J.F.T. Spencer, Canad. J. Chem., 1966, 44, 993). Biosynthetic studies strongly indicate that initially a homopolymer of D-mannuronic acid is formed (D.F. Pindar and C. Bucke, Biochem.J.,

1975, 152, 617), followed by partial C-5 epimerisation to L-guluronic acid, thus resulting in alginate (A. Haug and B. Larsen, Carbohydrate Res., 1971, 71, 297).

The common biosynthetic route of the extracellular heteropolysaccharides involves the pyrophosphate of the C_{55} iosprenoid alcohol, undecaprenol, which acts as a glycosyl carrier lipid (GCL). The repeating unit oligosaccharide is built by sequential transfer of glycosyl groups from sugar nucleotides, for example

```
         O⁻  O⁻
         |   |
GCL-O-P-O-P-O ← Gal ← Man ← Gal
         ||  ||             ↑
         O   O            GlcA
```

The completed repeating oligosaccharide unit is then transferred to the growing polymer chain (F.A. Troy, *et al.*, J. biol. Chem.,1971, 246, 118).

Homopolysaccharides like dextrans and levans are produced from sucrose by a number of bacterial species. However,the synthesis of these polymers differs from that of other extracellular polysaccharides in that neither sugar nucleotides nor lipid-linked intermediates are involved. Dextran synthesis is catalysed by the extracellular glycosyl transferase dextransucrase that transfers the α-D-glucosyl group from a sucrose molecule to the 6-position of the non-reducing terminal glucose residue of a growing dextran chain. The mode of introduction of branches is not yet known in detail.

Analogous to dextrans, bacterial levans are produced extracellularly from sucrose by the action of the enzyme levan sucrase. The fructose moiety of sucrose is utilised resulting in a polymer composed of α-D-fructofuranosyl residues connected by (2 ⟶ 6)-linkages.

The term dextran is used on a large group of bacterial exocellular polysaccharides composed of α-D-glucopyranosyl residues with the preponderance of (1 ⟶ 6)-linkages. (For reviews see R.L. Sidebotham, Adv. carbohydrate Chem. Biochem., 1974, 30, 371; P.A.Sandford, *ibid.*, 1979, 36, 265). The proportion of such linkages may vary considerably, from *ca.* 50% to *ca.* 90% depending on the strain of the producing organism and on the conditions of growth.

A number of species within the genera *Lactobacillus*, *Leuconostoc* and *Streptococcus* synthesise dextrans when grown on sucrose-containing media. Dextrans synthesised by *Leuconostoc* have been most thoroughly investigated in view of the industrial importance of these polysaccharides. However, dextrans produced by some *Streptococcus* species are of increasing interest because of their role in the formation of dental plaque (R.J.Gibbons and S.B.Banghart, Arch. oral. Biol., 1967, 12, 11; A.T.Arnett and R.M.Mayer, Carbohydrate Res., 1975, 42, 339).

The NRRL 512 strain of *Leuconostoc mesenteroides* is used generally for the industrial production of dextran for clinical and other purposes. This dextran contains *ca.*95% of α-(1 ⟶ 6)- and *ca.*5% of α-(1 ⟶ 3)-linkages and has a molecular weight of 40-50 x 10^6. All of the (1 ⟶ 3)-linkages are involved in chain branching and 85% of the side chains constitute one or two glycosyl residues (O.Larm, *et al.*, Carbohydrate Res., 1971, 20, 39). Clinical dextran is obtained by depolymerisation of the native polymer and isolation of the fractions with MW 40,000 and 70,000. Studies on other dextrans have revealed the presence also of α(1 ⟶ 2)- and α-(1 ⟶ 4)-linkages (F.R.Seymour, *et al.*, Carbohydrate Res., 1977, 53, 153).

Xanthan gum, the extracellular polysaccharide produced by the bacterium *Xanthomonas campestris*, is of importance as an emulsion-stabilising and gelling agent because of its favourable technological properties. Some of these can be explained in terms of the high molecular weight (from 1 to 10 million) and a rod-like ordered structure (I.C.M.Dea, *et al.*, Carbohydrate Res.,1977,57, 249). A repeating pentasaccharide unit structure has been proposed for the Xanthan polysaccharide (P.E.Jansson, *et al.*, Carbohydrate Res., 1975, 45, 275; L.D.Melton, *et al.*, *ibid.*, 1976, 46, 245).

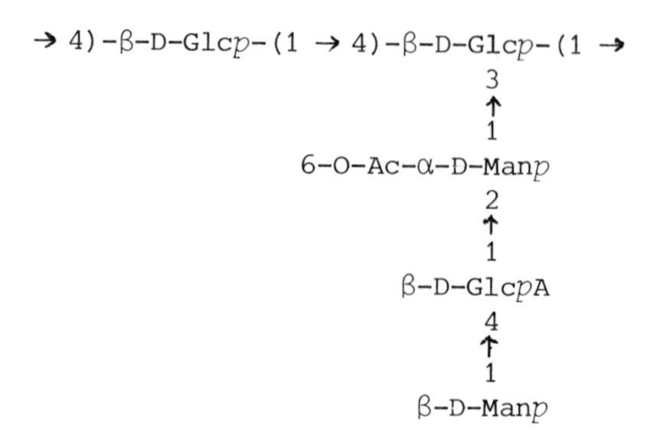

Every second terminal mannose residue carries a 4,6-pyruvic acid ketal group.

(d) Lipopolysaccharides have been reviewed by S.G.Wilkinson in "Surface Carbohydrates of the Prokaryotic Cell", ed. I.W.Sutherland, Academic Press, 1977, 97.

The lipopolysaccharides are cell wall components of Gram-negative bacteria, located in the outer membrane together with protein, lipoprotein and phospholipid. They constitute the specific surface antigens of the organism, they are toxins, and many of them serve as receptors for bacteriophages. The antigenic property resides in the polysaccharide portion of the molecule, whereas the toxicity is associated with the lipid moiety.

Numerous studies have shown that the chemical composition and structure of the lipopolysaccharides is highly complex. They are thought to consist of three structural regions (L. Lüderitz, *et al.*, Bacteriol.Rev.,1966,30,192).

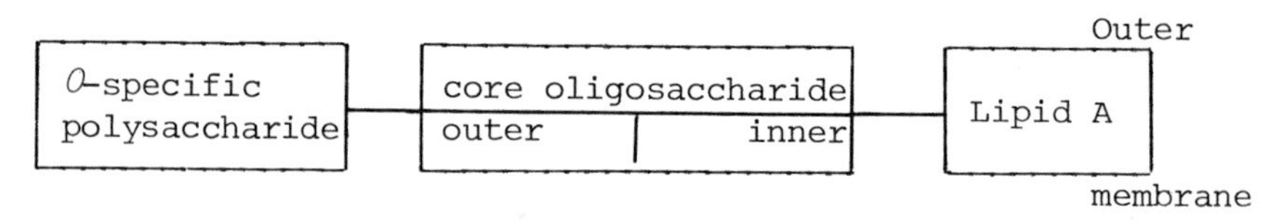

The *O*-specific polysaccharide consists of oligosaccharide repeating units and is distinguished by a great structural variability in accordance with the many immunological activities it can exert. The core oligosaccharide was originally found to be common for all *Salmonella* lipopolysaccharides and therefore called the "common core". It has been established later that there are several different but structurally closely related core oligosaccharides through which the serological R-specificities are expressed. The core can be divided into an outer region to which side chains may be attached, and an inner region, linked to lipid A.

Lipid A is a 2-acetamido-2-deoxy-D-glucopyranose-containing phospholipid that connects the polysaccharide to the outer bacterial membrane.

The polysaccharides of the enteric bacteria, mainly the *Salmonella* and *E.coli* have been investigated most extensively. The inner core hexasaccharide of the *Salmonella* lipopolysaccharides is composed of 3-deoxy-D-*manno*-octulosonic acid (KDO) and a heptose, usually L-*glycero*-D-*manno*-heptose (Hep).

```
                Hep                        KDO
                 ↓                          ↓
Outer core ⟶ Hep ⟶ Hep ⟶KDO ⟶ KDO ⟶ Lipid A
```

Alkali-labile, non-carbohydrate substituents have been omitted.

The *Salmonella* outer core oligosaccharide is a pentasaccharide for which the following structure has been proposed:

```
            α-D-GlcpNAc                             α-D-Galp
                 1                                      1
                 ↓                                      ↓
                 2                                      6
side  ⟶ 4)-α-D-Glcp-(1 → 2)-α-D-Galp-(1 → 3)-α-D-Glcp- inner
chain                                                    core
```

The uniformity of core composition is in marked contrast to the variability of the *O*-specific side chains. More often than not the latter contains several different sugars of which some may be unusual. The lipopolysaccharides have proved to be the source of a number of dideoxy sugars, amino dideoxy sugars and also *O*-methylated sugars, substances that have not been found previously in nature.

9. Glycoproteins

A comprehensive treatment of the subject is given in the two-volume monograph by Gottschalk (A. Gottschalk ed. "Glycoproteins: their Composition, Structure and Function", Elsevier, Amsterdam, 1972). For general reviews see R.G. Spiro, Adv. protein Chem., 1973, 27, 349; J.Montreuil, Adv. carbohydrate Chem. Biochem., 1980, 37, 157; W.J. Lennarz, ed. "The Biochemistry of Glycoproteins and Proteoglycans", (Plenum Press, New York, 1980). Glycoproteins are defined as proteins to which carbohydrate is covalently attached and they represent the most diverse group of naturally occurring macromolecules. The definition includes also the proteoglycans, but these polymers will be treated separately.

The glycoproteins are widely distributed in nature and have been isolated from a variety of sources, ranging from microorganisms to man. The carbohydrate content of glycoproteins may vary greatly from less than 1% to more than 80% of the total weight. They are involved in a wide range of biological functions, such as enzyme and hormone action, transport, lubrication and protection, receptor functions etc.

Isolation and purification of glycoproteins are usually complicated by the characteristic heterogeneity of the carbohydrate residues, thus leading to multiple components which have identical peptide chains, but differ in their saccharide moieties. This phenomenon was illustrated long ago by ion exchange chromatography of ribonuclease from bovine (T.H. Plummer, J. biol. Chem., 1968, 243, 5961), and porcine (V.N. Reinhold, *et al.*, *ibid.*, p. 6482) pancreatic juice by which means several members of a family of each enzyme were obtained and shown to differ only in their carbohydrate composition.

The amino acid composition of glycoproteins is not different from that of other proteins; the sugars most commonly encountered include D-mannose, D-galactose, D-glucose, L-fucose, 2-acetamido-2-deoxy-D-galactose, 2-acetamido-2-deoxy-D-glucose and the *N*- and *O*-acetyl and *N*-glycosyl derivatives of neuraminic acid (sialic acids). In addition D-xylose and uronic acids are found in the proteoglycans, some of which are also *N*- and *O*-sulphated. Because of the complex composition of most glycoproteins and the difference in stability of the peptide and the glycosidic bonds, and also of the various sugars, analysis and structural investigations of glycoproteins require the use of a variety of different methods (R.G. Spiro in Colowick and Kaplan, "Methods Enzymol", ed. V. Ginsburg, Academic Press, 1972, 28, 1).

The carbohydrate-peptide linkage is the unique structural feature of glycoproteins. This linkage involves C-1 of the most internal sugar residue and a functional group of an amino acid. Different types of carbohydrate-peptide linkage exist, and glycoproteins may be classified according to linkage type (R. Kornfeld and S. Kornfeld, Ann. rev. Biochem., 1976, 45, 217). The different glycopeptide bonds include one *N*-glycosidic (A) and three types of *O*-glycosidic linkage (B, C, D).

A. β-D-Glc*p*NAc-(1 → 4)-Asn

B. α-D-Gal*p*NAc-(1 → 3)-Ser (Thr)

D-Man*p*-(1 → 3)-Ser

D-Gal*p*-(1 → 3)-Ser (Thr)

α-L-Fuc*p*-(1 → 3)-Thr

C. β-D-Gal*p*-(1 → 5)-Hyl

D. L-Ara*f*-(1 → 4)-Hyp

D-Gal*p*-(1 → 4)-Hyp

A. The oligosaccharide moieties of the *N*-glycosylproteins may be classified into two main groups: the simple one which consisists of only mannose and 2-acetamido-2-deoxy-glucose units and the more complex one which in addition contains galactose, fucose and sialic acid. These groups both contain the same pentasaccharide core shown on the next page.

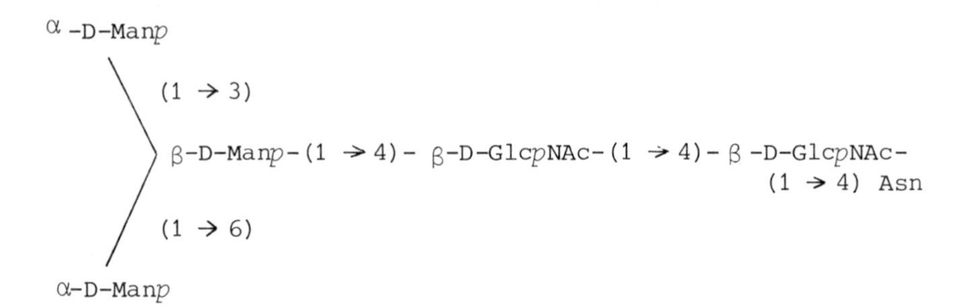

In the first group the pentasaccharide core is substituted exclusively by β-D-mannopyranose residues. Belonging to this group are the oligosaccharide moieties of hen ovalbumin (J. Conchie and I. Strachan, Carbohydrate Res., 1978, 63, 193), calf thyroglobulin unit A (S.Ito, *et al*., J. Biochem., 1977, 81, 621) and soybean lectin (H. Lis and N. Sharon, J. biol. Chem., 1978, 253, 3468).

In the second group the pentasaccharide is substituted by a variable number of residues of galactose, fucose, 2-acetamido-2-deoxyglucose and sialic acid. The disaccharide unit β-D-galactopyranosyl-(1 ⟶ 4)- β-2-acetamido-2-deoxy-D-glucopyranose (*N*-acetyllactosamine) is frequently found in this group and it has been proposed that these structures should be called the *N*-acetyllactosaminic type (J. Montreuil, Pure appl. Chem., 1975, 42, 431). This group can be exemplified by the oligosaccharide residues of α_1-acid glycoprotein (orosomucoid) (B. Fournet, *et al*., Biochem., 1978, 17, 5206), fetuin (B. Nilsson, *et al*., J. biol. Chem., 1979, 254, 4545) and human IgG glycopeptide (J. Baenziger and S. Kornfeld, J. biol. Chem., 1974, 249, 7270). The pentasaccharide depicted above does not occur in all the *N*-glycosylproteins. Some structures differ in that they contain a single 2-acetamido-2-deoxyglucose residue instead of two, as in the human myeloma IgM glycopeptide (F. Miller, Immunochem., 1972, 9, 217), and the glycan units of alveolar glycoprotein (S.N. Bhattacharyya and W.S. Lynn, J. biol. Chem., 1977, 252, 1172).

B. Glycoproteins with carbohydrate side chains attached through the hydroxyl group of serine or threonine can be divided into sub-classes according to the sugar residue involved. By far the largest sub-class is that with the linkage

α-D-Gal*p*NAc-(1 $\rightarrow$ 3)-Ser (Thr)

Usually the inner core structure

β-D-Gal*p*-(1 $\rightarrow$ 3)- α-D-Gal*p*NAc-(1 $\rightarrow$ 3)-Ser(Thr)

is found in this group, which is also termed the mucin type. It includes blood group substances and a number of other glycoproteins from a variety of sources. (For a review on the glycoproteins of mucus see J.R. Clamp ed. Brit.med. Bull., 1978, 34, 1). The term blood group substance was coined because the ABH and Lewis antigens were originally detected on the red cell membrane. However, these antigens have been found to be widely distributed in the body and occur as glycosphingolipids on the surface of both erythrocytes and other cells, as oligosaccharides in milk and urine and as *O*-glycosidically linked oligosaccharides attached to mucins secreted in the gastrointestinal and respiratory tracts. The enzymic basis for the ABH and Lewis group human blood group antigens has been reviewed (V. Ginsburg, Adv. Enzymol., 1972, 36, 131; S. Hakomori and A.Kobata in "The Antigens", ed. M. Sela, Academic Press, 1974, Vol. 2, p.79). The mucus glycoproteins are proteins possessing hundreds of attached oligosaccharide units, that each may contain from 2 to 15-20 sugar residues. This type of glycopeptide linkage is susceptible to cleavage by alkali by β-elimination. If the reaction is carried out in the presence of excess reducing agent, usually sodium borohydride, the intact oligosaccharide side chain will be released, the innermost sugar, *e.g.* 2-acetamido-2-deoxy-D-galactose being reduced to the corresponding alditol,thus stabilising the product from further alkaline degradation. Using this technique the oligosaccharide chains of a large number of glycoproteins from several sources have been investigated *e.g.* porcine submaxillary mucin (M.M.Baig and D. Aminoff, J.biol. Chem., 1972, 247, 6111), "anti-freeze" glycoprotein from antarctic fish (W.T.Shier, *et al.*, F.E.B.S. Letters, 1975, 54, 135), human gastric mucin (M.D. Oates, *et al.*, Carbohydrate Res., 1974, 34, 115) and human serum IgA_1 (J. Baenziger and S.Kornfeld, J. biol. Chem., 1974, 249, 7260, 7270). By treatment of two blood group active mucins with alkaline sodium borohydride eighteen reduced oligosaccharides were isolated and characterised, many of previously unknown structure. They all terminated in the disaccharide sequence β-D-galactopyranosyl-(1 ⟶ 3)-2-acetamido-2-deoxy-D-galactitol (L. Rovis, *et al.*, Biochm., 1973, 12, 5340, 5355). Some glycoproteins contain oligosaccharides linked through mannose to serine or threonine. In yeast mannan from a wild-type *Saccharomyces cerevisiae* the following structure has been established (T. Nakajima and C.E. Ballou, J. biol. Chem., 1974, 249, 7679).

α-D-Man*p*-(1 → 2)- α-D-Man*p*-(1 → 2)-α-D-Man*p*(1 → 3)-Ser

The glycoprotein mycodextranase secreted by *Penicillium melinii* (A.L. Rosenthal and J.H. Nordin, J. biol. Chem., 1975, 250, 5295) contains the glycopeptide structure shown below

α-D-Glc*p*-(1 → 2)-D-Man*p*-(1 → 3)-Ser

In plants the D-galactopyranosyl-*O*-serine linkage has been found in the cell wall glycoprotein extensin (D.Lamport *et al.*, Biochem. J., 1973, 133, 125; Y.P. Cho and M.J. Chrispeels, Phytochm., 1976, 15, 165). However, the anomeric configuration of the galactopyranosyl residue has not been established. In collagens isolated from annelid cuticle the same linkage has been shown to be present (R.G. Spiro and U.D. Bhoyroo, Fed. Proc., 1971, 30, 1223; L. Muir and Y.C. Lee, J. biol. Chem., 1970, 245, 502).

Finally the last structure of the B-type is that of a glycopeptide isolated from normal human urine (P.Hallgren, *et al.*, J. biol. Chem., 1975, 250, 5312).

β-D-Glc*p*-(1 → 3)-α-L-Fuc*p*-(1 → 3)-Thr

C. The carbohydrate-peptide linkage normally present in collagens and basement membranes is composed of β-D-galactopyranose or α-D-glucopyranosyl-(1 ⟶ 2)-β-D-galactopyranose linked *O*-glycosidically to the hydroxyl group of a hydroxylysine residue in the peptide backbone. These two sugar units occur in variable proportions and amounts in basement membranes and collagens from both vertebrate and invertebrate sources. In contrast to the *O*-glycosidic bond to the α-amino- β-hydroxy acids the linkage to the 5-hydroxyl group of hydroxylysine is stable to alkali, and the glycopeptides

α-D-Glc*p*-(1 → 2)- β-D-Gal*p*-(1 → 5)-Hyl and β-D-Gal*p*-(1 → 5)-Hyl

can be obtained from alkaline hydrolysates of collagen. (R.G.Spiro, Adv. protein Chem., 1973, 27, 349).

D. Hydroxyproline-linked oligosaccharides in which the linkage sugar is L-arabinofuranose are present in the cell wall of a wide range of plants. (For a review on plant glycoproteins see R.G. Brown and W.C. Kimmins in "International Review of Biochemistry. Plant Biochemistry II", ed. D.H. Northcote , University Park Press, 1977, p.183). Furthermore it has been established that D-galactopyranose is also linked *O*-glycosidically to hydroxyproline in the green alga *Chlamydomonas* (D.H. Miller, *et al*., Science, 1972, 176, 918) and in a glycoprotein from wheat endosperm (G.B. Fincher, *et al*., Biochem.J., 1974, 139, 535).

Despite numerous chemical studies on plant glycoproteins the anomeric configuration of the arabinofuranosidic and galactopyranosidic bond to hydroxyproline has not been unambiguously demonstrated. However, in one report on the hydroxyproline-linked arabinose oligomer from tobacco cells (Y. Akiyama and K. Kato, Agric. biol. Chem., 1977, 41, 79) the following configuration, based on n.m.r. measurements, is proposed

β-L-Ara*f*-(1 → 2)- β -L-Ara*f*-(1 → 4)-Hyp

Concerning the biosynthesis of glycoproteins of the *N*-glycosidic class it has been well established that the participation of lipid intermediates of the polyprenol phosphate type is required. (For a review on the role of lipids in glycoprotein synthesis see D.K. Struck and W.J. Lennarz in "The Biochemistry of Glycoproteins and Proteoglycans", ed. W.J. Lennarz, Plenum Press, 1980, p. 35).

10. Proteoglycans

For reviews see M.B.Mathews, "Connective Tissue. Macromolecular Structure and Evolution", Springer, 1975; L. Rodén in "The Biochemistry of Glycoproteins and Proteoglycans", ed. W.J. Lennarz, Plenum Press, 1980, p. 267).

The proteoglycans, previously termed mucopolysaccharides, differ from the typical glycoproteins in that they contain long polysaccharide chains attached to a polypeptide backbone at closely spaced intervals. These polysaccharides, also referred to as glycosaminoglycans, are composed of repeating disaccharide units in contrast to other glycoproteins where no true repeating sequences are found. Further, the proteoglycans contain a high proportion of uronic acid and/or sulphate groups and are thus highly charged polyanions. The proteoglycans are classified in five groups: Hyaluronic acid, chondroitin 4- and 6-sulphate, dermatan sulphate, heparin and heparan sulphate, and keratan sulphate.

(a) Hyaluronic acid occurs in various connective tissue, synovial fluid, umbilical cord and vitreous humor. It is chemically the simplest member of the family, being a polymer of the non-sulphated disaccharide unit

$$\rightarrow 4)-\beta\text{-D-Glc}p\text{A-}(1 \rightarrow 3)-\beta\text{-D-Glc}p\text{NAc-}(1 \rightarrow$$

The molecular weight is in the 2 to 8 million range, which is far higher than that of the other glycosaminoglycans. Although no conclusive evidence exists that hyaluronic acid is attached covalently to protein, it is nevertheless classified together with the other proteoglycans because of its similarity to these. The polyelectrolyte character of hyaluronic acid, the large volume it occupies in solution and its high molecular weight all contribute to its viscoelastic properties which make it well suited as joint lubricant and shock absorber.

(b) Chondroitin 4- and 6-sulphate occur in cartilage, skin, bone, arterial wall and cornea of the eye. The basic disaccharide repeating unit is

$$\rightarrow 4)-\beta\text{-D-Glc}p\text{A-}(1 \rightarrow 3)-\beta\text{-D-Gal}p\text{NAc-}(1 \rightarrow$$

in which the 2-acetamido-2-deoxy-β-D-galactose residue may be esterified with sulphate at position 4 or 6 and thus give rise to chondroitin 4- or 6-sulphate, respectively. Even if the product usually contains about one sulphate group per disaccharide, considerable deviations from a 1:1 ratio have been noted. Individual polysaccharide chains may also contain sulphate groups in both 4- and 6-positions. Each glycosaminoglycan chain contains between 30 and 100 such disaccharide units, corresponding to a molecular weight of 15-50,000 and the chain is linked to a serine residue of the peptide backbone through two galactose and one xylose unit (E.L. Stern, *et al.*, J. biol.Chem., 1971, 246, 5707).

→4)- β-D-Glc*p*A-(1 → 3)- β-D-Gal*p*-(1 → 3)-β -D-Gal*p*-(1 → 4)-
β-D-Xyl*p*-(1 → 3)-Ser

(c) Dermatan sulphate is present in skin, tendon, heart valve and arterial wall. Initially this product is presumably identical to that formed during biosynthesis of chondroitin sulphate. However,the modification of this polymer is not restricted to the introduction of *O*-sulphate groups, but includes also formation of L-iduronic acid residues by C-5 epimerisation of D-glucuronic acid units (A. Malmström, *et al.*, J. biol. Chem., 1975, 250, 3419). The D-glucuronic acid content may vary widely - from negligible amounts to more than 90% of the total uronic acid. Pig skin dermatan sulphate has the relatively simple repeating unit structure

→ 4)- α-L-Ido*p*A-(1 → 3)-β -D-Gal*p*NAc-(1 →

with ester sulphate at position 4 of the 2-acetamido-2-deoxy-galactose residue. However, other dermatan sulphates are more complex with extensively hybridised polysaccharide chain and also with both 4- and 6-*O*-sulphated 2-acetamido-2-deoxy-galactose groups. Over sulphated regions may be due to the occurrence of 2-*O*-sulphated L-iduronic acid units.

(d) Heparan sulphate and heparin. Heparan sulphate occurs in lung, blood vessel walls and on cell surfaces, whereas heparin is located in the mast cells of various tissues, such as lung, liver, skin and intestinal mucosa. These two glycosaminoglycans are thought to originate from the same polymerised disaccharide precursor:

→ 4)- β-D-Glc*p*A-(1 → 4)- α-D-Glc*p*NAc-(1 →

Further modification of this polymer proceeds as follows (U. Lindahl and M.Höök, Ann. Rev. Biochem., 1978, 47, 385).

1. Deacetylation of the 2-acetamido-2-deoxy-D-glucose.

2. *N*-Sulphation of the free amino group.
3. C-5 epimerisation of β-D-glucopyranosyluronic acid to α-L-idopyranosyluronic acid.
4. *O*-Sulphation at position 2 of the idopyranosyluronic acid.
5. *O*-Sulphation at position 6 of the *N*-sulphated 2-amino-2-deoxyglucose residues.

N-Deacetylation is a prerequisite not only to *N*-sulphation, but also to C-5 epimerisation of the uronic acid, since only the *N*-sulphated polymer is recognised as substrate by the epimerase (U.Lindahl, *et al.*, Fed. Proc., 1977, 36, 19). It should be noted that different disaccharide units of the polymer may remain at various levels of completion. It follows that the L-iduronic acid: D-glucuronic acid ratio increases with increasing sulphate content. Thus in general terms low-sulphated, D-glucuronic acid rich polysaccharides are classified as heparan sulphate whereas high-sulphated L-iduronic acid rich species are designated heparin. It should be emphasised that the former lacks anti-coagulant activity.

In similarity to chondroitin sulphate the glycosaminoglycan chains of heparin and heparan sulphate are linked to protein by means of the sequence:

→4)-β-D-Glc*p*A-(1→3)-β-D-Gal*p*-(1→3)-β-D-Gal*p*-(1→4)-β-D-Xyl*p*-(1→3)-Ser

Heparin is synthesised by mast cells and is stored within the cytoplasmic granules of these cells. However, it has been reported that in some tissues, such as bovine liver capsule, heparin occurs as single polysaccharide chains of molecular weight less than 1×10^4 (L. Jansson, *et al.*, Biochem. J., 1975, 145, 53). Other sources, for example rat skin (A.A. Horner, J. biol. Chem., 1971, 246, 231) and peritoneal mast cells (R.W. Yurt, *et al.*, J. biol. Chem., 1977, 252, 518) contain heparin with a molecular weight approaching 1×10^6 which apparently consists of several polysaccharide chains bound to a protein core structure.

Macromolecular heparin may be a proteoglycan with a unique polypeptide backbone consisting of serine and glycine in equimolar proportions (H.C. Robinson, *et al.*, J. biol. Chem., 1978, 253, 6687). Most of the serine residues are substituted by glycosaminoglycan chains. The occurrence of single-chain forms of heparin is attributed to cleavage of the polypeptide bound chains by specific endo-glycosidases at a few sites, and this yields products of a molecular weight range similar to that of commercial heparin (S. Ögren and U.Lindahl, Biochem.J., 1976, 154, 605).

(e) Keratan sulphate, that occurs in cartilage, invertebral disc and cornea of the eye, is classified as a proteoglycan although it is devoid of uronic acid. Instead it contains galactose linked to 2-acetamido-2-deoxyglucose to give the repeating disaccharide sequence:

$\rightarrow$4)-β-D-Glc*p*NAc-(1 $\rightarrow$ 3)-β-D-Gal*p*-(1 $\rightarrow$

This basic unit is *O*-sulphated to various degrees at position 6 of both sugar residues. Other sugars that appear to be components of certain keratan sulphates are mannose, fucose, 2-acetamido-2-deoxygalactose and sialic acid. Thus keratan sulphate bears some resemblance to the typical glycoproteins. It has been shown (F.J.Kieras, J. biol. Chem., 1974, 249, 7506; H.W. Stuhlsatz, *et al.*, Z. physiol.Chem., 1971, 352, 289) that bovine corneal keratan sulphate has the 2-acetamido-2-deoxy-D-glucose-asparagine glycopeptide linkage, while keratan sulphate from bovine cartilage has the 2-acetamido-2-deoxy-D-galactose-serine linkage.

Chapter 25

TETRAHYDRIC ALCOHOLS, THEIR ANALOGUES, DERIVATIVES AND OXIDATION PRODUCTS

R.A. Hill

1. *Tetrahydric alcohols*

(a) *Tetritols*

Erythritol can be made in high yield from 4,6-*O*-benzylidene-D-glucose by sequential treatment with sodium periodate, sodium borohydride and acid hydrolysis (W. Baggett *et al*, J. chem. Soc. C, 1966, 212). It can be produced on a large scale by fermentation of alkanes using *Candida zeylanoides* (K. Hattori and T. Suzuki, Agr. biol. Chem., 1974, 38, 581, 1203).

D-Threitol has been found in the edible fungus *Armellaria mellea* (K. Kratzl, H. Silbernagel, and K.H. Bässler, Monatsch., 1963, 94, 106). L-Threitol has not been found naturally, but has been prepared from 1,3-*O*-benzylidene-L-arabinatol by treatment with sodium periodate followed by sodium borohydride and subsequent acid hydrolysis (A.S. Perlin, Methods Carbohydrate Chem., 1962, 1, 68). Racemic threitol has been synthesised by *trans* hydroxylation of *cis*-1,4-hydroxybut-2-ene (J.L. Bose, A.B. Foster, and R.W. Stephens, J. chem. Soc., 1959, 3314).

CH_2OH CH_2OH $\xrightarrow{PhCO_3H}$ O CH_2OH CH_2OH $\xrightarrow{H^+}$

CH$_2$OH
-OH
HO-
CH$_2$OH

+

CH$_2$OH
HO-
-OH
CH$_2$OH

Neutron diffraction examination of the erythritol crystal structure has shown the positions of the protons and indicates that two sets of hydrogen bonds lock the molecules in a 3-D array (A. Shimada *et al*, Acta Crystallog. Sect. A, 1975, 31: Abst. 10.3-13).

The ^{13}C- n.m.r. spectrum of erythritol has been studied (W. Voelter *et al*, Angew. Chem. internat. Edn. 1970, 9, 803) and the effects of the formation of anhydro derivatives on the chemical shifts of erythritol and threitol reported (R.G.S. Ritchie *et al*, Canad. J. Chem. 1975, 53, 1424). The conformations of erythritol and threitol in solution have been studied using ^{13}C- n.m.r. spectroscopy and this shows that the preferred planar zig-zag conformations are adopted (G.W. Scnarr, D.M. Vyas, and W.A. Szarek, J. chem. Soc. Perkin I, 1979, 496).

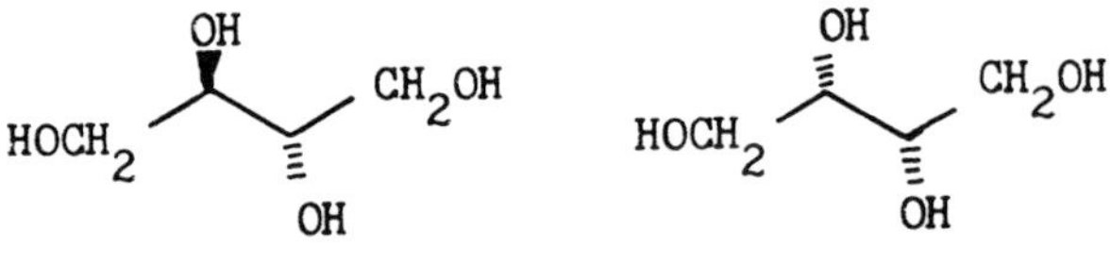

OH
HOCH$_2$
CH$_2$OH
OH

erythritol

OH
HOCH$_2$
CH$_2$OH
OH

threitol

These results are in agreement with ^{1}H-n.m.r. investigation of threitol using europium shift reagents in deuterium oxide. The proton coupling constants support the planar zig-zag arrangement, three hydroxyl groups can complex with the Eu^{3+} ion.

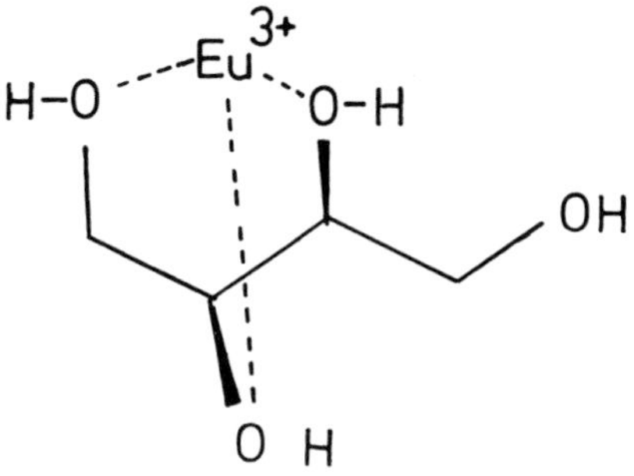

Erythritol does not complex strongly with Eu^{3+} (S.J. Angyal, D. Greeves, and J.A. Miller, Austral. J. Chem., 1974, 27, 1447).

Acid-catalysed dehydration of erythritol and threitol produces their 1,4-anhydro derivatives, *cis*- and *trans*-3,4-dihydroxytetrahydrofuran respectively. B.G. Hudson and R. Barker (J. org. Chem., 1967, 32, 3650) have studied the mechanism of this reaction.

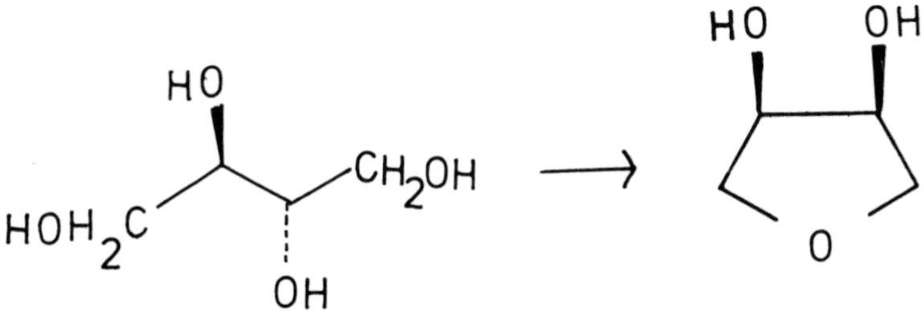

Acid-catalysed dehydration of erythritol also gives *trans*-2,5-bis(1,2-hydroxyethyl)-1,4-dioxan (A.H. Haines and A.G. Wells, Carbohydrate Res., 1973, 27, 261).

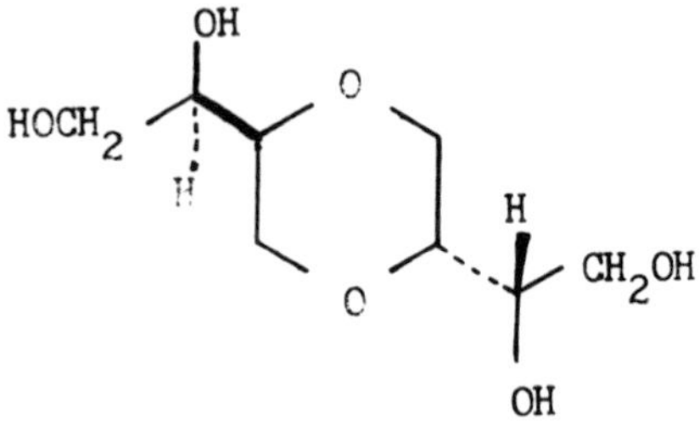

(b) Derivatives of the tetritols

Electrochemical hypochlorination of butadiene gives a mixture of dichlorobutanediols (D.A. Ashurov, *et al*, J. gen. Chem. USSR, 1978, 48, 808). 1,4-Dichlorobutane-2,3-diol, 2,4-dichlorobutane-1,3-diol, and 3,4-dichlorobutane-1,2-diol have been prepared in this way; the mechanism of their formation is discussed.

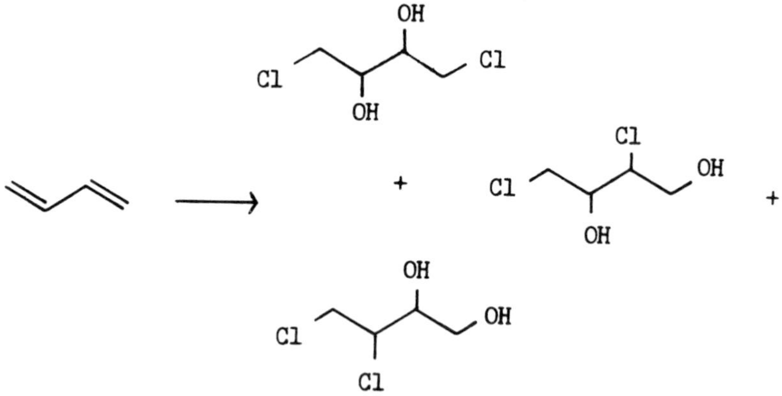

trans-Hydroxylation of *trans*-dihalobut-2-enes produces the corresponding *meso*-dihalobutane-2,3-diols. The route via the acetonide gives good yields (M.S. Malinovskii, V.G. Dryuk, and V.I. Avramenko, Zh. org. Khim., 1968, 4, 1725).

Cl / CH=CH / Cl $\xrightarrow{RCO_3H}$ epoxide

$\xrightarrow{Me_2CO}$ CH_2Cl acetonide CH_2Cl $\longrightarrow$ CH_2Cl–OH–OH–CH_2Cl

cis-Hydroxylation of *trans*-dihalobut-2-enes produces the corresponding (±)-*threo*-dihalobutane-2,3-diols. This is used in a high yielding synthesis of (±)-dithiothreitol, (±)-*threo*-1,4-dimercaptobutane-2,3-diol (G.I. Slepko *et al*, Izv. Akad. Nauk. SSSR Ser. Biol., 1970, 919).

Br / CH=CH / Br $\xrightarrow{KMnO_4}$ CH_2Br–OH, HO–CH_2Br $\xrightarrow{AcCl}$ CH_2Br–OAc, AcO–CH_2Br

KSAc →

CH_2SAc / —OAc / AcO— / CH_2SAc → CH_2SH / —OH / HO— / CH_2SH

meso-2,3-Diaminobutane-1,4-diol and 2,3-diamino-2,3-dideoxy-erythritol are prepared from 1,4-diacetoxybutane-2,3-diol (R. L. Martin and B. E. Norcross, J. org. Chem., 1975, 46, 523).

CH_2Ac / —OH / —OH / CH_2OAc —MsCl→ CH_2OAc / —OMs / —OMs / CH_2OAc —NaN_3→

CH_2OAc / —N_3 / —N_3 / CH_2OAc —1. KOH; 2. H_2, Pd→ CH_2OH / —NH_2 / —NH_2 / CH_2OH

Various methyl ethers of erythritol and threitol have been synthesised by methylating protected tetritols by the Kuhn procedure (MeI/Ag_2O). 1-*O*-Methyl-D-erythritol, 1-*O*-methyl-L-threitol, 1,4-di-*O*-methylerythritol and 1,4-di-*O*-methyl-L-threitol have been synthesised (P. Nanasi and A. Liptak, Carbohydrate Res., 1973, 29, 201). 2,3-Di-*O*-methyl-erythitol and Threitol are made from the corresponding tartaric acids (A.C.

Cope and A.S. Mehta, J. Amer. chem. Soc., 1964, 86, 1268).

Acid catalysed methylenation of erythritol (using paraformaldehyde and conc. H_2SO_4) gives 1,3:2,4- and 1,4:2,3-di-*O*-methylene-erythritol but none of the 1,2:3,4-derivative (I.J. Burden and J.F. Stoddart, J. chem. Soc. Perkin I, 1975, 666).

CH_2OH–CH(OH)–CH(OH)–CH_2OH ⟶ [1,3:2,4-di-O-methylene-erythritol] + [1,4:2,3-di-O-methylene-erythritol]

An improved low hazard procedure for the laboratory preparation of erythritol tetranitrate using a mixture of nitric and sulphuric acid in dichloromethan has been described (C.D. Marken *et al*, Synthesis, 1977, 484).

Dithiothreitol and the corresponding *meso*-compound, Cleland reagents, quantitatively reduce disulphide bonds in proteins (W.W. Cleland, Biochemistry, 1964, 3, 480).

R-S-S-R + CH_2SH–CH(OH)–CH(OH)–CH_2OH ⟶ 2 RSH + [4,5-dihydroxy-1,2-dithiane]

Dithioerythritol was found to be the better reagent for the reduction of disulphide bonds in polypeptide chains (L.I. Slobin and S.J. Singer, J. biol. Chem., 1968, 243, 1777).

The chiral Cleland reagent, L-(-)-1,4-dithiothreitol was prepared from L-(+)-tartaric acid as it might reduce right and left handed helical disulphide bridges at different rates (M. Cormack and C.J. Kelley, J. org. Chem., 1968, 33, 2171).

CO_2H / OH / HO / CO_2H ⟶ CO_2Et / O / O / CO_2Et ⟶ CH_2OH / O / O / CH_2OH

1. pTsCl, 2. AcSH ⟶ CH_2SAc / O / O / CH_2SAc ⟶ CH_2SH / OH / HO / CH_2SH

Seeback *et al* (Angew. chem. Internat. Edn., 1969, 8, 982) have used a similar approach to prepare chiral 1,4-bis(dimethylamino)-2,3-dimethoxybutanes from tartaric acid.

CO_2Et / OH / HO / CO_2Et —1. $HNMe_2$, 2. Me_2SO_4→ $CONMe_2$ / OMe / MeO / $CONMe_2$ —$LiAlH_4$→ CH_2NMe_2 / OMe / MeO / CH_2NMe_2

These ethers may be used as chiral solvents for the asymmetric synthesis of alcohols by the reaction of an alkyl lithium or a Grignard reagent with an aldehyde.

$$RCHO \quad + \quad R'Li \longrightarrow RR'C(OH)H$$

(c) Homologous tetrahydric alcohols

2-*C*-Methylerythritol has been isolated from *Convolvulus glomeratus* and the racemic compound synthesised from citraconic acid (T. Anthonsen *et al*, Acta chem. Scand., 1976, 30B, 91). The naturally occurring isomer has been shown to have the absolute stereochemistry 2*S*,3*R* (S.W. Shah *et al*, ibid, 1976, 30B, 903).

$$MeO_2C\text{-}C(Me)\text{=}CH\text{-}CO_2Me \xrightarrow[2.\ LiAlH_4]{1.\ KMnO_4} HOCH_2\text{-}C(Me)(OH)\text{-}CH(OH)\text{-}CH_2OH$$

citraconic acid

The chemical and economic aspects of the synthesis of penterythritol have been reviewed (R.T. Thampy, Chem. Process Eng. Bombay, 1969, 3, 28). A more general work on pentaerythritol has appeared (E. Berlow, R.H. Booth, and J.E. Snow, ACS Monograph No 136, 1958). The ^{13}C-n.m.r. of penterythritol is discussed (W. Voelter *et al*, Angew. Chem. internat. edn., 1970, 9, 803).

Hexane-1,2S,5S,6-tetraol has been synthesised from mannitol (C.C. Deane and T.D. Inch, Chem. Comm., 1969, 813).

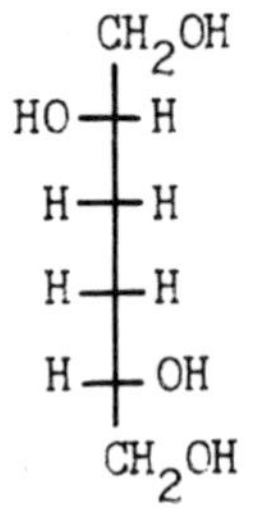

$$\begin{array}{c} CH_2OH \\ HO-C-H \\ H-C-H \\ H-C-H \\ H-C-OH \\ CH_2OH \end{array}$$

meso-Hexane-1,2,5,6-tetraol has been synthesised in low yield (R. Brettle and D.W. Latham, J. chem. Soc. C, 1968, 906).

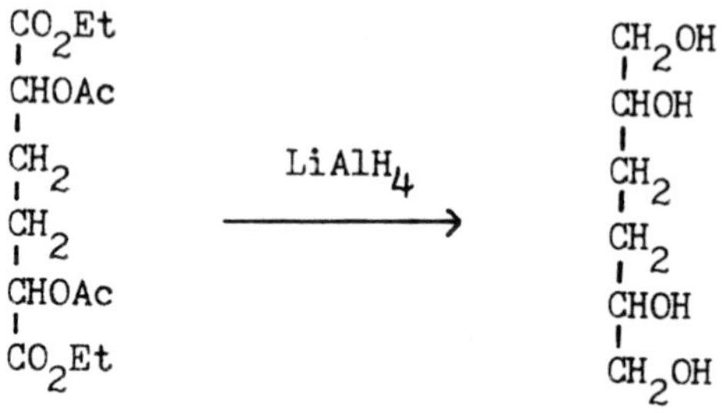

$$\begin{array}{l} CO_2Et \\ | \\ CHOAc \\ | \\ CH_2 \\ | \\ CH_2 \\ | \\ CHOAc \\ | \\ CO_2Et \end{array} \xrightarrow{LiAlH_4} \begin{array}{l} CH_2OH \\ | \\ CHOH \\ | \\ CH_2 \\ | \\ CH_2 \\ | \\ CHOH \\ | \\ CH_2OH \end{array}$$

2. *Trihydroxyaldehydes and trihydroxyketones*

A simple synthesis of L-erythrose from D-gulono-1,4-lactone has been reported (L.M. Lerner, Carbohydrate Res., 1969, 9,1).

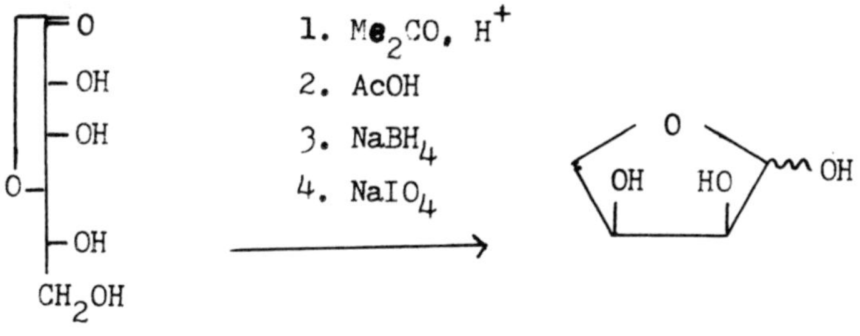

1. Me_2CO, H^+
2. AcOH
3. $NaBH_4$
4. $NaIO_4$

L-Threose is obtained from L-sorbose by treatment with silver carbonate on Celite (S. Morgenlie, Acta chem. Scand., 1972, 26, 2146).

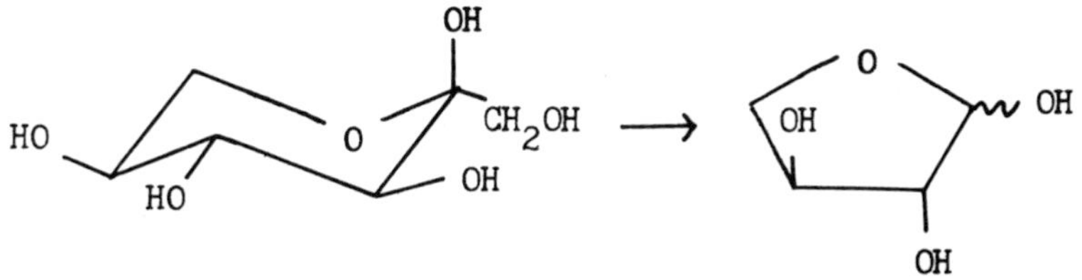

Both D-threose and D-erythrose are formed from 2,3-*O*-isopropylidene-D-glyceraldehyde after separation by preparative glc of the mixed products fromed in the first step (D.J. Walton, Canad. J. Chem., 1967, 45, 2922).

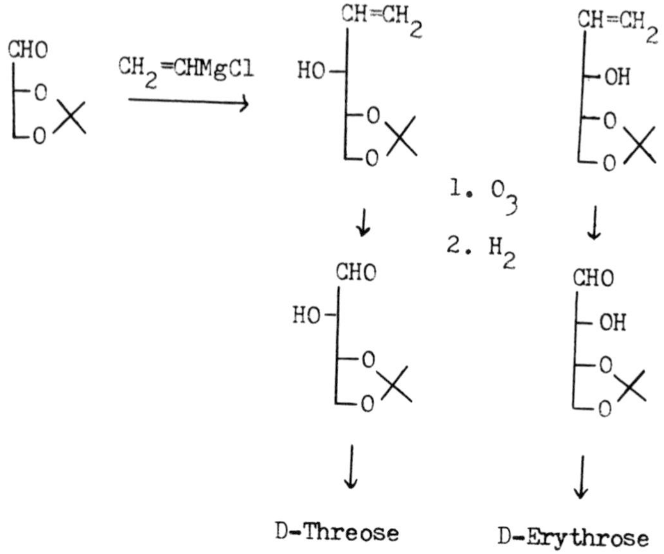

2,3-Di-*O*-methyl-L-threose and 2,3-di-*O*-methyl-D-erythrose are prepared by periodate treatment of 2,3-di-*O*-methyl-D-glucitol and 2,3-di-*O*-methyl-D-mannitol respectively (G.G.S. Dutton, K.B. Gibney and P.E. Reid, Canad. J. Chem., 1969, 47, 2494). The ^{1}H- n.m.r. spectra of erythrose and threose are discussed (K. Horitsu and P.A.J. Gorin, Agr. biol. Chem., 1977, 41, 1459).

L-Erythulose can be made in 23% yield by dehydrogenation of ethrythritol by *Acetobacter suboxydans* (H.J. Haas and B. Matz, Ann. Chem., 1974, 342). An improved total synthesis of ($\pm$)-erythulose starting from 1,4-dihydroxybut-2-yne has been reported (T. Ando, S. Shioi, and M. Nakagawa, Bull. chem. Soc. Japan, 1972, 45, 2611).

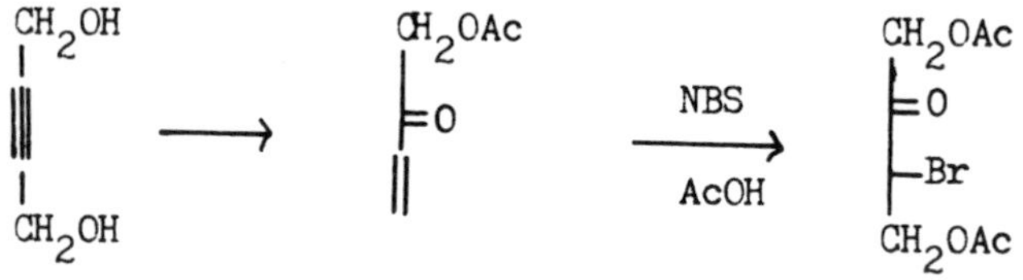

$$\xrightarrow{AgOAc} \begin{array}{l} CH_2OAc \\ \vert{=}O \\ \vert{-}OAc \\ CH_2OAc \end{array} \xrightarrow{Ba(OH)_2} \begin{array}{l} CH_2OH \\ \vert{=}O \\ \vert{-}OH \\ CH_2OH \end{array}$$

Oxidation by mercuric acetate of erythritol and D-threitol gives DL- and D-erythulose respectively (L. Stankovic and M. Fedoronko, Chem. Zvesti, 1971, 25, 441).
A correlation between the structures and ORD curves of D- and L-erythulose (glycero-tetrulose) has been established (T. Sticzay *et al*, Carbohydrate Res., 1968, 6, 418).
The four isomeric 2,4-di-*O*-methyl tetroses are prepared by periodate oxidation of known methylated sugars (G.G.S. Dutton and K.N. Slessor, Canad. J. Chem., 1964, 47, 614

L-Erythulose-1-phosphate has been isolated from *Propionibacterium pentosaceum* fed on erythritol (E.J. Wawszkiewicz, Biochemistry, 1968, 7, 683).

The products from the acetylation of D-erythulose include 1,2,3-tri-*O*-acetyl and β-D-erythrofuranose and acetylated dimers of D-erythulose (R. Andersson, O. Theander, and E. Westerlund, Carbohydrate Res., 1978, 61, 501).

$$\begin{array}{l} CH_2OH \\ \vert{=}O \\ HO{-}\vert \\ CH_2OH \end{array} \longrightarrow \text{(furanose ring with O, OAc, AcO, OAc)} + \textbf{dimers}$$

L-Deosamine, the enantiomer of the widely distributed antibiotic component has been synthesised (H.H. Baer and C.-W. Chiu, Canad. J. Chem., 1974, 52, 122).

L-Deosamine

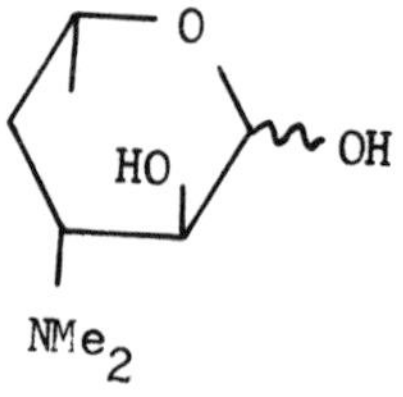

3. *Dihydroxydiones and their derivatives*

1,4-Dihydroxybutane-2,3-dione can be prepared by permanganate oxidation of but-2-yne-1,4-diol. The mechanism of this process has been studied (L.I. Simandi and M. Jaky, J. chem. Soc. Perkin II, 1973, 1861).

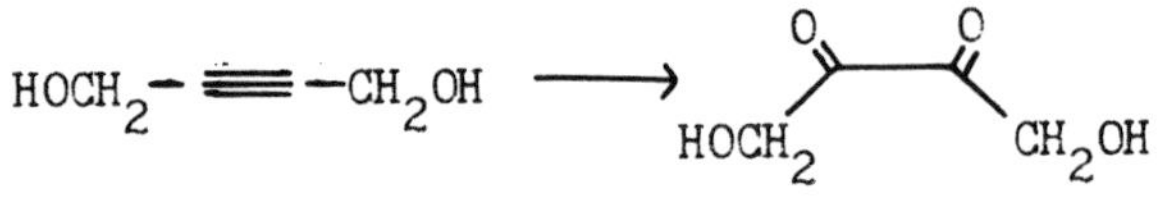

$$HOCH_2-C\equiv C-CH_2OH \longrightarrow HOCH_2COCOCH_2OH$$

2,5-Dihydroxyhexane-3,4-dione has been prepared from 2,5-dihydroxyhex-3-yne.

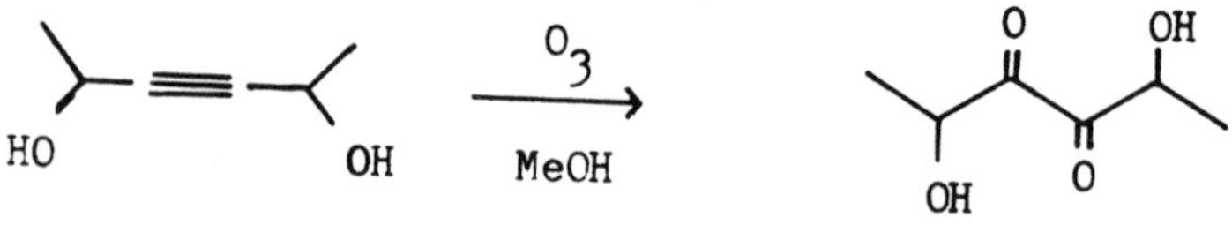

$$\xrightarrow[MeOH]{O_3}$$

Treatment of this compound with acid converts it into furaneol, a flavour component of strawberries and pineapples (L. Re, B. Maurer, and G. Ohloff, Helv. chim. Acta, 1973, 56, 1882). Furaneol can also be formed by hydrolysis of 2,5-dibromohexane-3,4-dione (G. Büchi, E. Demole, and A.F. Thomas, J. org. Chem., 1973, 38, 123).

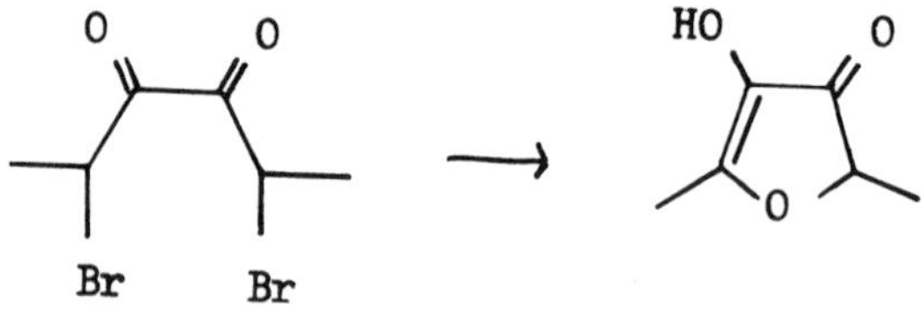

1,4-Dibromopentane-2,3-dione is produced when the parent dione is treated with bromine (C. Rappe and A. Norström,

Acta chem. Scand., 1968, 22, 1853).

1,4-Dibromobutane-3,4-dione reacts with triethyl phosphite in a double Perkow reaction (M.L. Honig and M.L. Sheer, J. org. Chem., 1973, 38, 3434).

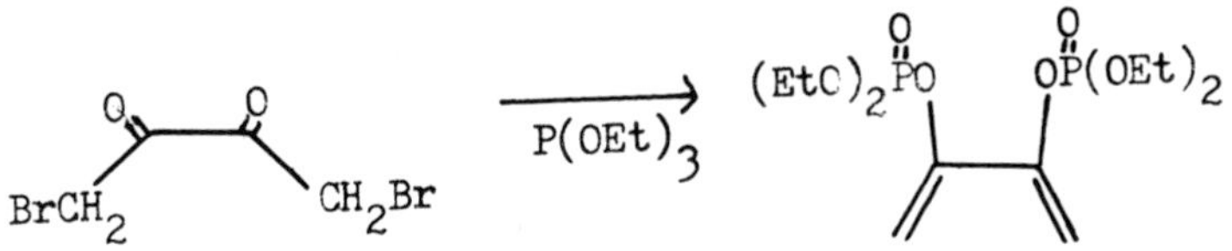

$BrCH_2$ CH_2Br $\xrightarrow{P(OEt)_3}$ $(EtO)_2\overset{O}{\overset{\|}{P}}O$ $O\overset{O}{\overset{\|}{P}}(OEt)_2$

1,4-Dibromobutane-3,4-dione is the starting material for the synthesis of 1,4-dinitro-1,3-butadiene-2,3-dicarboxylic dianhydride, a peptide coupling agent (B. Weinstein, Pept. Proc. Am. Pept. Symp. 5th, 1977, 539).

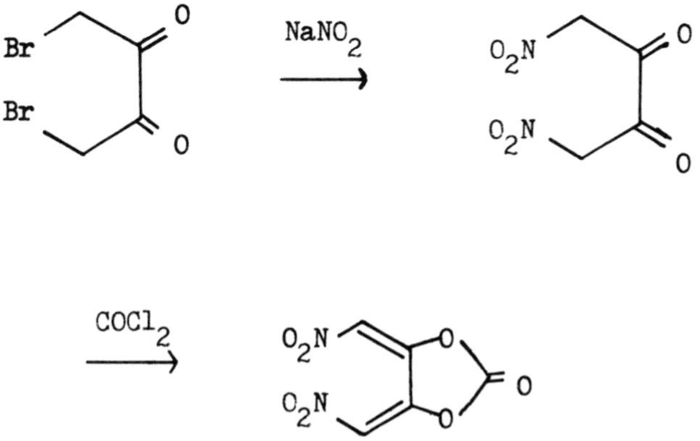

Br Br O O $\xrightarrow{NaNO_2}$ O_2N O_2N O O

$\xrightarrow{COCl_2}$ O_2N O_2N O O O

4. *Tetra-ones*

Nonane-2,4,6,8-tetrone is made by the reaction of the dianion of acetylacetone with the anion of methyl acetoacetate. Acid treatment of the tetraone gives 6-acetyl-5-methyl resorcinol (T.P. Murray and T.M. Harris, J. Amer. chem. Soc., 1972, 94, 8253).

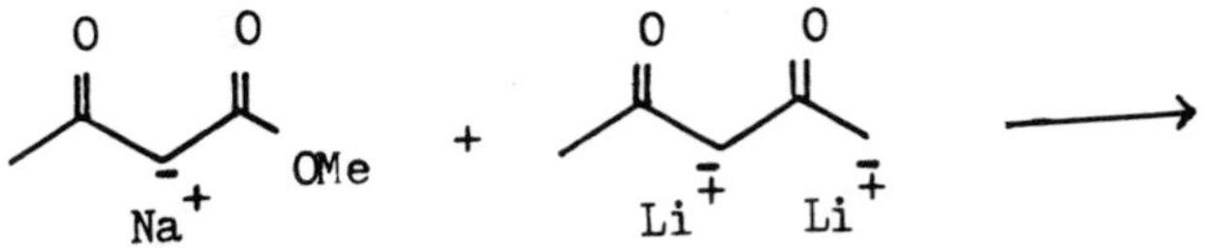

O O OMe Na^+ + O O Li^+ Li^+ $\longrightarrow$

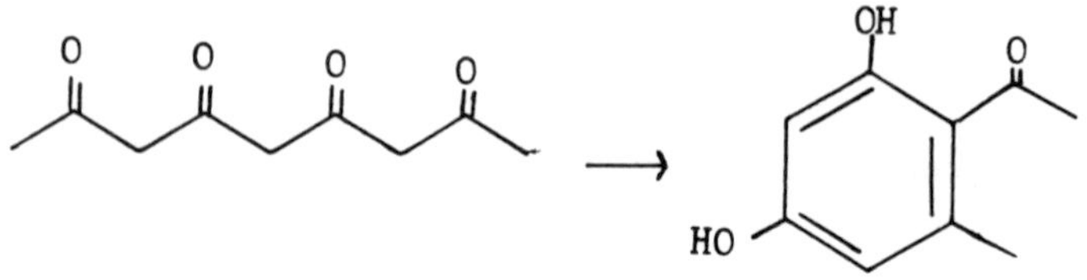

OH O O O O HO

Dimerisation of the dianion of acetylacetone gives decane-2,4,7,9-tetraone (K.G. Hampton and J.J. Christie, J. org. Chem., 1975, 26, 3887).

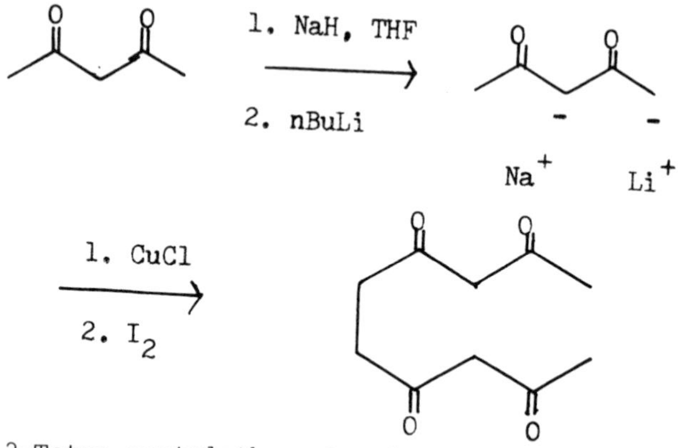

1. NaH, THF
2. nBuLi
Na^+ Li^+
1. CuCl
2. I_2

1,1,2,2-Tetra-acetylethane has been shown by neutron diffraction studies to exist in the solid state as a di-enol (L.F. Power, K.E. Turner, and F.H. Moore, J. cryst. mol. Struct., 1975, 5, 59).

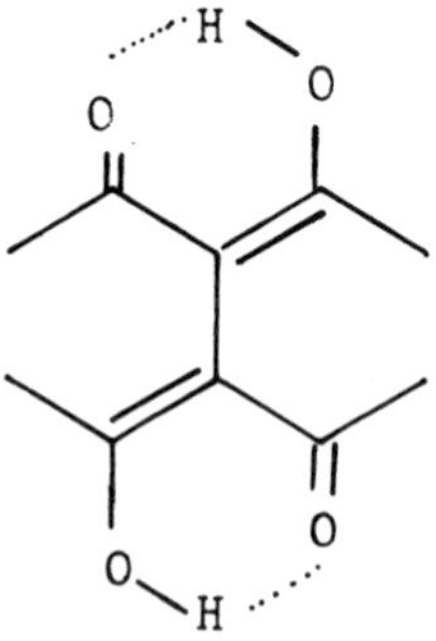

H O O O O H

5. *Trihydroxycarboxylic acids*

L-Threonic acid is prepared in high yield by oxidation of ascorbic acid with excess alkaline hydrogen peroxide (H.S. Isbell and H.L. Frush, Carbohydrate Res., 1979, 72, 301).

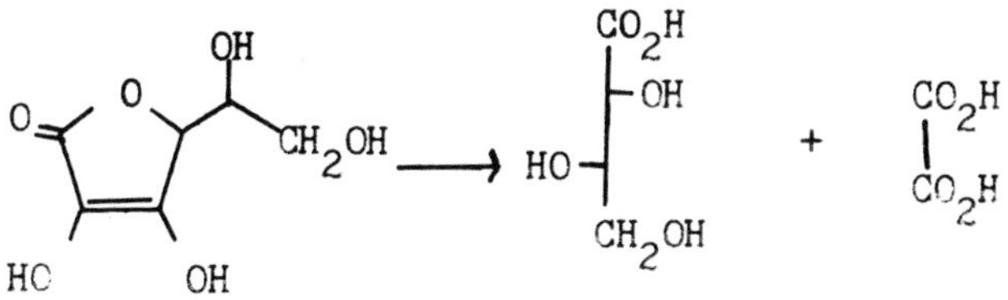

D-Erythronolactone is converted into D-erythrose by treatment with bis(3-methyl-2-butyl)borane followed by hydrogen peroxide (T.A. Giudici and A.L. Fluharty, J. org. Chem., 1967, 32, 2043).

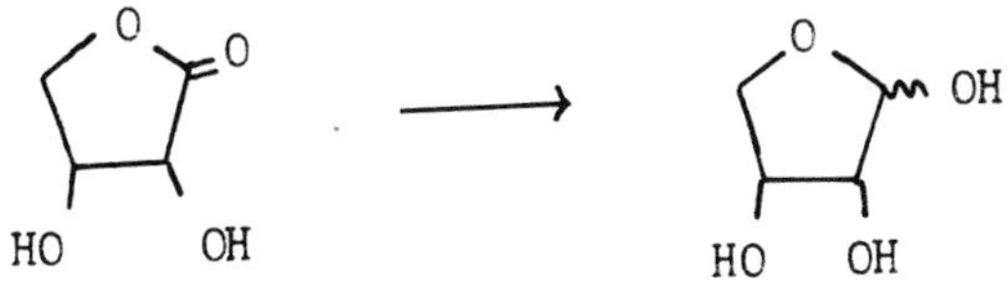

Condensation of copper glycinate and benzyloxyacetaldehyde gives a mixture of 2-amino-4-benzyloxy-3-hydroxybutanoic acids. These are resolved into the optically active *erythro* forms by Takadiastase and the optically active *threo* forms by using Vogler's method (K. Okawa *et al*, Bull. chem. Soc. Japan, 1969, 42, 2720).

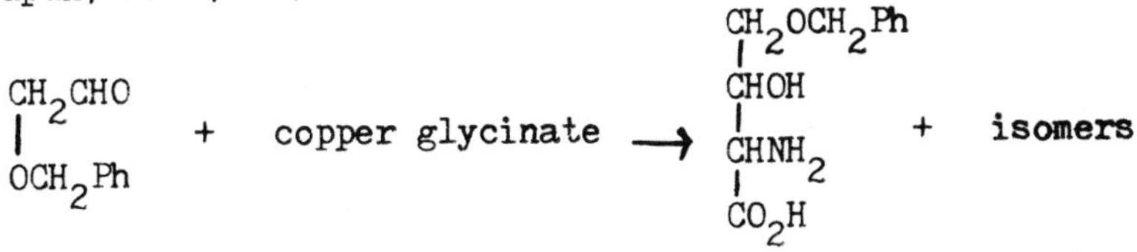

$$\begin{array}{c} CH_2CHO \\ | \\ OCH_2Ph \end{array} + \text{copper glycinate} \longrightarrow \begin{array}{c} CH_2OCH_2Ph \\ | \\ CHOH \\ | \\ CHNH_2 \\ | \\ CO_2H \end{array} + \textbf{isomers}$$

The alarm pheromone of the sea anemone *Anthopleura elegantissima* has been identified as 4-amino-4-deoxy-L-threonic acid, three of the four possible stereoisomers have been synthesised (J.A. Musich and H. Rapoport, J. Amer. chem. Soc., 1978, 100, 4865).

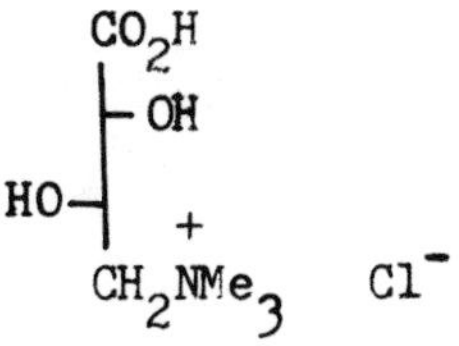

Eritadenine has been isolated from *Lentinus edodes*, and has been synthesised from the corresponding amino compound (T. Kamiya *et al*, Tetrahedron, 1972, 28, 899).

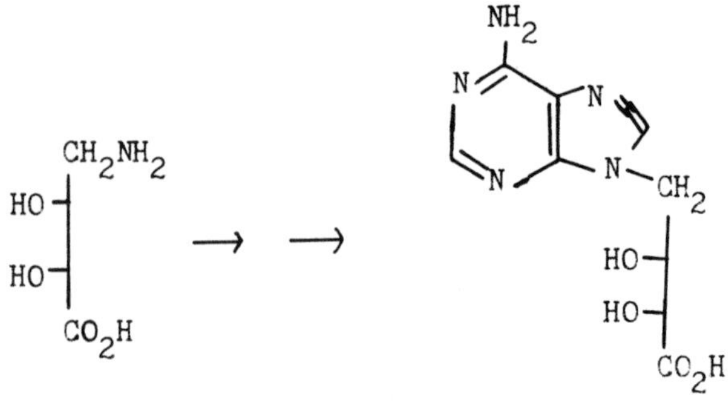

Negamycin and its antipode were synthesised by S. Shibahara *et al* (J. Amer. chem. Soc., 1972, 94, 4353). A total synthesis of racemic negamycin has also been reported (W. Streicher, H. Reinshagen and F. Turnowsky, J. Antibiot., 1978, 31, 725).

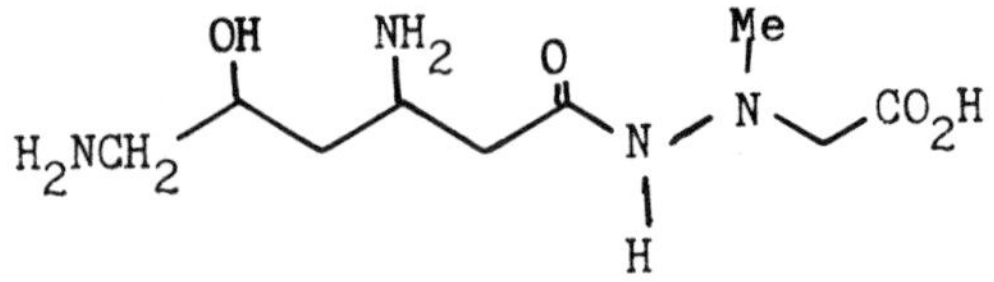

Threo-4-hydroxy-β-lysine is obtained from hydrolysis of the antitubercular peptides, tuberactinomycin A and N (T. Wakamiya, T. Shiba, and T. Kaneko, Bull. chem. Soc. Japan, 1972, 45, 3668).

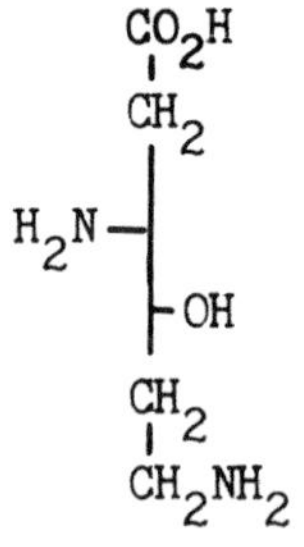

6. *Dihydroxyoxocarboxylic acids*

Treatment of the triacetate shown below with sodium ethoxide produces methyl 5,6-dihydroxy-4-oxohexanoate (I. Dyong *et al*, Ber., 1977, 110, 1175).

CH_2OAc / OAc / OAc / CO_2Me —NaOEt→ CH_2OH / OH / =O / CH_2 / CH_2 / CO_2Me

7. *Hydroxydioxocarboxylic acids*

Ethyl 3-hydroxy-2,4-dioxopentanoate has been synthesised (N. Sugiyama *et al*, Bull. chem. Soc. Japan, 1967, 40, 2594).

CO_2Et —SO_2Cl_2→ Cl, CO_2Et —1. KOAc, 2. HCl→ OH, CO_2Et

8. *Dihydroxyalkanecarboxylic acids*

(a) Dihydroxysuccinic acids

The isolation, identification and culture of organisms producing L-(+)-tartaric acid from glucose has been reviewed (K. Yamada *et al*, Ferment. Advan. Pap. Int. Ferment. Symp. 3rd, 1968, 541, Ed. D. Perlman, Academic Press, New York, 1969). Dimethyl 2,3-dimethoxybutanedioate and dimethyl 2-chloro-3-methoxybutanedioate are obtained by cleavage of cyclobutenes W. Kirmse, F. Scheidt, H.-J. Vater, J. Amer. chem. Soc., 1978, 100, 3945).

Cl, OMe —1. O_3, 2. CH_2N_2→ CO_2Me / Cl / OMe / CO_2Me

OMe, OMe (cyclobutene) → 1. $NaIO_4$ 2. CH_2Cl_2 → CO_2Me, OMe, OMe, CO_2Me

The crystal structure of *meso*-tartaric acid has been reported (G.A. Bootsma and J.C. Schoone, Acta Cryst., 1967, 22, 522). The crystal chemistry of tartaric acid and its simple and complex salts has been reviewed (G.A. Kiosse, Krist. Strukt. Neorg. Soedin., 1974, 103).

Di-isopropyl 2,3-di-*O*-methyl-L-(+)-tartrate has been used as a solvent for the asymmetric synthesis of alcohols by the reaction of dialkyl cadmium reagents and ketones (H.-J. Bruer and R. Haller, Tetrahedron Letters, 1972, 5227).

$CO_2{}^iPr$, OMe, MeO, $CO_2{}^iPr$

(b) Dihydroxyglutaric acids

2,4-Dimercaptoglutaric acids have been prepared. A mixture of diasteriomeric dimethyl 2,4-dibromoglutarates is converted into the 1,2-dithiolane as shown. This is then separated into crystalline racemic compound and syrupy *meso* compound. Reduction with zinc and hydrochloric acid gives 2,4-dimercaptoglutaric acid (M.-O. Hedblom, Tetrahedron Letters, 1970, 5159).

MeO_2C, CO_2Me, Br, Br → KSAc → MeO_2C, CO_2Me, SAc, SAc

→ KOH, I_2 → HO_2C, CO_2H, S-S → Zn, HCl → HO_2C, CO_2H, SH, SH

(c) Dihydroxyadipic acid

Kolbe electrolysis of the monomethyl ester shown below gives dimethyl 2,5-diacetoxyadipate (R. Brettle and D.W. Latham, J. chem. Soc. C, 1968, 906).

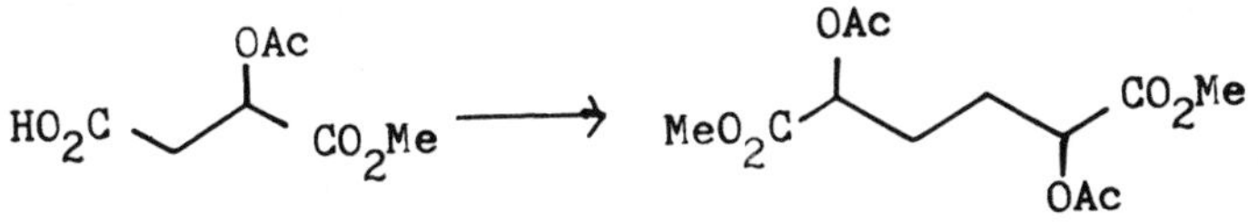

Dimethyl *threo*-3,4-dimethoxyadipate has been synthesised from D-(-)-tartaric acid as outlined below. The adipate is an intermediate in the synthesis of 1R,7R-*exo*-brevicomin (K. Mori, Tetrahedron, 1974, 30, 4223).

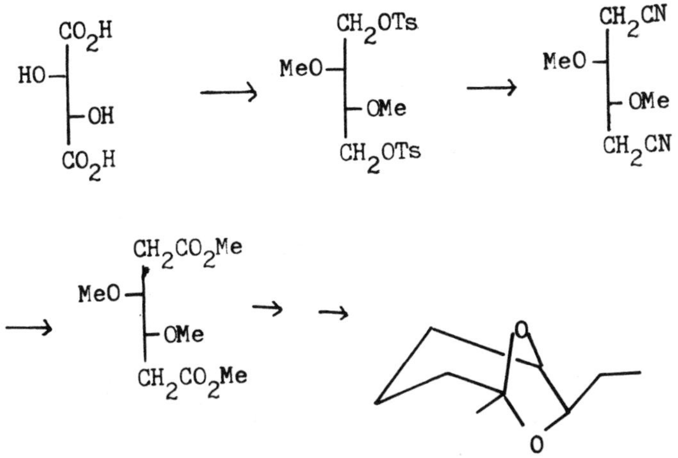

Treatment of *racemic* 2,5-dibromoadipic acid with sodium carbonate yields the dilactone shown below (R.B. Sandin, W.J. Rebel, and S. Levine, J. org. Chem., 1966, 31, 3879).

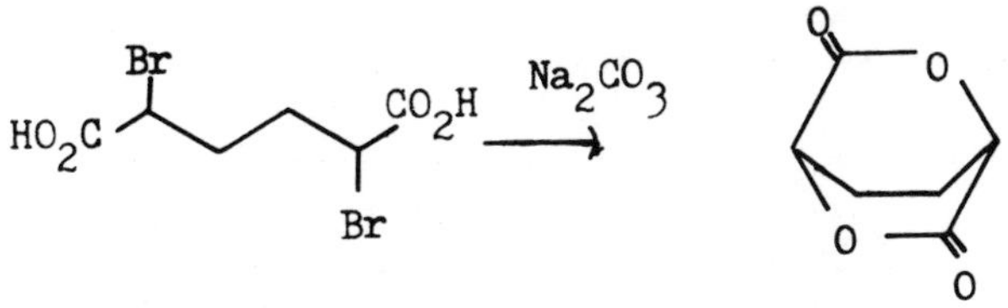

(d) Dihydroxypimelic acids

The production and biosynthesis of 2,6-diaminopimelic acid has been reviewed (K. Nakayama, Microbial Prod. Amino Acids, 1972, 369). A laboratory synthesis of 2,6-diaminopimelic acid starting from glutaraldehyde has been reported (A. Arendt *et al*, Rocz. Chem., 1974, 48, 883). The racemate has been resolved with papaine and aniline (A. Arendt, *ibid*, 1974, 48, 635). Various methyl substituted 2,6-diaminopimelic acids have been synthesised (B. Cavalleri *et al*, Farmac. ed. Sci., 1974, 29, 257).

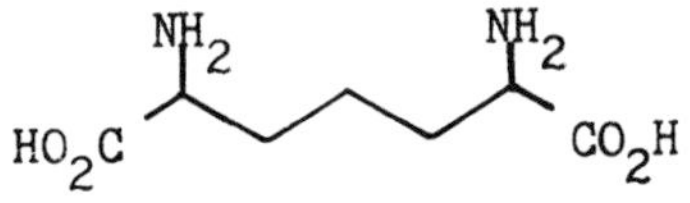

NH_2 NH_2
HO_2C CO_2H

9. *Dioxocarboxylic acids*

Diethyl 2,3-diacetylsuccinate has been synthesised from the sodium salt of ethyl acetoacetate and ethyl 2-chloroaceto-acetate. The diacetylsuccinate showed no evidence of enol-isation. Treatment with strong acid gave the surprising product, diacetyl maleate. Diacetyl maleate and diacetyl fumarate have been synthesised by other routes and their structures proven (S.C. Airy, C. Martin, and J.M. Sullivan, J. org. Chem., 1979, 44, 1891).

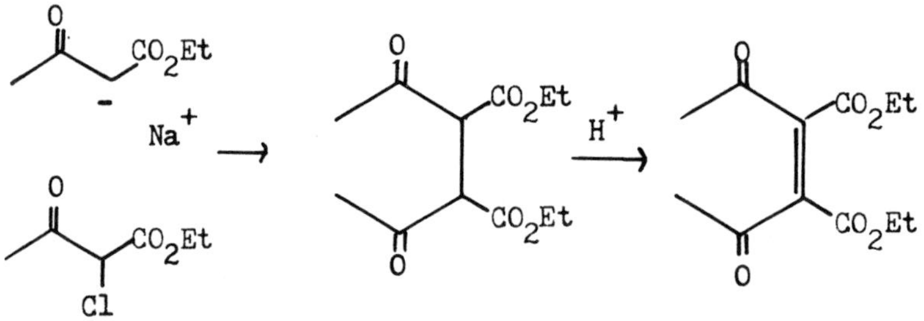

O CO_2Et - Na^+ → O CO_2Et CO_2Et O $\xrightarrow{H^+}$ O CO_2Et CO_2Et O

O CO_2Et Cl

10. *Hydroxycarboxylic acids*

The production of citric acid by fermentation has been reviewed (L.B. Lockwood, Microb. Technol. (2nd Edn.), 1971, 1, 355, Ed. H.J. Peppler and D. Perlman, Academic Press, New York and L.M. Miall, Econ. Microbiol., 1978, 2, 47). The use of citric acid in pharmaceutical preparations and in cosmetics has been reviewed (E.F. Bouchard and E.G. Merritt, Kirk-Othmer Encycl. Chem. Technol., 3rd Edn., 1979, 6, 150, Ed. M. Grayon and D. Eckroth, Wiley, New York). A more general review of citric acid has been published (L.R. Roberts, Encycl. Chem. Process. Des., 1979, 8, 324, ed. J.J. McKetta and W.A. Cunningham, Dekker, New York).

Trimethyl 1-hydroxybutane-1,1,4-tricarboxylate has been synthesised by the action of fuming nitric acid on trimethyl 1,1,4-butanetricarboxylate (R. Brettle and D.P. Cummings, J. chem. Soc. Perkin I, 1977, 2177).

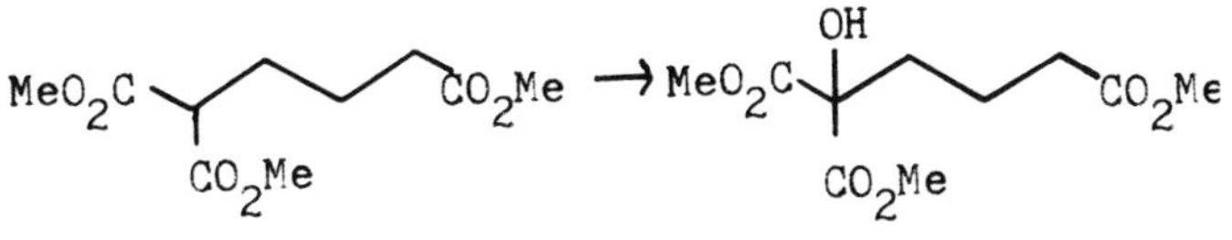

The isomers of homoisocitric acid have been synthesised from the corresponding cyclohexenes as shown below. 1R-Hydroxybutane-1,2S,4-tricarboxylic acid has been shown to be an intermediate in the α-aminoadipic acid pathway in lysine biosynthesis (K. Chilina *et al*, Biochemistry, 1969, 8, 2846).

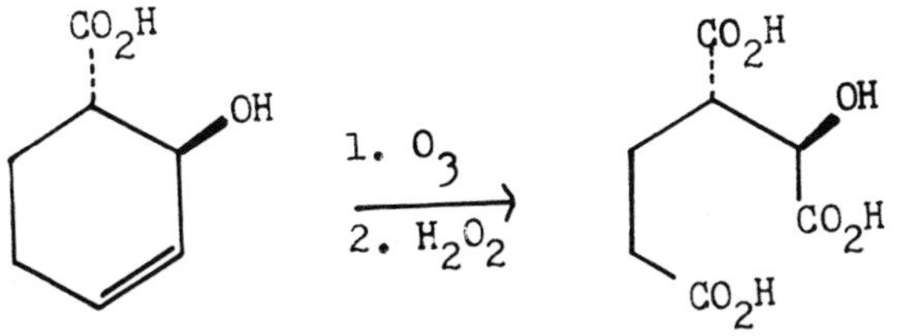

The vitamin K dependent formation of γ-carboxyglutamic acid (4) has been reviewed (R.E. Olson and J.W. Suttie, Vitam. Horm., 1977, 35, 59 and J. Stenflo and J.W. Suttie, Ann. Rev. Biochem., 1977, 46, 157). The crystal structure of γ-carboxyglutamic acid has been determined (K.A. Satyshur and S.T. Rao, Acta Cryst., 1979, 35B, 2260).
L-γ-Carboxyglutamic acid has been synthesised by the route outlined below. The protected pyroglutamate (1) is treated with Bredereck"s reagent, and the resulting compound (2) reacts with trichloroethyl chloroformate to give compound (3). The latter is deprotected to give L-γ-carboxyglutamic

acid (4) (S. Danishefsky *et al*, J. Amer. chem. Soc., 1979, 101, 4385).

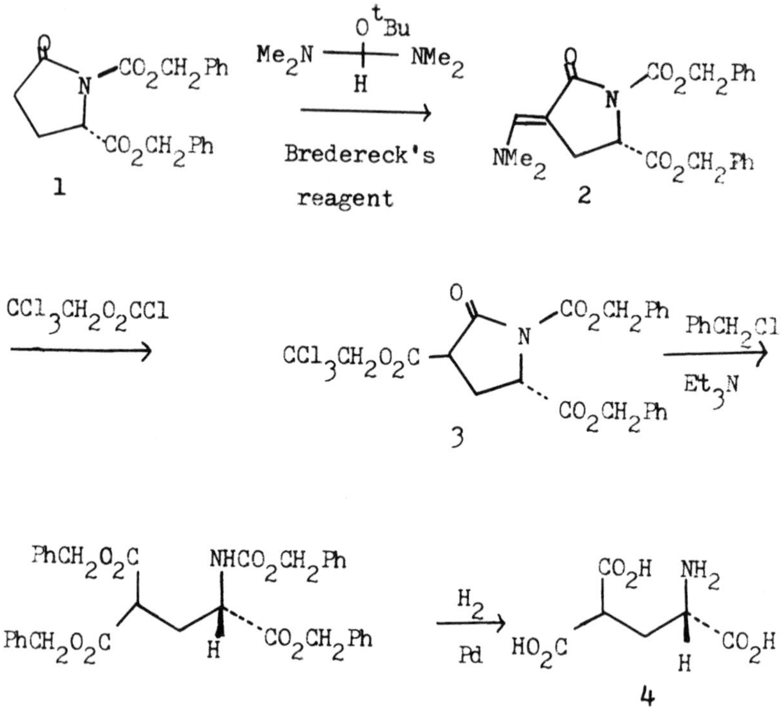

Me$_2$N–CH(O^{t}Bu)–NMe$_2$

Bredereck's reagent

1

2

$CCl_3CH_2O_2CCl$

3

$PhCH_2Cl$

Et_3N

H_2

Pd

4

Racemic γ-carboxyglutamate has been resolved as di-t-butyl-N-benzyloxycarbonyl- γ-carboxyglutamate using (-)-ephedrine and quinine (W. Märki *et al*, Helv. chim. Acta, 1977, 60, 798).

11. *Oxotricarboxylic acids*

Triethyl 3-oxobutane-1,1,4-tricarboxylate has been prepared as shown below (T. Kato, H. Kimura, and K. Tanji, Chem. pharm. Bull., 1978, 26, 3880).

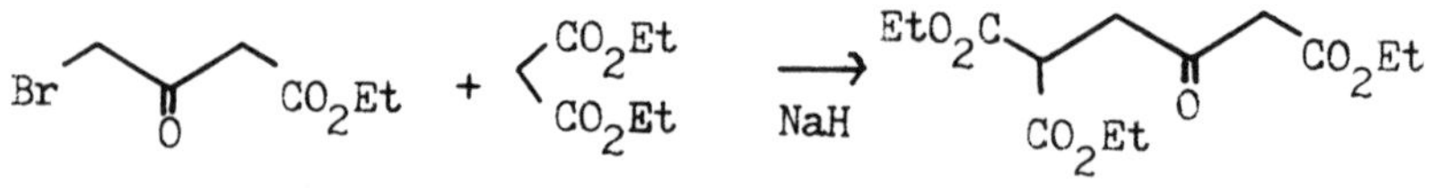

NaH

12. *Tetracarboxylic acids*

The synthesis and the chemistry of tetracarboxylic acids has been reviewed (B.I. Zapadinskii, B. I. Liogonkii, and A.A. Berlin, Russ. chem. Rev., 1973, 939).

(a) Alkanetetracarboxylic acids

(i) Carboxyl groups attached to two different carbon atoms

Tetraethyl ethane-1,1,2,2-tetracarboxylate has been synthesised from diethyl bromomalonate by treatment with sodium diethyl phosphite (K.H. Takemura and T.J. Tuma, J. org. Chem., 1969, 34, 252).

$$Br\text{-}CH(CO_2Et)_2 \xrightarrow{(EtO)_2PONa} (EtO_2C)_2CH\text{-}CH(CO_2Et)_2$$

This compound has also been made by indirect electrochemical oxidation of diethyl malonate (D.A. White, J. electrochem. Soc., 1977, 124, 1177). Alkyl substituted malonates have been oxidised electrochemically to give the corresponding tetracarboxylates (H.G. Thomas, M. Streukens, and R. Peek, Tetrahedron Letters, 1978, 45).

$$R\text{-}CH(CO_2Et)_2 \longrightarrow (EtO_2C)_2\underset{R}{C}\text{-}\underset{R}{C}(CO_2Et)_2$$

The crystal structure of tetramethyl ethane-1,1,2,2-tetracarboxylate has been determined (J.P. Schaefer and C.R. Costin, J. org. Chem., 1968, 33, 1677).

(ii) Carboxyl groups attached to four different carbon atoms

All the isomers of pentane-1,2,3,4-tetracarboxylic acid have been synthesised from the corresponding isomer of 3-methylcyclohexene-4,5-dicarboxylic acid (R.I. Galeeva and V.N. Okinokov, Khim. Vysokomol. Zh. org. Khim., 1973, 9, 666).

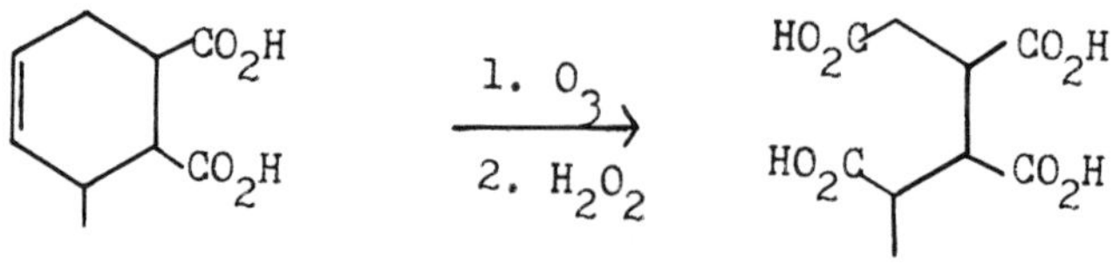

The dianhydride of butane-1,2,3,4-tetracarboxylic acid has been shown to be the 1,2:3,4-anhydride rather than the 1,3:2,4-anhydride (R. Nagao *et al*, Acta Cryst., 1971, 27B, 569).

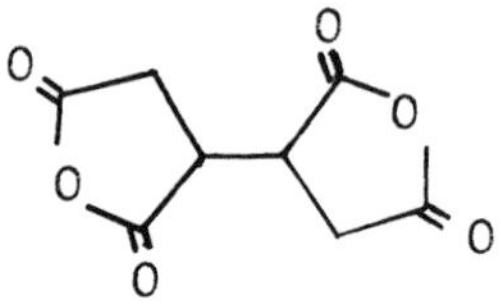

(b) Alkenetetracarboxylic acids

Tetraethyl ethenetetracarboxylate reacts with hydrazine to give 4,5,5-tricarbethoxy-3-pyrazolidinone whereas with phenyl hydrazine it gives the uncyclised adduct (J.M. Patterson *et al*, J. org. Chem., 1978, 43, 3039).

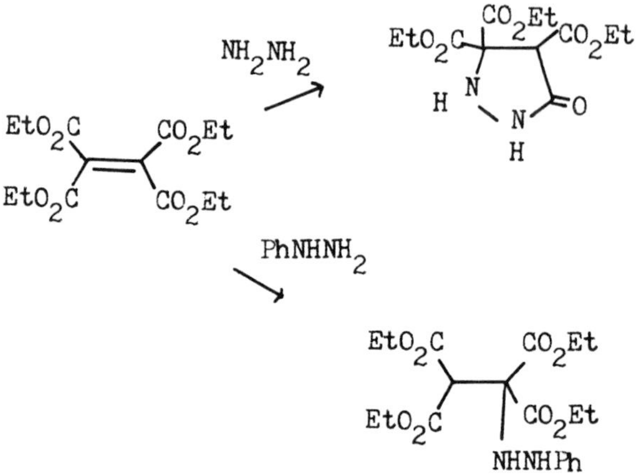

Guide to the Index

This index is constructed in a similar manner to the volume indexes of the first edition of the Chemistry of Carbon Compounds. However, to make the index easier to use, more descriptive entries have been made for the commonly occurring individual, and groups of chemicals.

The indexes cover primarily the chemical compounds mentioned in the text, and also include reactions and techniques, where named, and some sources of chemical compounds such as plant and animal species, oils, etc.

Chemical compounds have been indexed alphabetically under the names used by authors, editing being restricted to ensuring uniformity of entries under the same heading. In view of the alternative nomenclature that can often be used, a limited amount of cross-referencing has been done where it is considered to be helpful, but attention is particularly drawn to Convention 2 below.

For this and the succeeding volumes, the indexing conventions listed below have been adopted.

1. *Alphabetisation*

(a) The following prefixes have not been counted for alphabetising:

n-	*o-*	*as-*	*meso-*	D	*C*
sec-	*m-*	*sym-*	*cis-*	DL	*O-*
tert-	*p-*	*gem-*	*trans-*	L	*N-*
	vic-				*S-*
		lin-			*Bz-*
					Py-

Some prefixes and numbering have been omitted in the index, where they do not usefully contribute to the reference.

(b) The following prefixes have been alphabetised:

Allo	Epi	Neo
Anti	Hetero	Nor
Cyclo	Homo	Pseudo
	Iso	

(c) A letter by letter alphabetical sequence is followed for entries, firstly for the main entry, followed by the descriptive entry. The only exception to this sequence is the placing of plural entries in front of the corresponding individual entries to prevent these being overlooked by a strict alphabetical sequence which could lead to a considerable separation of plural from individual entries. Thus "butanes" will come before *n*-butane, "butenes" before 1-butene, and 2-butene, etc.

2. *Cross references*

In view of the many alternative trivial and systematic names for chemical compounds, the indexes should be searched under any alternative names which may be indicated in the main body of the text. Only a limited amount of cross-referencing has been carried out, where it is considered that it would be helpful to the user.

3. *Esters*

In the case of lower alcohols esters are indexed only under the acid, e.g. propionic methyl ester, not methyl propionate. Ethyl is normally omitted e.g. acetic ester.

4. *Derivatives*

Simple derivatives are not normally indexed if they follow in the same short section of the text.

5. *Collective and plural entries*

In place of "– derivatives" or "– compounds" the plural entry has normally been used. Plural entries have occasionally been used where compiunds of the same name but differing numbering appear in the same section of the text.

6. *Main entries*

The main entry of the more common individual compounds is indicated by heavy type. Multiple entries, such as headings and sub-headings over several pages are shown by "–", e.g., 67–74, 137–139, etc.

INDEX